The SCIENCE *of the* EARTH

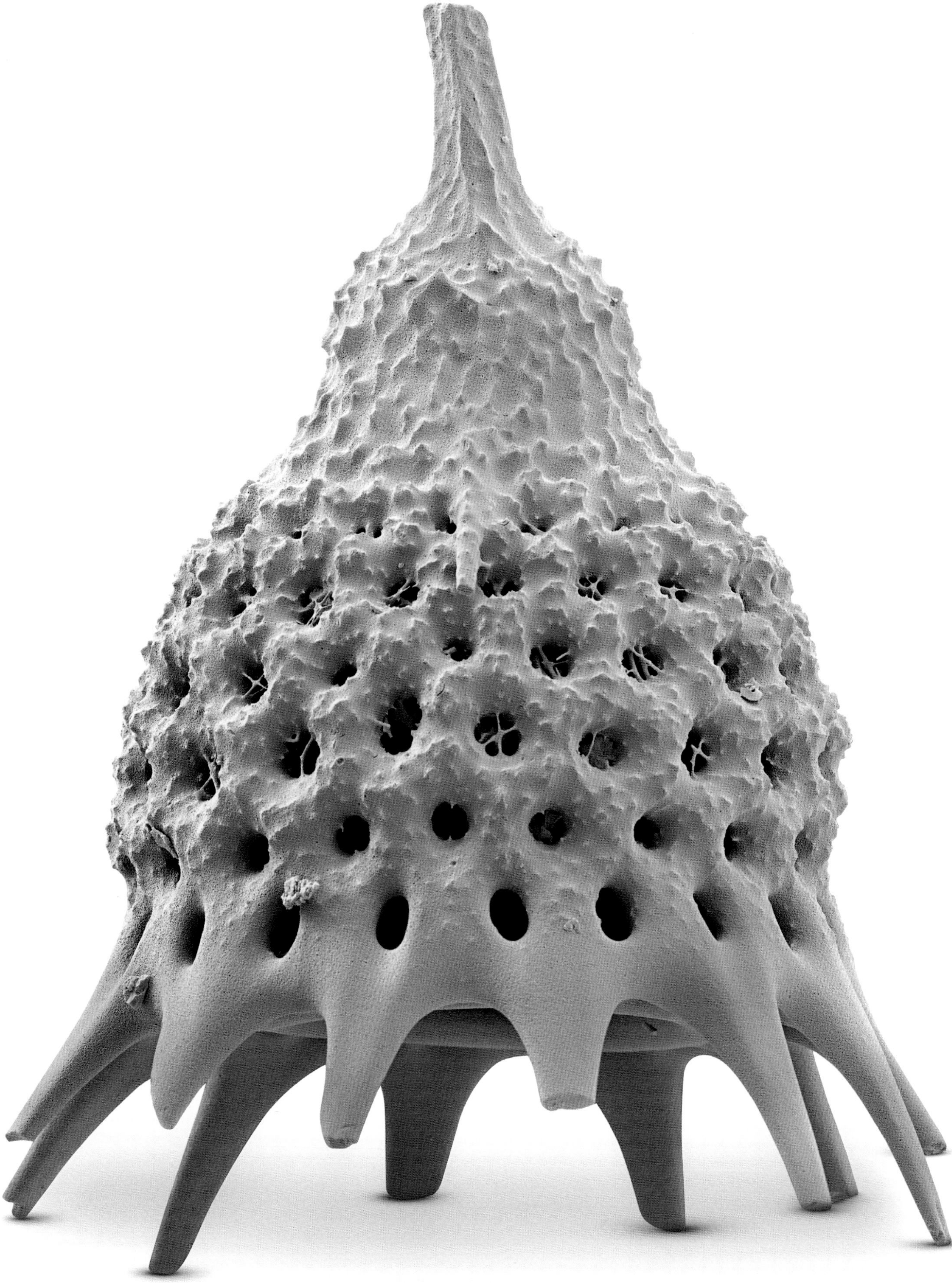

DK SMITHSONIAN

The SCIENCE *of the* EARTH

THE SECRETS OF OUR PLANET REVEALED

DK Penguin Random House

DK LONDON

Senior Editors Peter Frances, Rob Houston
Editors Polly Boyd, Jemima Dunne, Sarah MacLeod, Cathy Meeus, Steve Setford
Senior US Editor Megan Douglass
Senior Production Controller Meskerem Berhane
Managing Editor Angeles Gavira Guerrero
Associate Publishing Director Liz Wheeler
Publishing Director Jonathan Metcalf

Senior Art Editors Sharon Spencer, Duncan Turner
Illustrator Phil Gamble
Senior Jacket Designer Akiko Kato
Jacket Design Development Manager Sophia MTT
Managing Art Editor Michael Duffy
Art Director Karen Self
Design Director Phil Ormerod

DK DELHI

Senior Editor Dharini Ganesh
Project Editor Priyanjali Narain
Senior Picture Researchers Deepak Negi, Surya Sankash Sarangi
Senior Jacket Designer Suhita Dharamjit
Senior Jacket Coordinator Priyanka Sharma
Picture Research Manager Taiyaba Khatoon
Senior Managing Editor Rohan Sinha
Editorial Head Glenda R Fernandes

Senior Art Editor Mahua Mandal
Project Art Editor Anjali Sachar
Art Editors Nobina Chakravorty, Debjyoti Mukherjee
DTP Designer Vikram Singh
Production Editor/Senior DTP Designer Vishal Bhatia
Managing Art Editor Sudakshina Basu
Preproduction Manager Balwant Singh
Production Manager Pankaj Sharma
Design Head Malavika Talukder

First American Edition, 2022
Published in the United States by DK Publishing
1745 Broadway, 20th Floor, New York, NY 10019

22 23 24 25 26 10 9 8 7 6 5 4 3 2 1
001-326803-Oct/2022

A catalog record for this book is available from the Library of Congress.
ISBN: 978-0-7440-6007-2

Printed in the United Arab Emirates

For the curious
www.dk.com

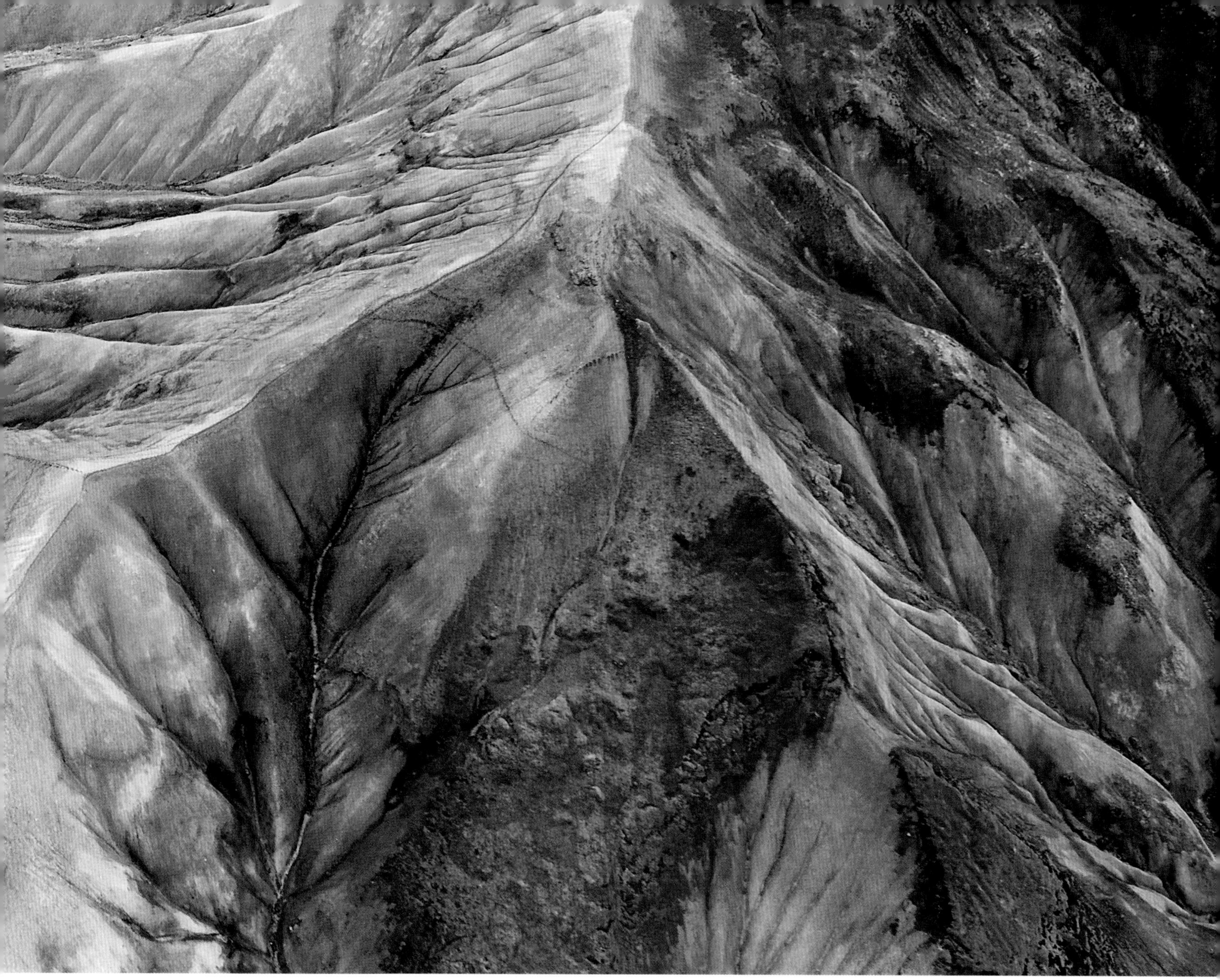

Contributors

Philip Eales studied physics and remote sensing at University College, London. As well as writing about Earth and space science, he runs a computer graphics company specializing in the visualization of astronomical and geographical data and phenomena.

Gregory Funston is a Canadian paleontologist who studied at the University of Alberta and is now a postdoctoral fellow at the University of Edinburgh. His research on dinosaurs and mammals has taken him to fossil sites around the world.

Derek Harvey is a naturalist with a particular interest in evolutionary biology, who studied Zoology at the University of Liverpool. He has taught a generation of biologists, and has led student expeditions to Costa Rica, Madagascar, and Australasia.

Anthea Lacchia is a writer and journalist based in Ireland. She mainly writes about science and nature, and has a PhD on the study of fossil ammonoids—extinct relatives of squid and cuttlefish.

Dorrik Stow is a geologist, oceanographer, and author of more than 300 papers and books. He is Emeritus Professor at Heriot-Watt University in Edinburgh, Distinguished Professor at the China University of Geoscience in Wuhan, and a Leverhulme Emeritus Fellow.

Half-title page Geode with amethyst and calcite crystals
Title page Colored scanning electron micrograph of the fossil radiolarian *Anthocyrtidium ligularia*
Above Rhyolite and geothermal deposits at Laugahraun lava field, Iceland
Contents page Flies, borer beetles, and worker ants preserved in amber from the Dominican Republic

Consultants

David Holmes is a geographer who studied at Leeds University, UK, and has a first degree in physical geography and a master's degree in environmental science. He is a Fellow of the Royal Geographical Society in London and is the author of several well-known geography school textbooks.

Cally Oldershaw is former curator of gemstones at the Natural History Museum and Chair of the Gemmological Association of Great Britain. She is an Earth science educational consultant and lecturer, and the author of numerous books and articles.

Douglas Palmer is an Earth science writer who has written and contributed to numerous books, especially on paleontology, which he originally lectured on at Trinity College Dublin. He also works at Cambridge University's Sedgwick Museum.

Kim Dennis-Bryan is a zoologist who began her career studying fossil fish at London's Natural History Museum, before becoming an Open University lecturer specializing in natural sciences. She has written for and consulted on many science books, including DK's *Animal*, *Ocean*, and *Prehistoric Life*.

SMITHSONIAN

Established in 1846, the Smithsonian Institution—the world's largest museum and research complex—includes 21 museums and galleries and the National Zoological Park. The total number of artifacts, works of art, and specimens in the Smithsonian's collections is estimated at 155.5 million, the bulk of which is contained in the National Museum of Natural History, which holds more than 126 million specimens and objects. The Smithsonian is a renowned research center, dedicated to public education, national service, and scholarship in the arts, sciences, and history.

contents

living planet

directory

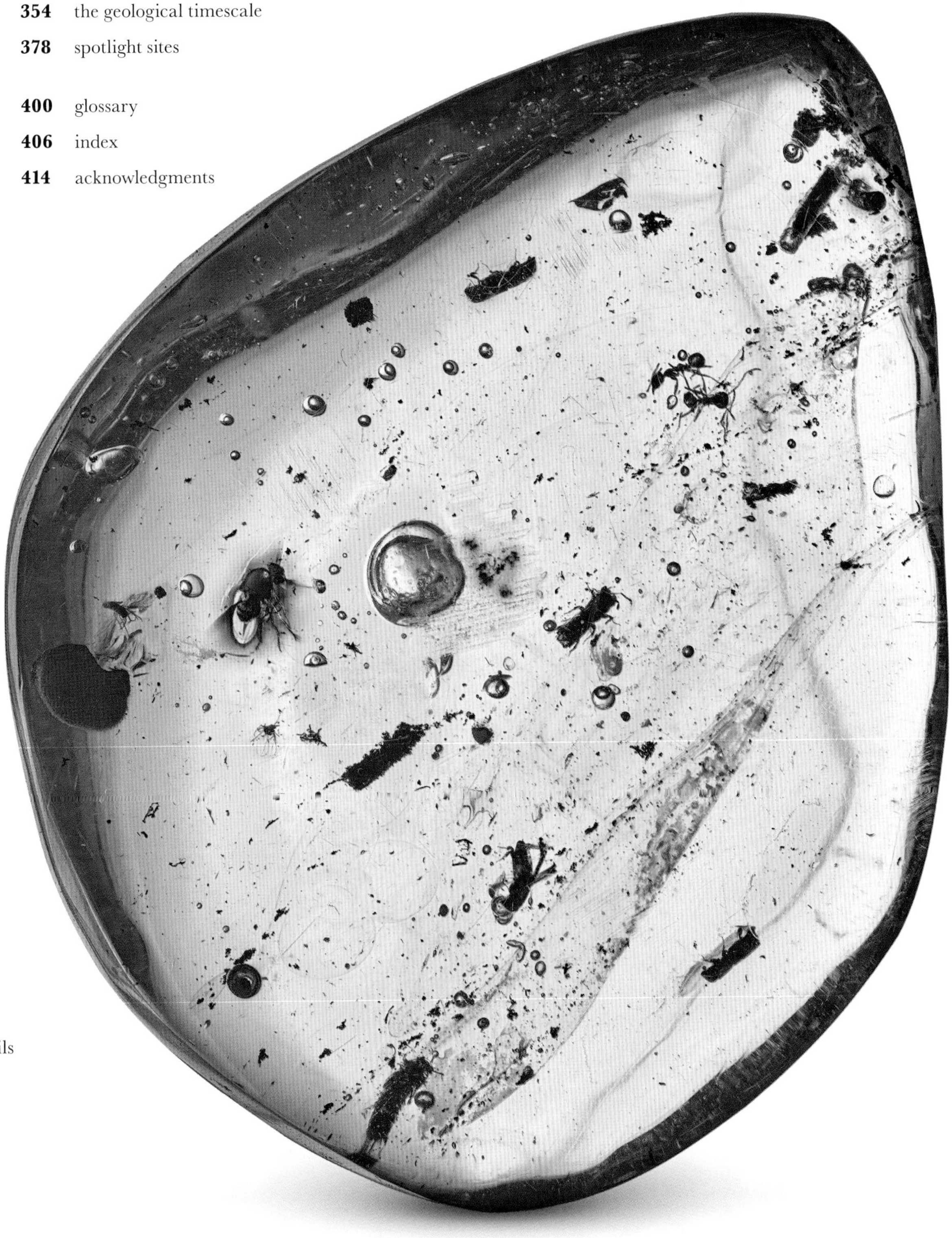

foreword

Initially molten, later frozen, once watery and landless, a planet without a sky or clouds—it's hard to imagine our Earth without sunshine or rain, our beautiful spinning sphere entirely without life. That is not because of the shortcomings in our scientific understanding of its history, it's down to the complexity of factors involved and, critically, our perception of time. Time is tricky to get a measure of.

The fiery ball we now call home formed 4.6 billion years ago. Life appeared at least 3.77, maybe even 4.41, billion years ago. All life descends from the "Last Universal Common Ancestor" (LUCA), an organism from which every living species alive today has evolved—yes, every plant, animal, and fungal organism is related to the one single-celled life form that would not have looked too different from a small, modern-day bacterium. We have no fossils of LUCA, but we think that it appeared in the hot waters of deep-sea ocean vents 4.28 billion years ago. It is a big idea, complete with very big numbers! So, how do we begin to gather what we know about Earth's complex past and use it to appreciate how the world exists now?

This remarkable and beautiful book offers us that chance. By patiently breaking the big ideas into smaller features about the fundamentals of physics, chemistry, and biology, we can figure out the processes that once formed our world and continue to shape it today. From crystals to tornadoes and fossils to volcanoes, each page offers exciting insights into what makes Earth work. It is the most exciting story ever told, because without its myriad chapters and all its twists and turns, we simply would not be here today ... and it's been quite a journey.

CHRIS PACKHAM
NATURALIST, BROADCASTER, WRITER,
PHOTOGRAPHER, AND CONSERVATIONIST

SANDSTONE RIDGES ON NORWAY'S VARANGER PENINSULA

planet Earth

Earth came into existence about 4.6 billion years ago as a hot, rocky body orbiting the newly formed sun. Its position in the solar system makes it unique among the planets in having liquid water on its surface. The continued outward transfer of heat from its interior has also kept it geologically active, driving constant motion and recycling of its outer layers.

interstellar origins

With three exceptions, every element on Earth was forged long ago in the furnace of a star. The exceptions—hydrogen, helium, and lithium—were created when the universe was born. The heat and pressure in a star's core are thought to be high enough to fuse lighter elements to create heavier ones, up to iron with an atomic weight of 56. Elements heavier than iron, such as lead, are created by the even more powerful forces unleashed when a dying star explodes as a supernova. The blast scatters all the elements created by the star into interstellar space, seeding the next generation of stars.

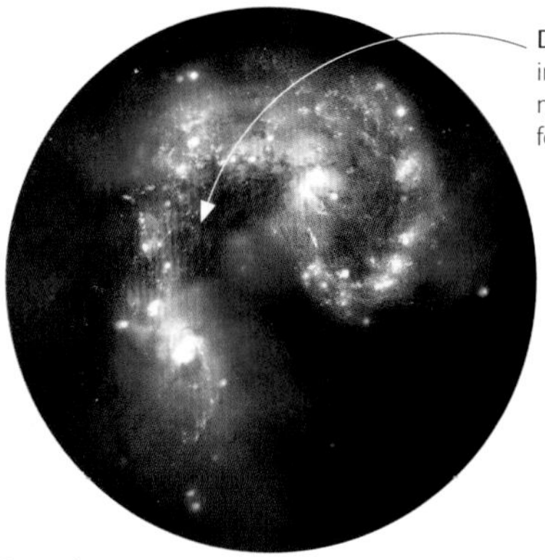

Star trigger
When galaxies collide, new waves of star formation are triggered. The Antennae galaxies have been colliding for about 100 million years, throwing clouds of gas and dust together.

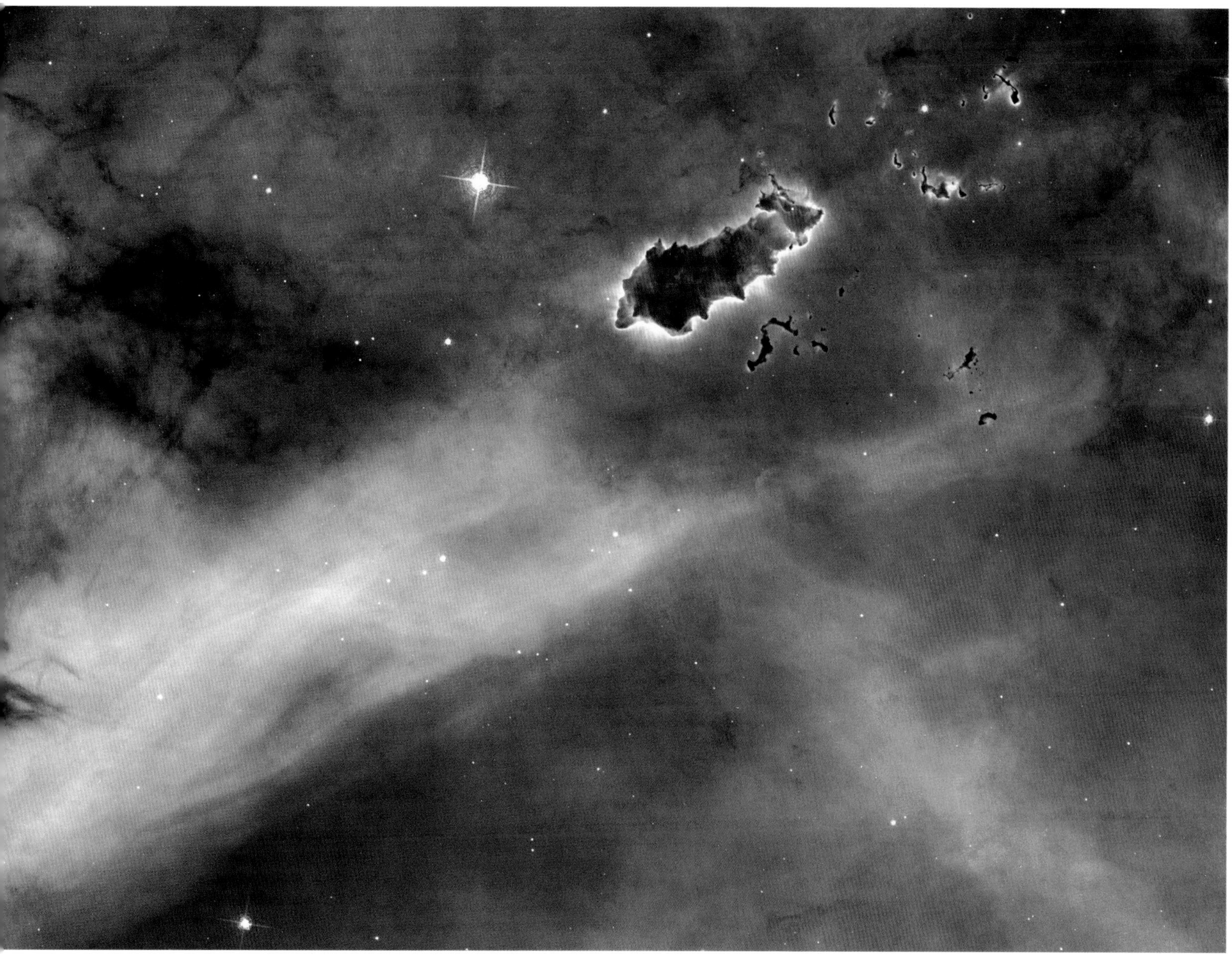

THE MILKY WAY

Our home galaxy, the Milky Way, is a barred spiral galaxy containing 100–400 billion stars. Its two major spiral arms extend from the ends of a central bulge densely populated with old stars. As the galaxy rotates, the interstellar gas and dust become compressed, producing areas of star formation. Our solar system lies in a minor arm called the Orion Spur, a little over halfway between the galaxy's core and its outer arm.

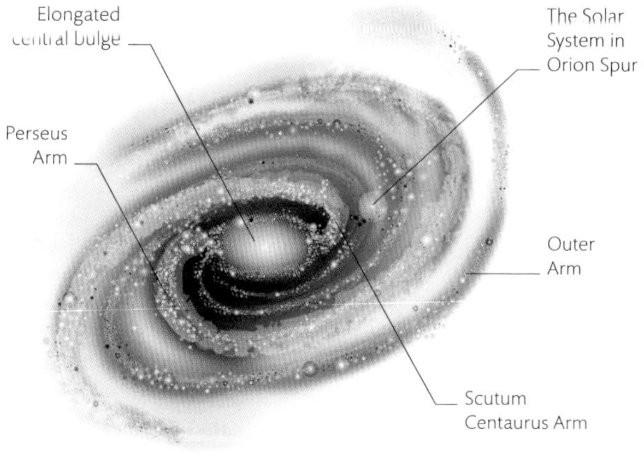

STRUCTURE OF THE MILKY WAY

Molecular clouds

The material expelled from stars accumulates as clouds of gas and dust. Consisting mostly of hydrogen molecules, these are known as molecular clouds. The Keyhole Nebula (above, left) is an example of a molecular cloud. Molecular clouds can contract through their own gravity or an external trigger to form dense globules, such as "the Caterpillar" (above, right), that give birth to new stars.

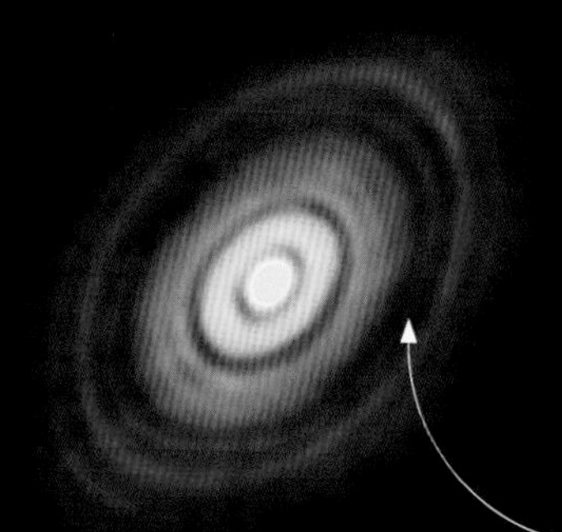

Protoplanetary disk
A disk of dust, called a protoplanetary disk, can be seen swirling around the young star HL Tauri in this image captured by a radio telescope. HL Tauri is 450 light-years from Earth in the constellation Taurus. Bright and dark circles within the disk show where dust is coalescing to form planets.

The sun
Accounting for 99 percent of the mass of the solar system, the sun is a yellow dwarf star, about halfway through its life. It has a diameter of about 865,000 miles (1.4 million km). The sun is the most important source of light and heat for life on Earth. Its energy comes from nuclear fusion in its core, where the temperature reaches 27 million°F (15 million°C).

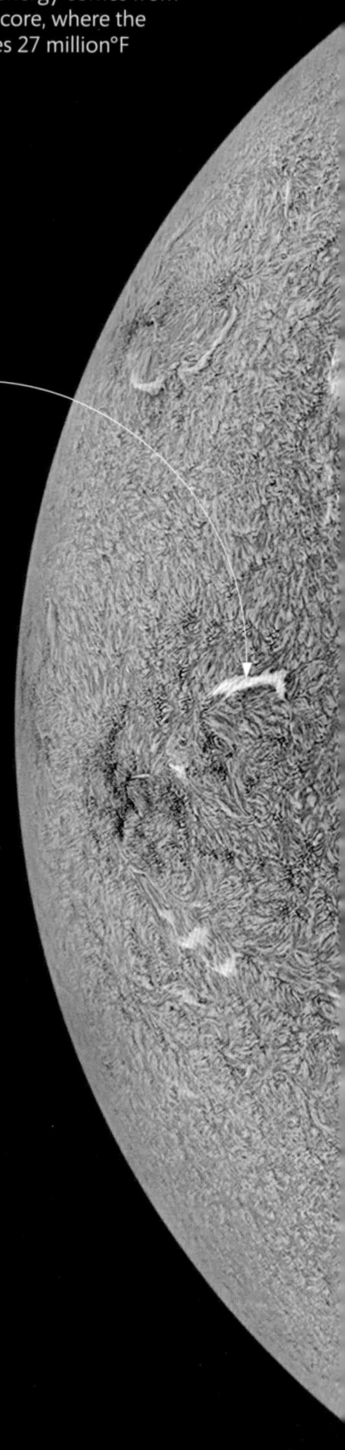

the formation of the solar system

About 4.6 billion years ago, a giant cloud of interstellar gas and dust started to collapse in on itself, perhaps triggered by the shock waves from a star exploding nearby. As the cloud collapsed, it started to spin and heat up, forming a flat, swirling disk of gas and dust. Once the heat and pressure at the core of the nebula were high enough, hydrogen atoms started to fuse into helium atoms, releasing a huge amount of energy. A new star, our sun, was born at the heart of the rotating disk. Gravity drew the remaining materials in the disk together, creating asteroids, comets, planets, and moons.

HOW THE PLANETS FORMED

The material in the disk around the young sun included dust, gas, and in the cold outer reaches, fragments of ice. This material collided and sometimes stuck together in a process called accretion. Fragments of dust clumped together to form pebbles, pebbles came together to form rocks, and rocks grew into planetary building blocks called planetesimals. Some of these objects grew large enough for gravity to shape them into spheres, and they became planets or moons.

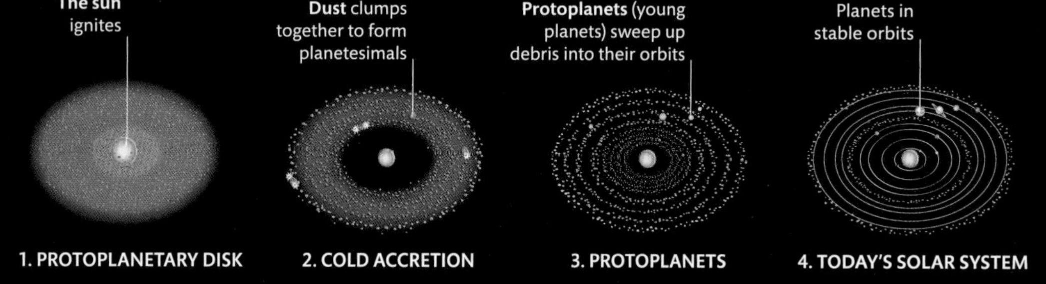

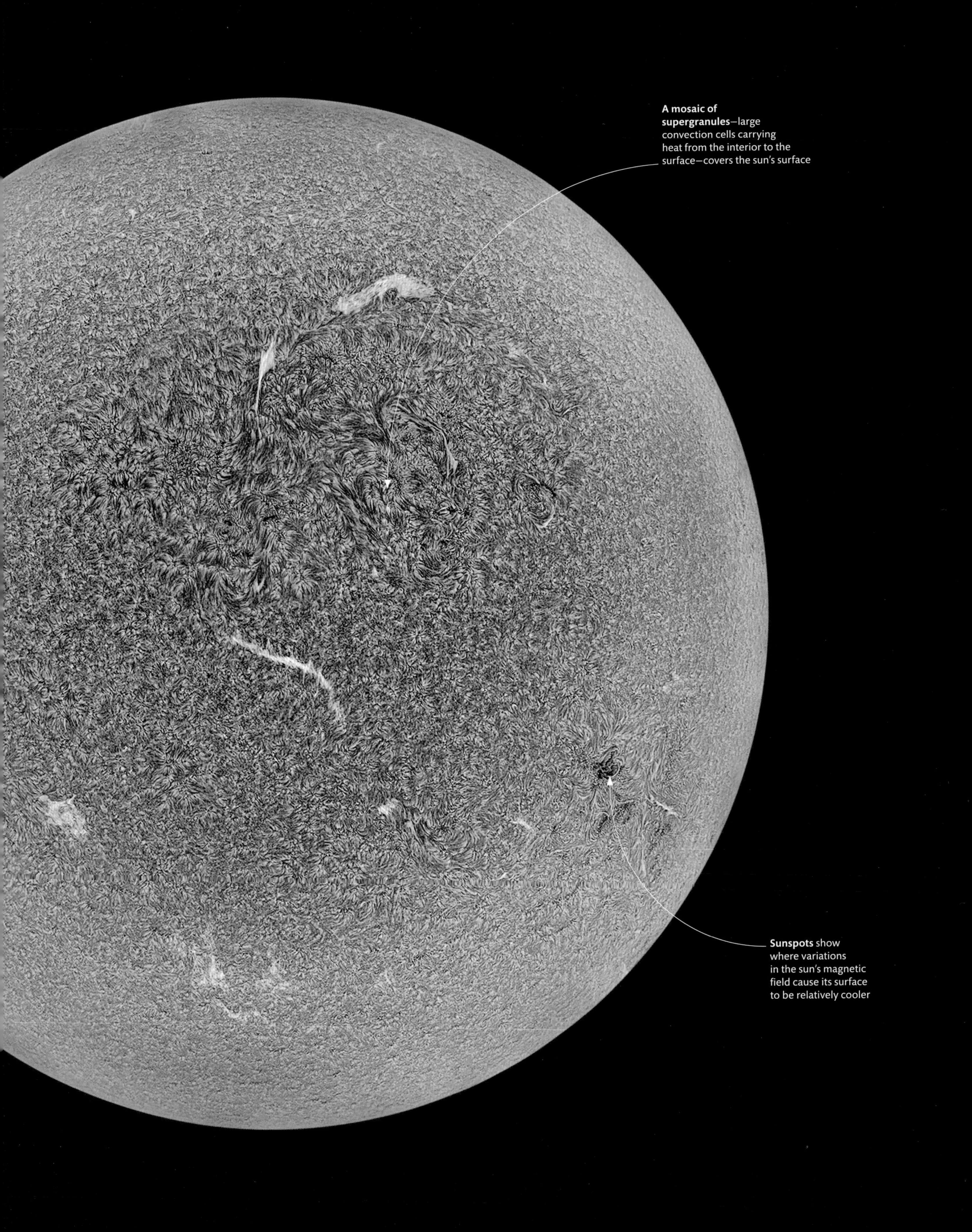
A mosaic of supergranules—large convection cells carrying heat from the interior to the surface—covers the sun's surface
Sunspots show where variations in the sun's magnetic field cause its surface to be relatively cooler

Chondrite micrometeorite is less than 2 mm (1/16 in) across

Micrometeorite
Shown here is a scanning electron micrograph of a tiny meteorite that was found on a beach on the east coast of the US. This specimen belongs to a class of meteorites called chondrites that are made up of the same primitive material that originally formed the rocky planets.

meteorites

Each year, about 77,000 tonnes (70,000 metric tons) of extraterrestrial material enter Earth's atmosphere. While most of it consists of microscopic dust particles, a significant proportion of larger objects fall to the surface to become meteorites. Today, most meteorites come from the asteroid belt between Mars and Jupiter, where collisions between the asteroids can send fragments of rocks toward Earth. Among these are meteorites containing carbon and even complex organic molecules, which may have played a role in the origin of life. Like all planets, Earth was showered with objects when the solar system was young. The moon bears the scars of this bombardment, but the continual recycling of Earth's crust through plate tectonics has erased all but the most recent impact craters from our planet's surface.

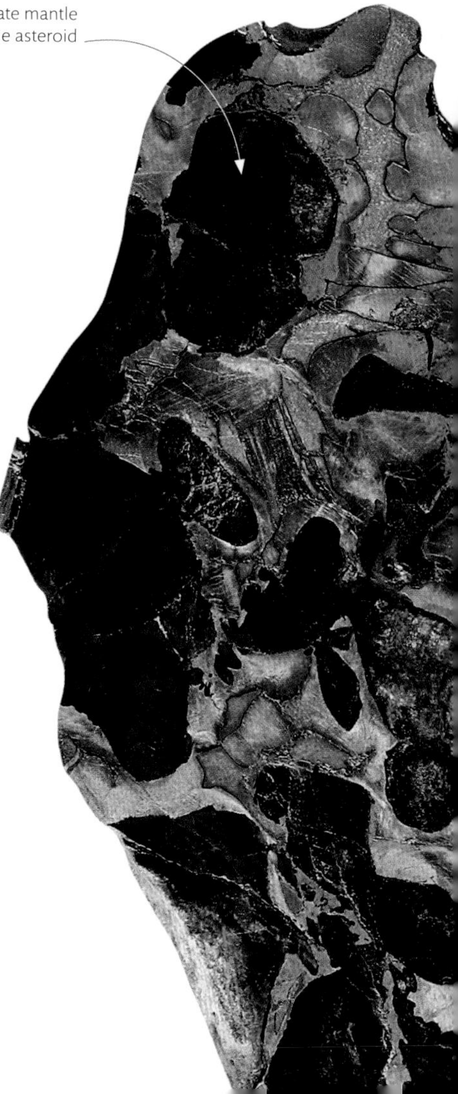

Fragments of rock suggest this fragment formed at the boundary between the metallic core and the silicate mantle of the asteroid

IMPACT CRATERS

If an asteroid or meteoroid is large enough and traveling fast enough, it leaves a circular crater when it hits the ground and explodes. The kinetic energy of the speeding rock is instantly converted into heat that may be sufficient to vaporize the meteorite and some of the surface rock. A part of the meteorite may survive as melted rock or broken fragments (breccia), and some of the debris thrown out settles into a halo of ejecta. The rock layers beneath the crater may be fractured, uplifted, or overturned.

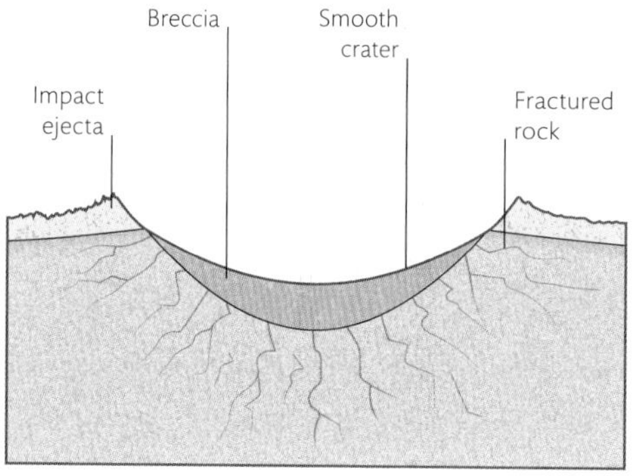

Olivine (magnesium iron silicate) crystals embedded in iron-nickel metal

Core remnant

These cut and polished sections of stony-iron meteorites found in Seymchan, Russia, show crystals of iron and nickel that were formed in the molten core of a planetesimal or a differentiated asteroid—one that has segmented into a core, mantle, and crust. Some meteorites are rocky, with a few showing signs of melting in their parent bodies. Others have a chemical composition similar to the sun's and are thought to be representative of the solar nebula from which the planets formed.

Crystals of nickel and iron formed in the hot core of an asteroid

The distinctive texture of iron-nickel alloy crystals, which appears when a cross section of an iron meteorite is cut, polished, and etched with weak acid, is called the Widmanstätten pattern

Erosion and resurfacing have obliterated most of Earth's many impact craters, leaving only the youngest. Old, stable rocks and lack of vegetation make central Australia a good place to find them. One of the few remnants of Earth's bombardment by comets and asteroids, the Tnorala impact feature, also known as Gosse Bluff, is located about 125 miles (200 km) west of Alice Springs in Australia's Northern Territory.

spotlight Tnorala

Tnorala was created in the early Cretaceous period, 142 million years ago, when an asteroid or comet 2,000 ft (600 m) across, hurtling at 25 miles per second (40 km/s), slammed into the plains of central Australia, leaving behind a 14-mile- (22-km-) wide puncture in Earth's crust. Today, erosion has removed almost all traces of the outer rim of this impact crater, but the central uplift remains in the form of a circular ring of sandstone hills about 3 miles (5 km) across. Although eroded flat, the crater's rim is an outer circle of darker rock seen in satellite images. The land surface of Tnorala now lies about 1.25 miles (2 km) lower than it was at the time of impact. The collision left evidence including deformed and fragmented rocks; impacted and melted quartz grains; and, most compellingly, shatter cones in the rock, left by shock waves radiating from the impact.

Tnorala is a sacred site for the Western Aranda Aboriginal people, whose folklore also attributes its creation to an object falling from the sky. In their story, a baby in a basket falls to Earth, dropped by a group of celestial women dancing along the Milky Way. The rocks were pushed up when the basket landed.

Sandstone ring
Composed of a circle of sandstone hills, Tnorala rises 600 ft (180 m) above the surrounding plain. It is one of Australia's most studied impact craters. Its impact origin was first proposed in the 1960s, based largely on the abundance of shatter cones found in the area.

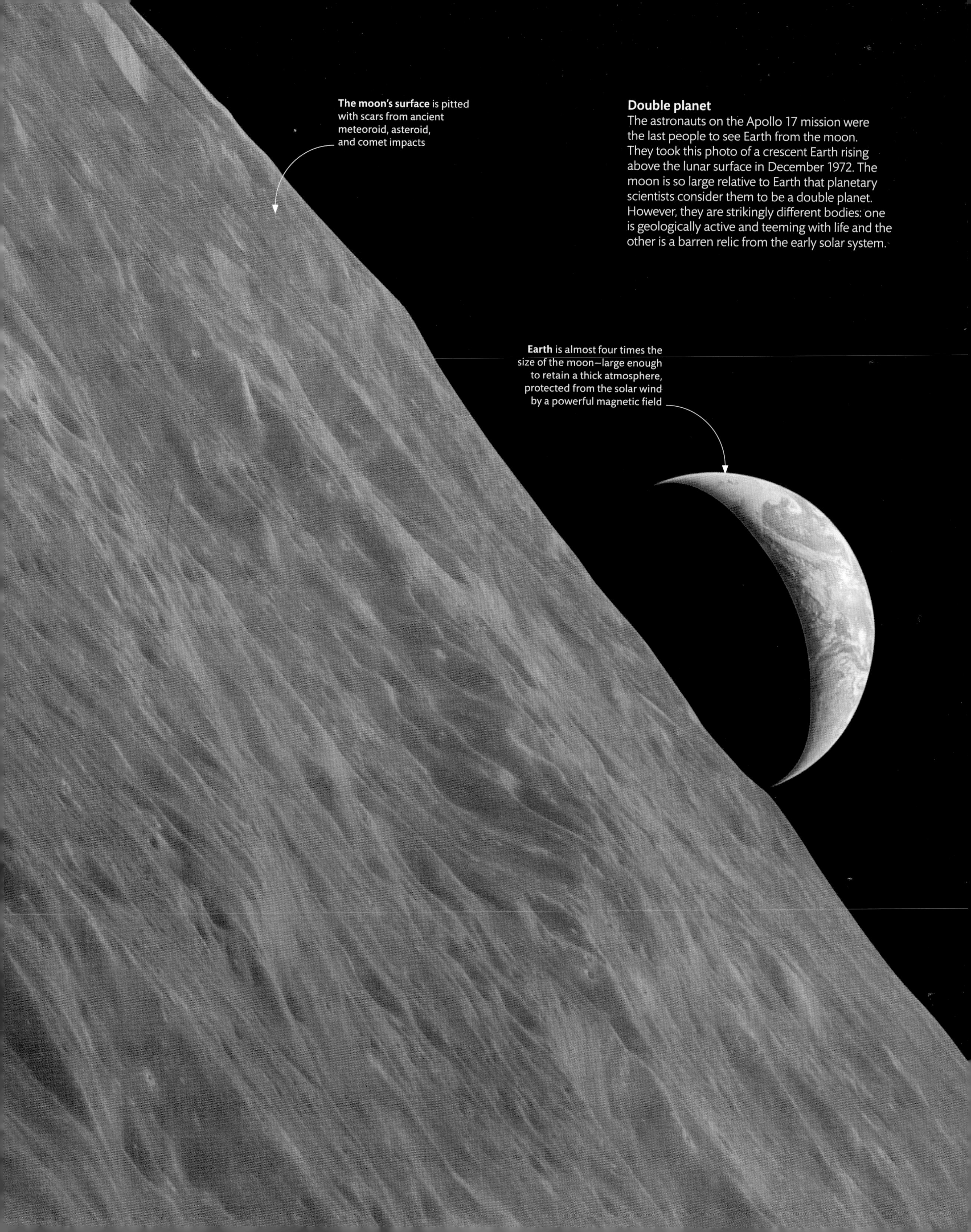

Double planet
The astronauts on the Apollo 17 mission were the last people to see Earth from the moon. They took this photo of a crescent Earth rising above the lunar surface in December 1972. The moon is so large relative to Earth that planetary scientists consider them to be a double planet. However, they are strikingly different bodies: one is geologically active and teeming with life and the other is a barren relic from the early solar system.

Moon rock
Rock samples from the moon collected by astronauts in NASA's Apollo program between 1969 and 1972 reveal that the moon's surface is similar in composition to Earth's mantle. Up to 4.4 billion years old, these specimens help date Earth's early history and the moon's origins.

Vesicles (holes) in the rock, created by gas bubbles in molten lava

This 3.5-billion-year-old rock sample brought back by astronauts aboard the Apollo 15 mission is similar to the basalt rocks found around Hawaii

the moon

It is thought that fewer than 100 million years after Earth formed, it was struck by a smaller planet about the size of Mars. This cataclysmic impact melted the smaller planet and a significant part of Earth's crust, and splashed the hot debris into space, where it coalesced and solidified to form Earth's only natural satellite. The young moon was bombarded by comets and asteroids, which have left craters all over its surface. Some of the largest impact basins have been flooded by lava, creating the dark "seas" seen from Earth. The far side of the moon is more mountainous. The moon is the solar system's largest satellite relative to the size of its parent planet, and has had a correspondingly large influence on the evolution of Earth and its life forms. As well as driving the ocean tides, the moon has stabilized Earth's rotation rate, axial tilt, and climate. It may also have enhanced Earth's protective magnetic field. It looms large in our night sky and was an early target for space exploration.

THE MOON'S ORBIT

The gravitational tug of the moon is the main cause of tides in Earth's oceans, but the moon also pulls at our planet's rocks, and vice-versa. This exchange of gravitational energy has slowed Earth's rotational period from about 5 hours when the moon was young to 24 hours today. It has also increased the diameter of the moon's orbit from 80,000 miles (128,000 km) to 238,000 miles (383,000 km) and tidally locked the moon—which means it has the same rotation period as Earth—so only one side, called the near side, is visible from Earth.

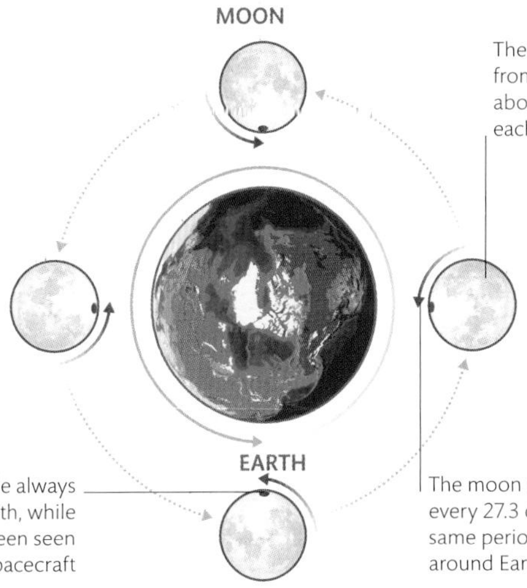

The moon is receding from Earth at a rate of about 1½ in (3.8 cm) each year

The moon's near side always faces toward Earth, while the far side has been seen only from spacecraft

The moon rotates once every 27.3 days—the same period as its orbit around Earth

our place in the solar system

Once the planets acquired enough mass, their own gravity started to compress them, transforming gravitational energy into heat energy. This, along with the internal heat from radioactive decay, was enough to melt their interiors, allowing heavier elements to sink toward the core and lighter elements to float toward the surface in a process known as planetary differentiation. This left Earth with a solid crust and a large, partly liquid iron core capable of generating a magnetic field powerful enough to prevent the stream of charged particles known as the solar wind from stripping away its atmosphere. Earth is uniquely structured and positioned to allow life to thrive.

Hostile world
The surface of early Earth would have been a hellish place, caught between protoplanetary debris raining down from above and heat rising up from the layers below.

SOLAR SYSTEM ORBITS

The solar system is a family of bodies held in position by the gravitational pull of the sun. The bodies are grouped into rocky inner planets (Mercury, Venus, Earth, and Mars), a belt of tiny asteroids, the gas and ice giants (Jupiter, Saturn, Neptune, and Uranus), the Kuiper Belt of distant icy bodies that includes the dwarf planet Pluto, and the more distant Oort Cloud of comets. The astronomical unit (AU) is used to compare distances in the solar system: 1 AU is the average distance between Earth and the sun. Earth is the only planet within the solar system's habitable zone (shown in green below)—it is neither too hot nor too cold, so liquid water is stable on its surface.

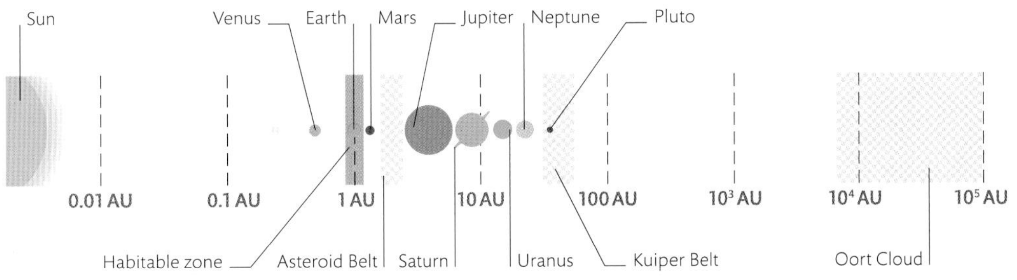

Ice-filled crater
Mars has enough water to flood its entire surface. However, liquid water cannot exist on Mars due to the low atmospheric pressure and the extremely low temperature. Mars lies beyond the outer edge of the solar system's habitable zone. Its water is trapped as ice, mostly below the surface, but sometimes also on the floors of craters, such as Korolev Crater, near the poles.

After shedding its leaves in winter, a beech tree bears leaf buds shielded by overlapping scales, so that the leaves are ready to develop in spring

In beeches, pollen from male flowers is dispersed by the wind when conditions for spreading are ideal—that is, when the weather is cool and dry

Long, woody branches spread the leaves across a wide area

Seasonal adaptation
Life has adapted to the seasonal variation of sunlight. Some trees have developed tough evergreen needles that can survive the freezing cold of the winter. Other plants, such as the beech tree shown here, have evolved broad, thin leaves that maximize exposure to sunlight in summer. The trees grow vigorously in the summer months and become dormant in the winter.

THE PLANET'S TILT

The planetary collision that created the moon (see pp.20–21) left Earth's axis of rotation tilted to the plane of the planet's orbit around the sun. This would have made Earth unstable, but the gravitational pull of the moon steadied Earth's motion.

Our planet's axial tilt has changed only by a few degrees in a million years, allowing a relatively stable climate to develop, with a predictable annual variation of sunlight received in the northern and southern hemispheres. This variation gives us the seasons.

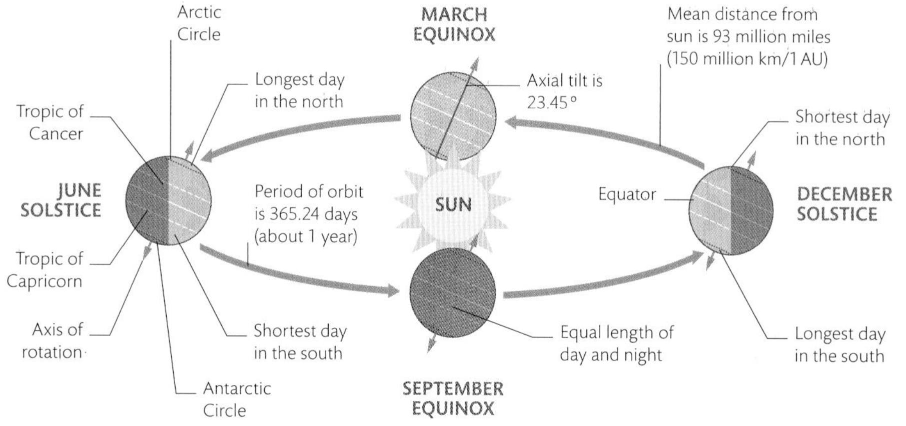

days and seasons

Earth rotates on its axis once every 23 hours and 56 minutes. It also revolves in its orbit, and so it takes a little longer for the sun to return to the same point in the sky, giving us the length of one day, which we split into 24 hours. At the equator, we experience 12 hours of daylight and 12 hours as night, but as we move north or south of the equator, the tilt of Earth's rotational axis causes the days to be longer in summer and the nights to be longer in winter. This daily and annual variation in energy received from the sun is what drives Earth's seasons.

Midnight Sun

As we move north of the Arctic Circle or south of the Antarctic Circle (66.5° North and South), there is at least one day in summer when the sun does not set, and one day in winter when it does not rise. The multiple-exposure image here shows the sun above the Arctic Circle dipping toward the horizon close to midnight. It does not quite set and instead, rises immediately after midnight. The North and South poles experience 6 months of daylight followed by 6 months without seeing the sun.

Foucault's pendulum in Paris
This replica of Foucault's original apparatus is installed in the Paris Panthéon. Using a system originally devised by Vincenzo Viviani, Foucault used a wire 220 ft (67 m) in length to suspend a 62 lb (28 kg) lead weight from the dome of the building.

Spiral of weather
This satellite view of a low-pressure system over Australia shows the influence of the Coriolis effect. Resulting from Earth's rotation, it causes storms and cyclones to spin: counterclockwise in the northern hemisphere and clockwise (as here) in the southern hemisphere.

history of Earth science

Earth's rotation

Although astronomers in India in the 5th century BCE, and in the Islamic world in the 10th century, believed that Earth rotates once a day, the accepted view in 16th-century Europe was that the heavens revolved around a fixed Earth. In 1543 Nicolaus Copernicus proposed a rotating Earth as part of his heliocentric model, but a century later critics were still rejecting his theory for lack of physical evidence of Earth's spin.

The search for proof of Earth's rotation centered on experiments to test whether a moving object deflects slightly sideways—a phenomenon now known as the Coriolis effect. It was not until the late 18th century that a measurable effect was confirmed, when weights were observed to deflect by a few centimeters when dropped from towers more than 500 ft (150 m) tall in height.

Much clearer evidence for Earth's rotation was provided by French physicist Jean Bernard Leon Foucault in 1851. He built a long pendulum with a special mount that allowed it to rotate freely. If Earth is not rotating, the pendulum would be expected to swing in a fixed plane, conforming to the principle of conservation of angular momentum. However, if Earth is rotating, angular momentum would be conserved by the plane of the swing rotating (precessing) relative to Earth's surface.

For a pendulum at the north or south pole, the swing plane would complete a full 360-degree rotation in 24 hours. At lower latitudes the effect is less, so the precession is slower, falling to zero at the equator. At the latitude of Paris, where Foucault conducted his experiment, his pendulum took almost 32 hours to complete one rotation, precessing clockwise at about 11 degrees per hour. Demonstrations of his experiment in different locations across Europe and America attracted huge public interest.

Although small on a human scale, the effects of rotation are highly significant on a planetary scale, giving rise to the circulation patterns of Earth's atmosphere and oceans.

> "You are invited to come to see the Earth turn, tomorrow, from three to five, at Meridian Hall of the Paris Observatory."
>
> LÉON FOUCAULT, INVITATION CARD, FEBRUARY 3, 1851

RADIOMETRIC DATING

Certain Earth materials, including the ancient zircon below, can be aged by radiometric dating. This uses the slow but clocklike decay of radioactive isotopes, such as uranium-235 to lead, to provide an absolute scale against which we can measure geological time. The ratio of uranium-235 to lead in a rock therefore tells us its age. The isotopes of other elements can tell us about conditions when some rocks were formed. For instance, the ratio of oxygen-18 to oxygen-16 tells us about the presence of liquid water, and the ratio of carbon-12 to carbon-13 about the presence of life.

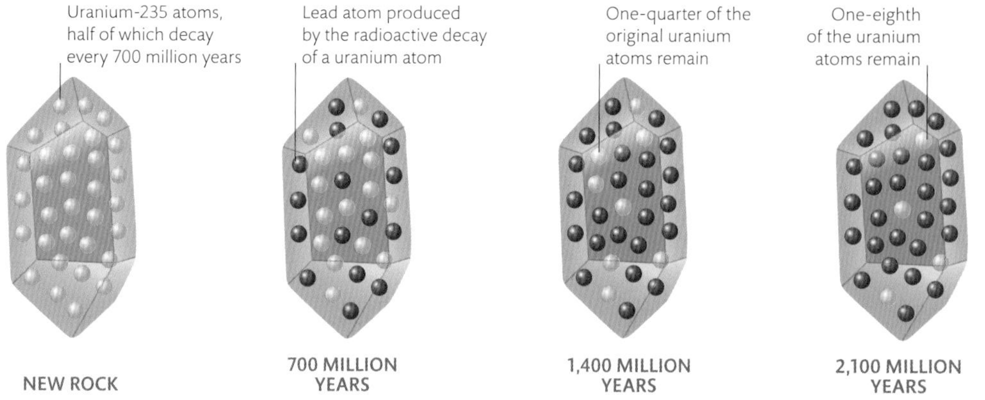

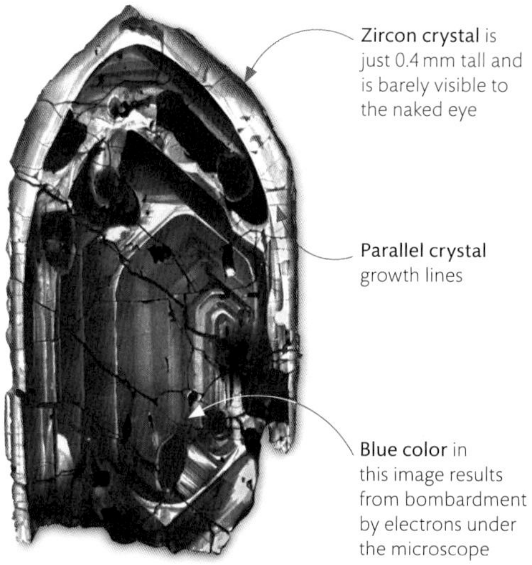

Extraterrestrial water
Some of Earth's water will have arrived on comets that frequently impacted the young planet. Comets are composed of about half ice and half dust. They lose some of their ice, which is about 80 percent water, through spectacular geyser-like eruptions as they warm up on entering the inner solar system. The Rosetta spacecraft captured this in close-up as it approached Comet 67P/Churyumov-Gerasimenko in 2014.

Traces of the first ocean
The oldest materials yet found on Earth are crystals of the mineral zircon, formed 4.4 billion years ago. They contain oxygen isotopes that indicate the presence of liquid water at that time. Zircon is as tough as diamond and a long-lived geological record keeper.

origin of the oceans

As Earth settled into layers in the process of differentiation (see p.22), volatile materials were expelled from its interior by intense volcanic activity, producing an atmosphere of nitrogen, carbon dioxide, and water vapor. The surface quickly cooled to form a solid crust, allowing water vapor to condense and pool, forming the first ocean. There is evidence from ancient crystals of zircon found in the Jack Hills of Western Australia that surface water was present at least 4.4 billion years ago, only 160 million years after the Earth formed. Some meteorites contain 15–20 percent water, and early Earth is thought to have had a similar composition, providing enough water for the first ocean.

Old at heart
The oldest rocks are found in the stable centers of some continents—remnants of Earth's early cratons. This metamorphic rock from an island in the Acasta River in Canada is about 4 billion years old.

During this rock's long history, bands were produced by metamorphosis under high heat and pressure

the continents form

During the process of differentiation (see p.22) Earth's interior shaped itself into distinct layers. A metallic core formed beneath a silicate-rich mantle and a lightweight crust. This primordial crust was uniform, but it, too, began slowly differentiating, into two types—oceanic and continental crust. The crust, fused to the uppermost layer of the mantle, floated on hot, mobile material beneath. The heat from below caused the outer layers to move and fracture into tectonic plates. Some of these plates were dragged down into the mantle, while new crust formed where the plates were pushed apart. This repeated recycling of the crust caused the concentration of lighter elements in some locations, creating the thick, buoyant continental crust that now covers about 40 percent of Earth's surface.

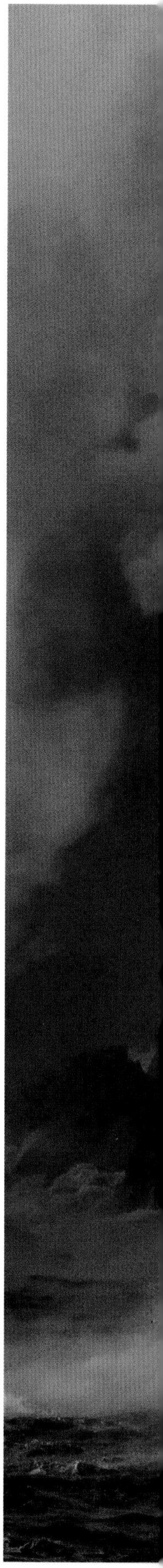

Waterfall of lava
Lava pours into the Pacific Ocean, vaporizing the water at Kamokuna lava delta on Hawaii's Big Island. The island is gradually growing as hot, runny lava flows from the Kilauea volcano into the sea, adding material to the island's broad flanks. This process, continuing today in Hawaii, was also part of the process of early continent building.

HOW THE FIRST CONTINENTS FORMED

About 4 billion years ago, Earth's plates began to move, and some of the primordial crust was dragged down into the mantle. The crust melted and released water, causing the mantle to melt, producing magma rich in light elements that rose up to form volcanic islands. Plate motion pushed the islands together to form cratons—masses of light rock. Weathering and sedimentation further concentrated the lighter materials on the cratons. When plate motion brought cratons together, they formed larger masses of continental crust.

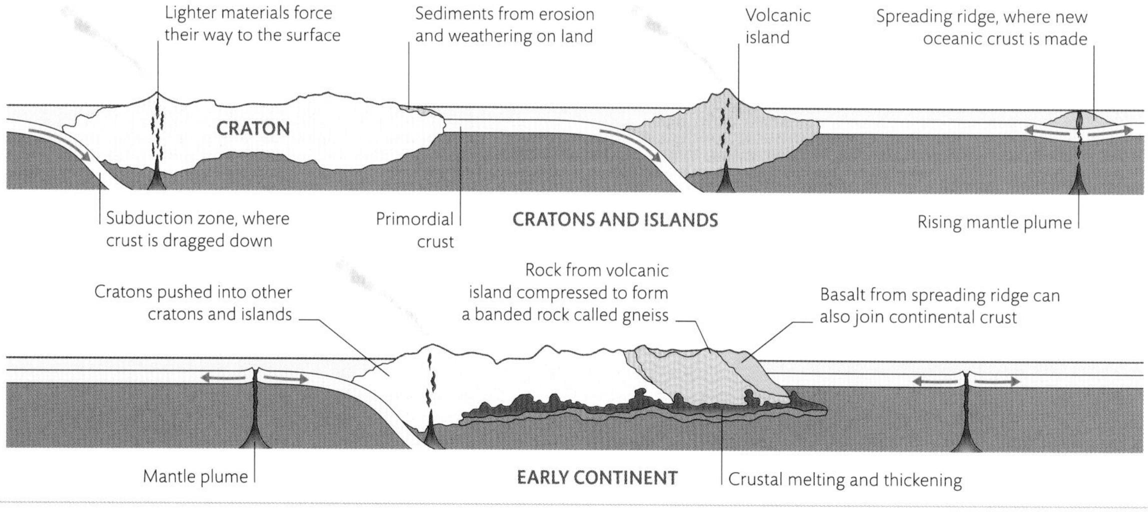

the ages of **Earth**

If Earth's history were to be compressed into one day, the dinosaurs would appear for about an hour at 10:40 p.m., and our human ancestors would not arrive until two minutes to midnight. Instead of hours, minutes, and seconds, geological time is split into eons, eras, and periods. These intervals do not mark equal slices of time, but what can be interpreted in the rock record. The Hadean eon (derived from the Greek word for "beneath the Earth") is the first; unfortunately, there are almost no surviving rocks from this time. Archean (ancient) rocks contain sparse evidence of bacteria, but as rocks get younger, through the Proterozoic (earlier life) to the Phanerozoic eon (visible life), they begin to contain more and more evidence of the development of continents, oceans, their environments, and the evolution of living organisms.

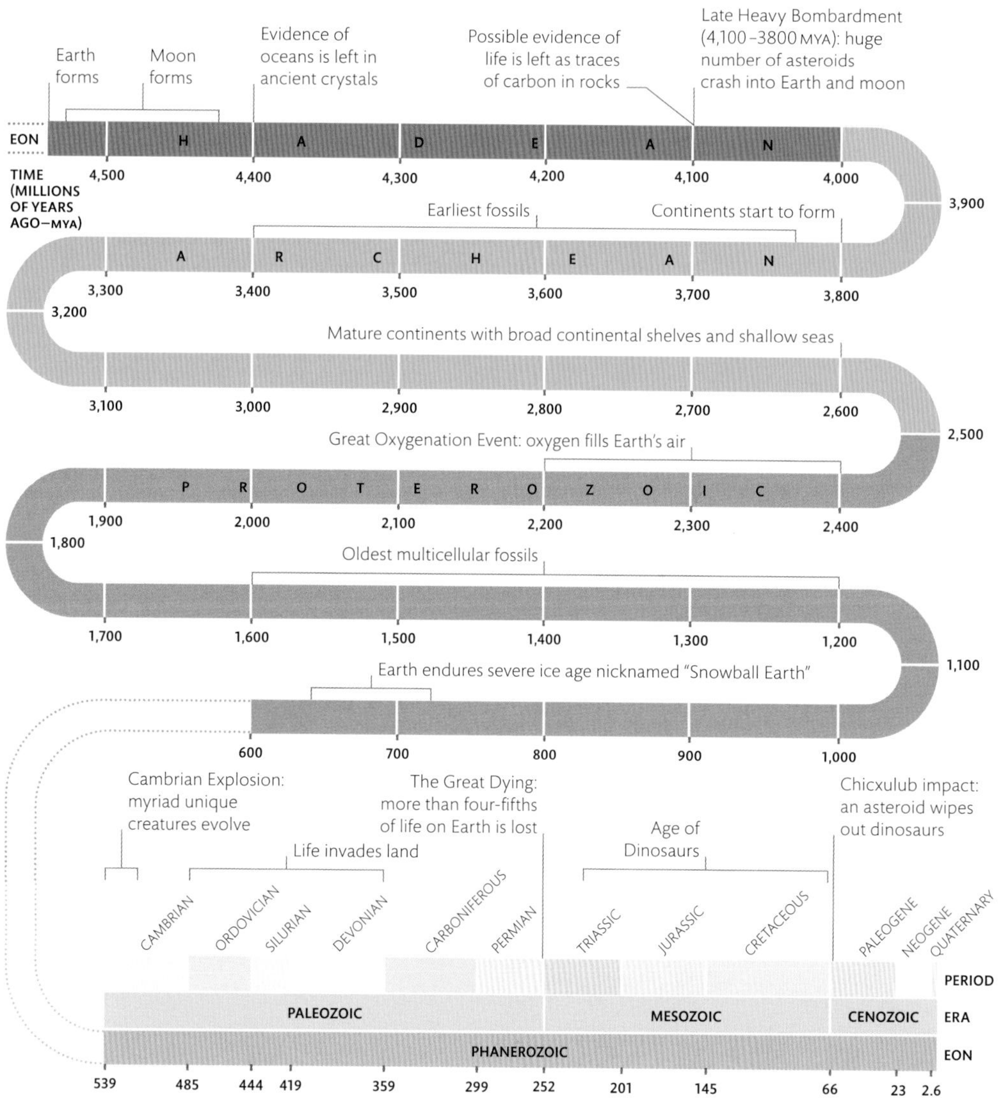

The story of our planet
The geological timescale provides a chronology of major events in Earth's history since it formed around 4.6 billion years ago. In this chart, the most recent eon, the Phanerozoic, is divided into eras, which are further subdivided into periods. The timescale is marked by the appearance and extinction of new life forms.

Rock record
In some places, Earth's rock layers can be read like the pages of a history book. The layers of shale and clay at Bentonite Hills in Utah resulted from the deposition of mud, silt, fine sand, and volcanic ash in swamps and lakes during the Jurassic period. By studying rock layers like these, we can learn what life on Earth was like 145 million years ago.

Earth's materials

Earth's thin outer crust is made of a great diversity of rocks that have developed as a result of billions of years of geological activity. The natural building blocks of rocks—and all the solid bodies of the universe—are minerals. The crust is overlain and partly concealed by an even thinner layer of soil and vegetation, and by bodies of liquid and frozen water.

Marcasite crystals
The mineral marcasite is a polymorph of pyrite, meaning it has the same chemical formula as pyrite but a slightly different crystal structure. Unlike pyrite, marcasite tarnishes and breaks down quickly when exposed to air.

Spear-shaped marcasite crystals have a different shape than pyrite crystals and belong to a different crystal system

Pyrite crystals
Pyrite is found in almost every environment on Earth. Its crystals can be shaped as cubes, octahedrons, or 12-sided pyritohedrons. This cubic crystal is from Guangxi Province in China. Pyrite's name is derived from the Greek word for "fire," and it gives off sparks when struck hard with a hammer.

crystal structure

Minerals are inorganic solids that occur naturally and are made up of chemical elements whose atoms are typically arranged in a systematic internal pattern. When free to grow in an open space, minerals develop regular geometric shapes called crystals, which can often be seen with the naked eye. These are a reflection of a mineral's internal crystal structure, the regular arrangement in three-dimensional units of its atoms or ions—ions are atoms or molecules that have an electric charge. One way of describing the symmetry of a crystal is through axes of symmetry: imaginary lines drawn through a crystal's center from the middle of opposite faces. Crystals are divided into seven groups, or crystal systems, each with a characteristic shape and symmetry. Most crystals do not achieve a perfect crystal shape because they form in restricted spaces in nature.

Cubes are the most symmetrical of all crystal shapes, with three equal axes of symmetry at right angles and four-fold symmetry (it looks the same at four points during a complete rotation)

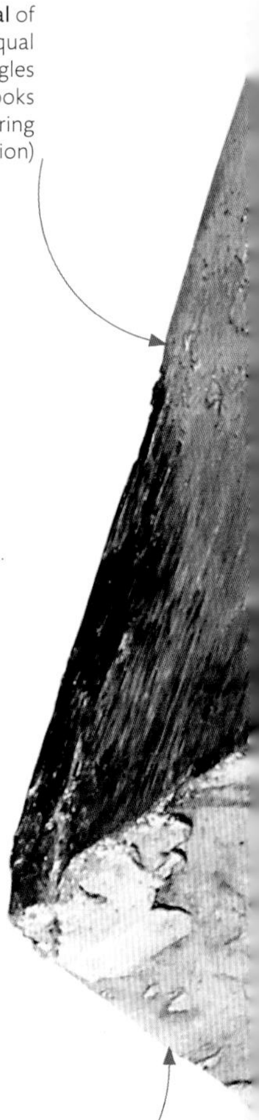

With its yellow color and metallic luster, pyrite is sometimes mistaken for gold, earning it the nickname "fool's gold"

CRYSTAL TWINNING

Two or more crystals of the same mineral sometimes grow symmetrically. These are called crystal twins. There are two main types: contact and penetration twins. In contact twins, there is a distinct boundary between the crystals, which appear as mirror images. In penetration twins, individual crystals appear to grow through each other. The world's most abundant minerals, feldspars, are often twinned. Quartz and spinel form contact twins, while orthoclase feldspar, pyrite, and fluorite form penetration twins. The type of twinning can help identify a mineral.

The contact between twins is sometimes called Carlsbad contact

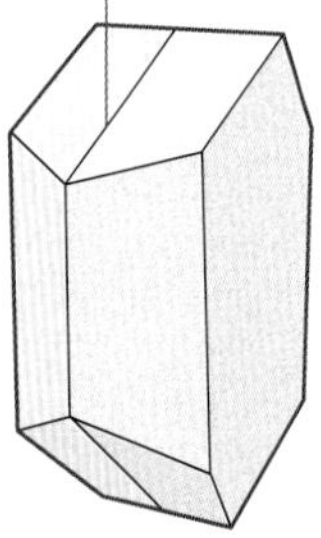

CONTACT TWINS

An irregular surface separates the two crystals

This is a Carlsbad twin, a form of penetration twinning in feldspars

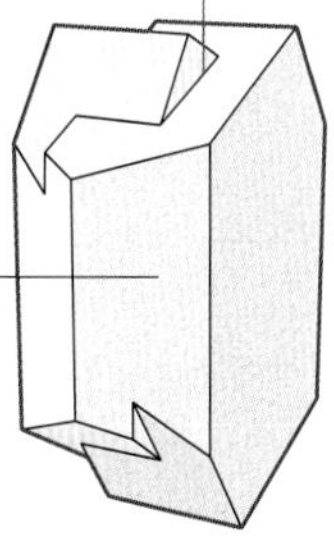

PENETRATION TWINS

Striations, or linear stripes or marks, on pyrite's cubic crystal faces are usually caused by two different crystal forms developing at the same time

Forms and faces

The term "form" describes a set of identical crystal faces that have symmetrical planes. Forms are referred to as "closed" or "open". In closed forms, such as cubes or octahedra, the faces are identical. In contrast, open forms have faces of different shapes and sizes. For example, a specimen that has mainly parallel prism faces (see right) has to terminate in a different type of face (here, a pyramid) to enclose the space. If a crystal consists of more than one set of faces, the habit is named for the faces that predominate (here, prismatic).

PRISMATIC
Quartz

CUBIC
Halite (rock salt)

OCTAHEDRAL
Cuprite

Appearance

Some habits are named for their general appearance, rather than for the crystal's forms and faces. This is the case with aggregates, which are crystals that grow together in a group rather than as separate, individual crystals, generally because they are imperfectly developed. In some aggregates, the crystals are microscopic. There are many different terms used to describe aggregates, which are highly variable in size and habit.

MASSIVE
Dumortierite

FIBROUS
Tremolite

RADIATING
Pyrophyllite

BLADED
Kyanite

NEEDLELIKE
Mesolite

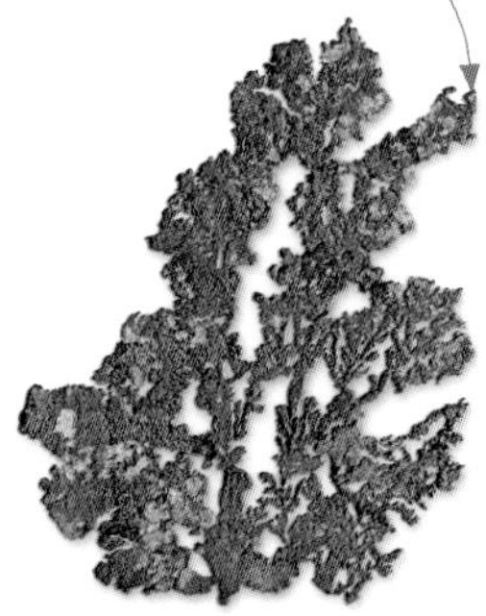

DENDRITIC
Copper

CONCENTRIC
Rhodochrosite

crystal habits

Habit refers to a crystal's external shape, including the forms and faces of individual crystals. The description can refer to a single, well-formed crystal or to a cluster of intergrown crystals, known as an aggregate. In nature, perfectly formed crystals are rare, because their development is limited by the size and shape of the cavities in which they grow, as well as by gravity. It is possible for a mineral to have a variety of habits depending on the conditions in which it forms.

BOTRYOIDAL
Malachite

TABULAR
Torbernite

Aquamarine
The blue variety of beryl, aquamarine forms deep inside Earth and is often associated with granite intrusions (see pp.134–135). Its crystals form six-sided prisms, with its faces running parallel to a central axis. This specimen is from the Jos Plateau in Nigeria.

Silvery white color is typical of freshly produced bismuth; over time it can develop an iridescent sheen

Brilliant, shiny luster gives diamond a sparkly look

ADAMANTINE
Diamond

Surface of mineral is rough and nonreflective

DULL
Hematite

Unreflective surface resembles dried mud, soil, or clay

EARTHY
Bentonite

Microscopic irregularities on the surface give the mineral a greasy look

GREASY
Chrysocolla

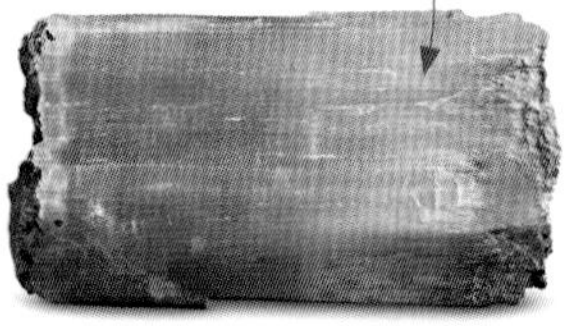

Slight shimmer given off by the mineral's parallel fibers

SILKY
Ulexite

Crystal prisms gleam like glass

VITREOUS
Quartz

reflecting light

A mineral is often described or identified by its luster—that is, its reflective qualities and the extent of its sheen. Minerals can be broadly referred to as metallic, submetallic, or nonmetallic. Those described as having metallic luster are reflective and opaque to light, like polished metal. Submetallic minerals are duller and less reflective, and minerals with nonmetallic luster are usually lighter in color, with many showing some degree of transparency or translucency. There are several types of nonmetallic luster, some of which are shown opposite.

Metallic luster
Bismuth, with its brilliant, highly reflective surfaces that gleam like metal, is described as having metallic luster. As with all similar minerals, the "polished metal" effect is produced when light striking the surface stimulates electrons, causing them to vibrate and emit a diffuse light.

REFRACTION AND INTERNAL REFLECTION

The extent of a crystal's transparency depends on what happens to light rays as they pass through the crystal. If the incident ray (incoming light) is at right angles to the crystal-air interface, all the light goes straight through the crystal (1). More often, the ray is at a different angle, and the light is partly refracted and partly reflected as it slows down and changes direction (2). As the angle of incidence increases, the proportion of reflected light also increases. The critical angle (3) is the angle of incidence above which total internal reflection occurs. When the angle of incidence is greater than the critical angle, all light is reflected (4).

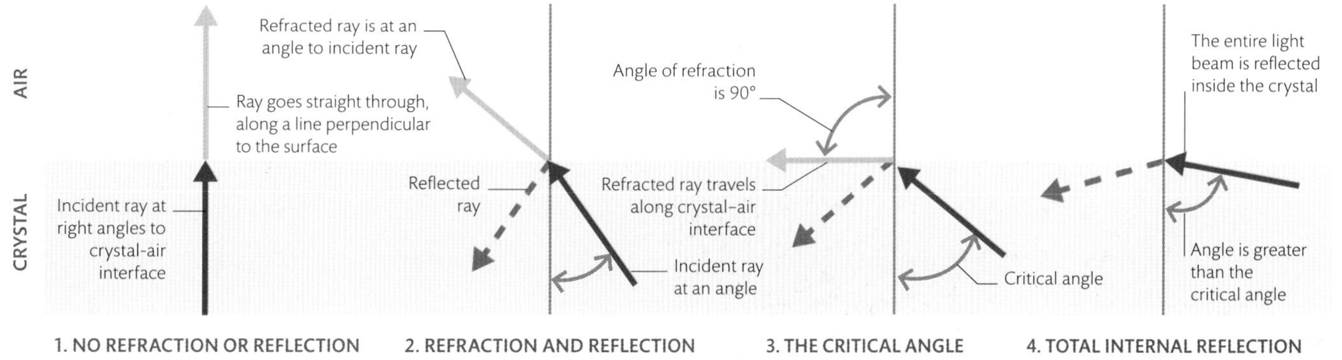

Talc
The softest of all minerals, talc has a value of 1 on the Mohs Hardness Scale and can be easily scratched by a fingernail. Cosmetics, including talcum powder, are made from ground-down talc.

Rough diamond
This 424-carat diamond was found in South Africa in 2019. Diamonds are the hardest naturally occurring substances; as such, they can scratch any other mineral, but only a diamond can scratch a diamond. These precious gems are pure carbon and form about 100 miles (160 km) below Earth's surface, under great pressure and intense heat. The diamonds we see at the surface have been carried upward by very deep-seated volcanic eruptions.

from hard to soft

Testing hardness is a very useful way to identify an unknown mineral. The hardness of a mineral refers to the extent of its resistance to scratching or abrasions, rather than its toughness or strength; very hard minerals can actually be quite brittle and break easily. A scratch mark on a mineral marks the place where atoms have been removed from its surface. The strength of the bonds that hold the atoms together can influence a mineral's hardness. For example, graphite is a relatively soft mineral, with weak bonds between its atoms, whereas diamonds, which are the hardest minerals, have strong atomic bonds. All minerals can be assigned a hardness value using one of several testing methods, including Mohs scale and the Knoop scale.

ASSESSING HARDNESS

The most widely used hardness test for minerals is called Mohs scale. This measures hardness relative to 10 standard minerals of increasing hardness, from 1 (as soft as talc) through to 10 (as hard as a diamond). To test a mineral's hardness, it can be scratched using a known mineral (a harder mineral will scratch a softer one, but not vice versa) or a common object of known hardness. For example, a fingernail, copper coin, and steel file will scratch minerals that are less than 2.5, 3.5, and 6.5 respectively. For the Knoop test, a load is applied to the surface of the sample using an indenting tool. The hardness value is based on the ratio of the applied load to the area of indentation.

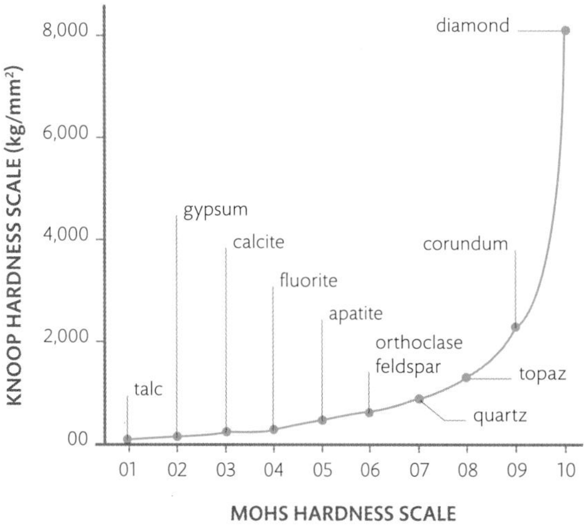

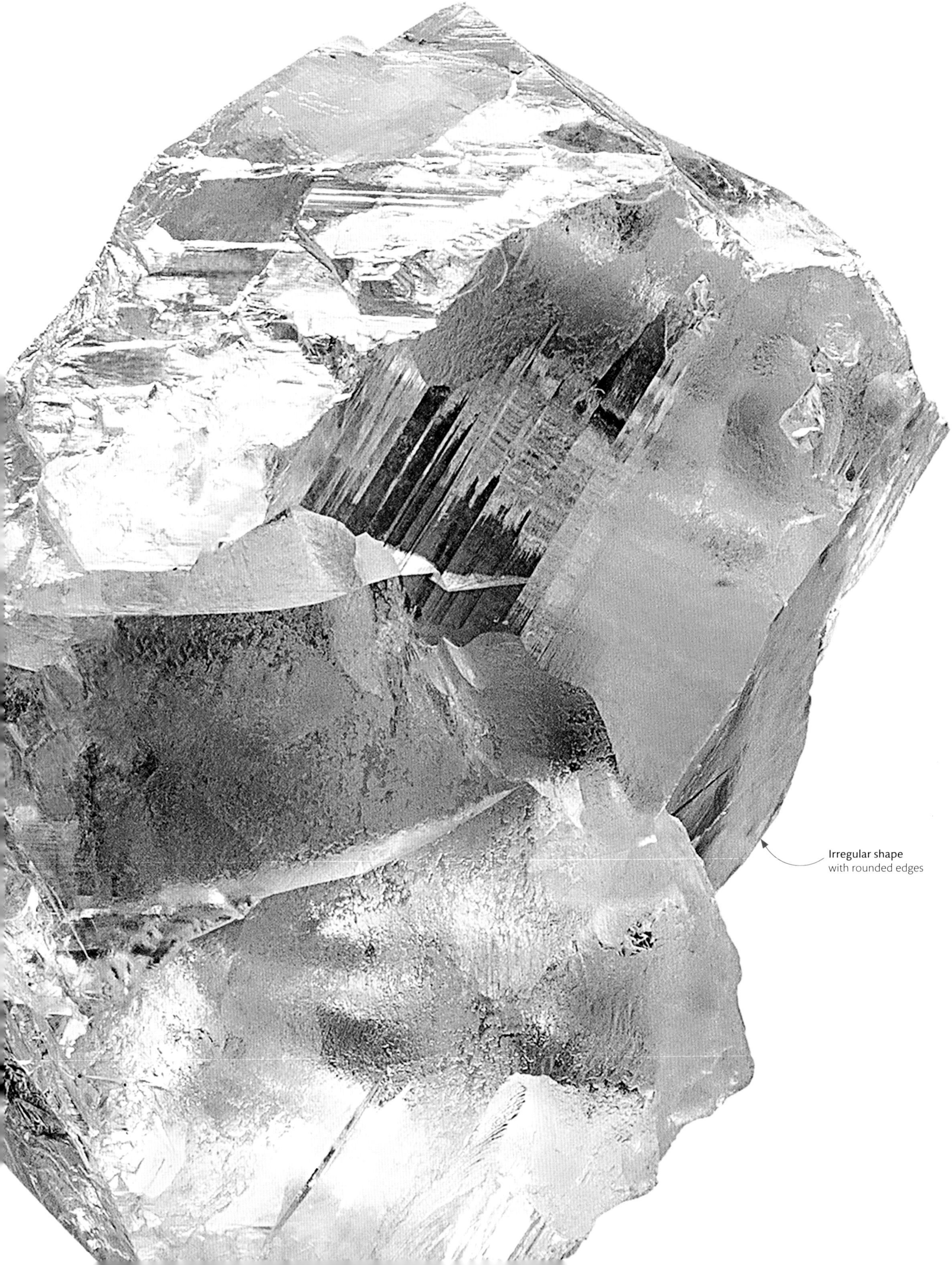
Irregular shape
with rounded edges

native elements

Most minerals are made up of a combination of chemical elements in compounds. However, some elements, known as native elements, occur in nature in a relatively pure form. Common native elements are divided into three groups: metals (gold, silver, platinum, copper, and iron), semimetals (arsenic and bismuth), and nonmetals (sulfur and carbon). Other native elements occur more rarely. Native elements occur in various rock types, and are often of economic value. The minerals gold, silver, platinum, osmium, and iridium are the main sources for the elements with the same name.

SILVER

PLATINUM

GRAPHITE

SULFUR

NATURAL FORMS OF CARBON

Carbon is one of the most abundant chemical elements on Earth and exists in many forms. Diamond and graphite are both crystalline: diamond develops in high-temperature, high-pressure conditions and graphite in high temperatures. Amorphous carbon (which includes soot and coal) is noncrystalline and forms through incomplete combustion. Carbon is rare among elements in producing long, strongly bonded chains of atoms called polymers.

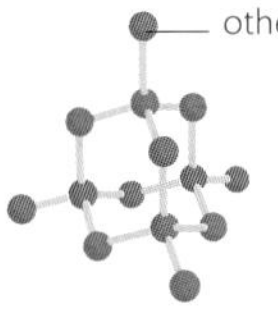

DIAMOND

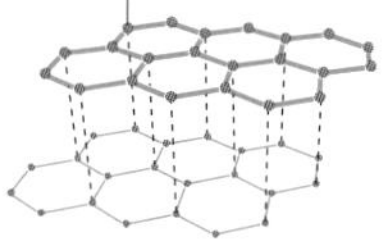

GRAPHITE

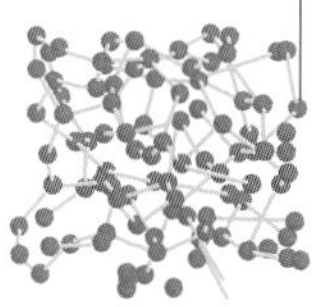

AMORPHOUS CARBON

Native copper
Probably the first metal to be worked by humans, copper is a bright, reddish gold metal that tarnishes to dull brown, and often forms green or black crusts when exposed to oxygen. Traditionally used to make coins along with silver and gold, copper conducts heat and electricity well, and is used to make many kinds of electrical equipment.

Green patches of tarnish caused by oxidation are visible on the surface
Distorted, branched appearance is typical for copper

Metal compound
Most metals combine with nonmetallic chemical elements or with semimetals to form a great variety of minerals. For example, tetrahedrite (shown here) is a compound of the metal copper with the chemical element sulfur and the semimetal antimony. Semimetals have properties that are intermediate between those of metals and nonmetals.

ABUNDANCE OF METALS IN EARTH'S CRUST

This diagram shows that the most abundant metals in Earth's crust are common, rock-forming elements that combine with oxygen to form silicate minerals such as quartz and feldspar. These elements include iron, aluminum, magnesium, titanium, and manganese. The rarest metals are elements that are depleted in the crust because they mainly reside deeper within Earth's interior. These rare metals include platinum and gold.

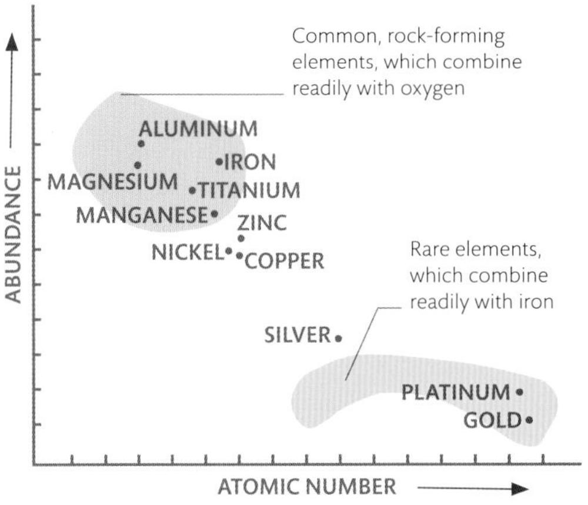

Native metal
Some metals can be found in nature in their elemental form, meaning they are not combined with other chemical elements. Known as native metals, this group includes gold, platinum, silver, mercury, and copper (see pp.44–45). Gold is found in igneous rocks or in river deposits as flakes or nuggets. This specimen was found in Queensland, Australia.

Smooth but bumpy surface and shiny, bright, golden yellow color

Uneven fracture of the crystals

Earth's metals

Metals occur in various forms in Earth's crust. They are typically defined as solid, shiny, and opaque, are usually crystalline, and comprise the main group of chemical elements in the periodic table. Not all metals feature in the table, however, as some form compounds with other metals or nonmetals—for example, steel is not included since it is a mix of iron, nickel, and chromium. Ores are rocks that contain valuable metals or metal compounds in amounts that are large enough for their extraction to be commercially worthwhile. Metals conduct electricity and heat well, and are both malleable and resilient, making them sought after for many uses.

FORTIFICATION AGATE

GRAPE AGATE

CARNELIAN

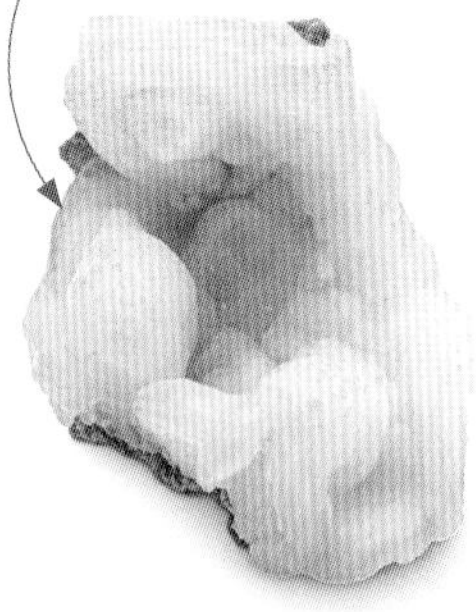

PINK CHALCEDONY

Microcrystalline quartz
In some varieties of quartz, described as microcrystalline (or cryptocrystalline), the crystals are too small to be seen with the naked eye and are visible only under high magnification. Microcrystalline quartz typically forms at low temperatures in volcanic settings.

AMETHYST

CITRINE

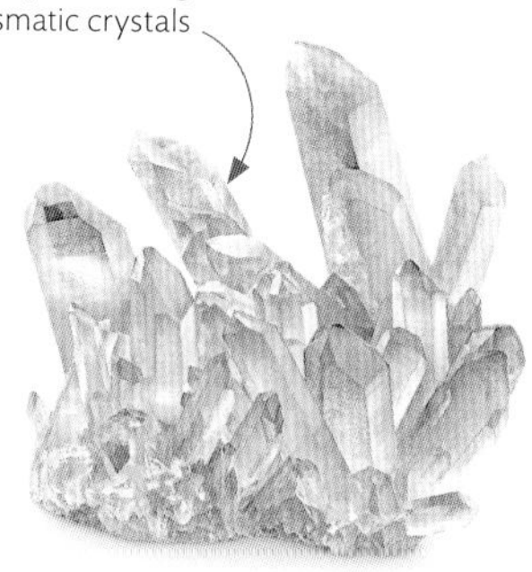

ROCK CRYSTAL

ROSE QUARTZ

Macrocrystalline quartz
Quartz varieties with crystals that are visible to the naked eye are described as macrocrystalline. The crystals are six-sided prisms and pyramids, often perfectly formed, although less well-developed crystals form imperfect shapes. Aggregates or crusts of macrocrystalline quartz often line rock cavities (see pp.56–57).

varieties of quartz

Found in most rock types, quartz is the second most common mineral in Earth's continental crust, after feldspars (see pp.50–51). There are many varieties, and although pure quartz is colorless, the mineral is found in a wide range of colors, which are determined by chemicals present in the rock. Quartz is fairly hard, due to the strength of its chemical bonds, and is the only silicate (oxygen- and silicon-based compound) to be made up entirely of silicon and oxygen. Together with feldspars and mica, quartz is one of the main components of granite. There are two forms of quartz: microcrystalline and macrocrystalline.

Smoky quartz
When quartz specimens, such as the smoky quartz shown here, have space to grow, they form beautiful, hexagonal prisms that terminate in six-sided pyramids. The brown hue of smoky quartz can range from light brown to nearly black, and is produced when the mineral is exposed to radiation underground.

CHRYSOPRASE

RED JASPER

GREEN JASPER

ONYX

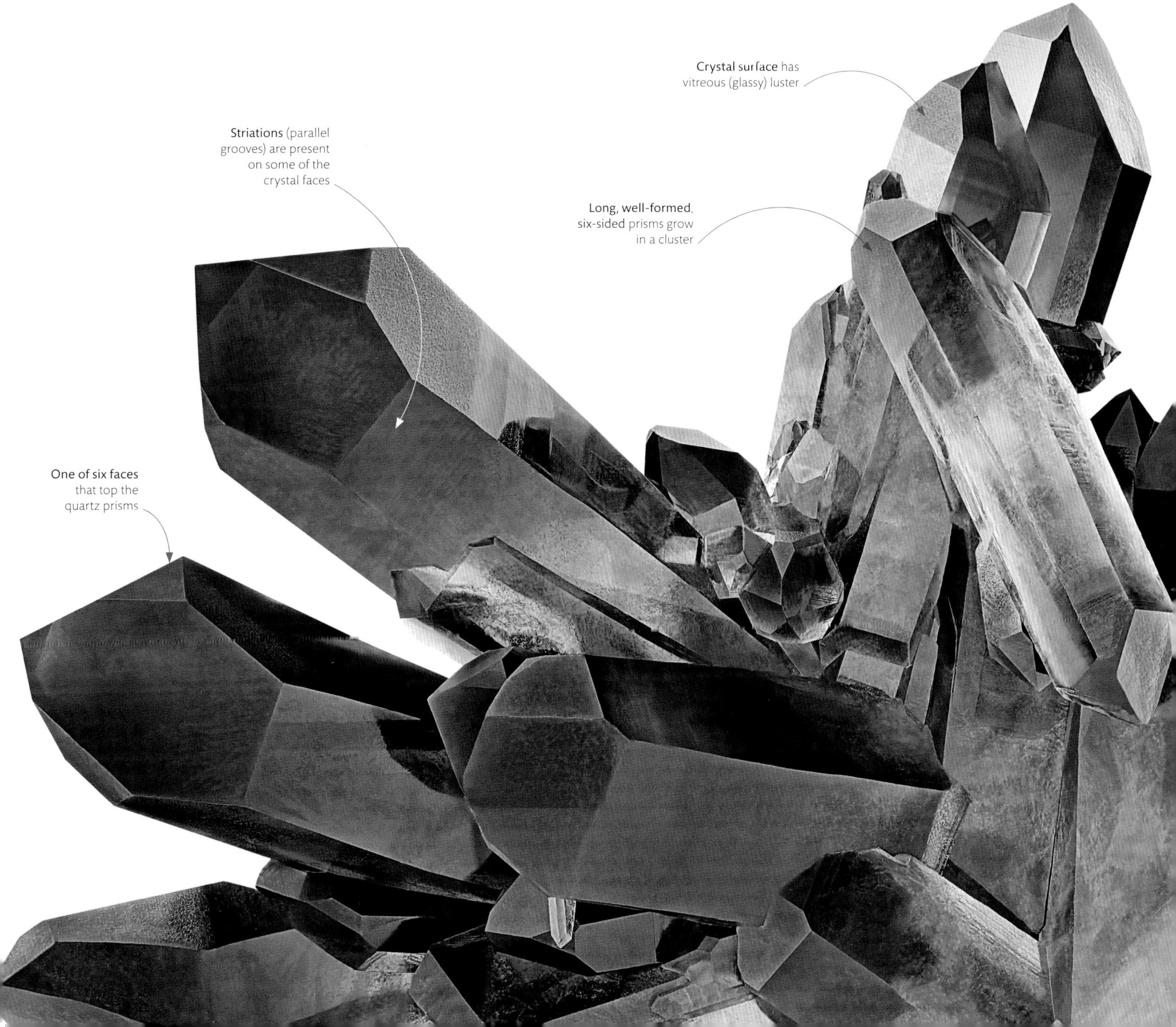

SILICATE GROUPS

All silicate minerals (including feldspars) share a basic unit: the silica tetrahedron, in which a silicon atom is surrounded by four oxygen atoms. These tetrahedrons, which can be visualized as pyramids or as a ball-and-stick model, occur singly or connected in arrays. The many silicate minerals are divided into groups according to how the tetrahedrons, and other elements, are linked together.

THE SILICA TETRAHEDRON

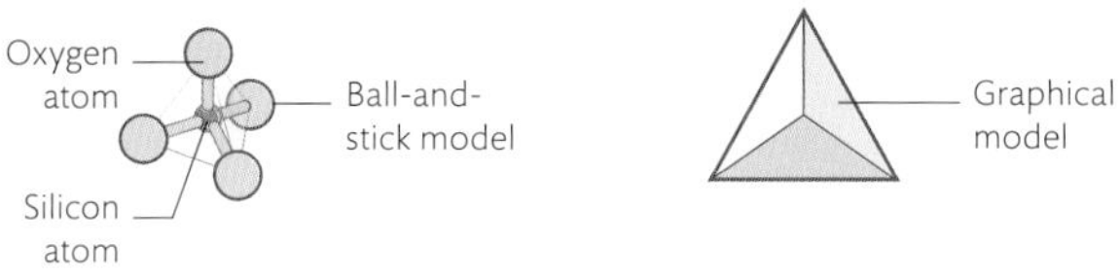

RINGS, CHAINS, AND FRAMEWORKS

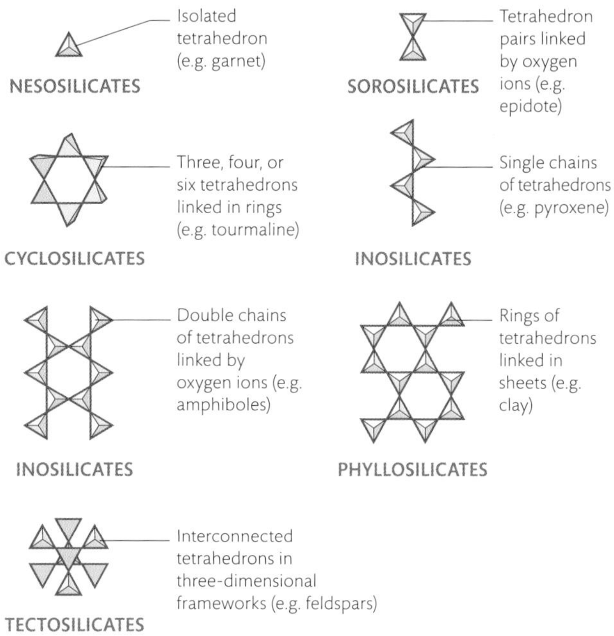

Intergrown, tabular white crystals

ALBITE

Blue-green amazonite is intergrown with other minerals

AMAZONITE

feldspars

The most abundant of all minerals, feldspars make up about two-thirds of Earth's crust. They belong to the silicates, which form the largest and most important class of minerals. Although there are many varieties of feldspar, they can be broadly divided into two groups: potassium-rich feldspars, and sodium- and calcium-rich (collectively known as plagioclase) feldspars. Potassium-rich feldspars are often found in igneous rocks such as granite, as well as in gneiss and sandstone. Plagioclase feldspars are common in igneous rocks such as gabbros, and in meteorites and moon rocks.

Outer crust does not show labradorescence as the twinning surface is not exposed

Labradorite
A plagioclase feldspar, labradorite is easily identified by its iridescent sheen. The source of this iridescence (or labradorescence) is the mineral's internal structure composed of lamellar twinning—alternating layers of two or more feldspars of different chemical compositions. When light penetrates the mineral, it reflects off these different twinning surfaces, producing flashes of color.

ORTHOCLASE

ANORTHITE

BYTOWNITE

OLIGOCLASE

Malachite
A secondary mineral, malachite is formed when copper-rich minerals are altered. This bright green mineral forms when carbonated water (water containing dissolved carbon dioxide) interacts with copper or when limestone interacts with a copper-rich fluid. It often forms round masses of crystals arranged in concentric bands.

altered minerals

Minerals are chemical compounds that can undergo reactions with other compounds to form new minerals called secondary minerals. When chemical processes change a mineral's chemistry or the properties of its crystals, that mineral is said to be altered. Minerals can be altered when they come into contact with chemicals dissolved in water. This can happen, for example, when existing minerals come into contact with oxygen-rich groundwater percolating down from the surface or with water that has been heated below ground by a body of magma.

Chalcopyrite
A copper iron sulfide, chalcopyrite is the main source of copper in the world. When it is exposed to air, its surface tarnishes to iridescent purple, yellow, blue, and green. When chalcopyrite reacts with some water solutions, it can be altered to other minerals such as malachite.

DIAGENESIS

Minerals can be altered into new minerals when sediments are deposited and buried and then consolidated into rock. This process, which takes place near Earth's surface, is known as diagenesis. It can also involve the progressive bonding together of grains of sediment, called cementation. For example, quartz can grow between sand grains, cementing them together.

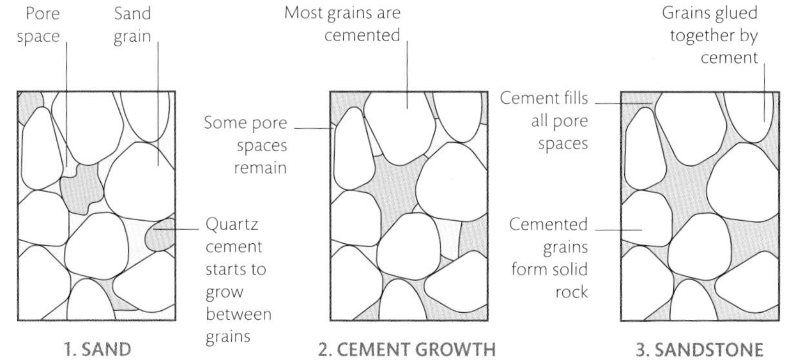

mineral associations

Minerals frequently occur in groups, or associations, and certain minerals are often found together. For example, malachite is often found in association with pyrite and chalcopyrite, and white gold with quartz. Some minerals form together in a specific environment. Ore minerals, for example, can develop in rocks surrounding cooling magma bodies in Earth's crust (see panel, below). Other mineral associations are related to the chemical composition of the fluids that precipitate them, like the minerals found lining crystalline cavities such as geodes (see pp.56–57). Knowing the environment of formation and the mineral associations helps with the identification of the minerals.

Minerals may be intergrown, showing they grew around the same time

Pyrite with sphalerite and quartz
This specimen shows colorless quartz with brassy colored pyrite and dark gray, metallic sphalerite.

HYDROTHERMAL VEINS

Mineral associations are often found in deposits in Earth's crust called hydrothermal veins. These form by precipitation of minerals from hot, mineral-bearing liquids moving through fissures inside rocks. The liquids can be released by cooling magma bodies called intrusions, or by heated groundwater. The deposits often include economically valuable ore minerals, with tin and tungsten formed close to the intrusion, and copper, zinc, and lead farther away.

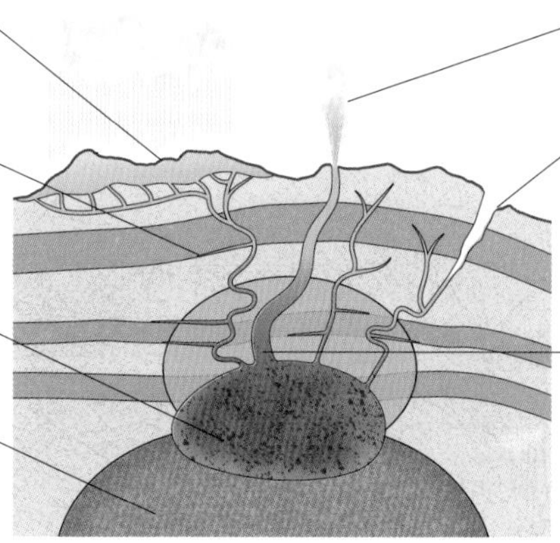

Prehnite with apophyllite
The minerals prehnite and apophyllite are frequently found together, lining cavities inside igneous rocks, such as basalt, or sometimes growing on granite. These two minerals are also found together in veins within rocks (see panel, below). Individual crystals of prehnite are rare. It more usually forms as botryoidal or rounded crystal aggregates.

Clear crystals of apophyllite

The rock on which the crystals grow is called the matrix

Apophyllite forms clusters of blocky, tabular crystals
Rare, pinkish orange prehnite crystals have a glassy luster

Clear quartz crystals are growing on top of the lining in the center of the cavity

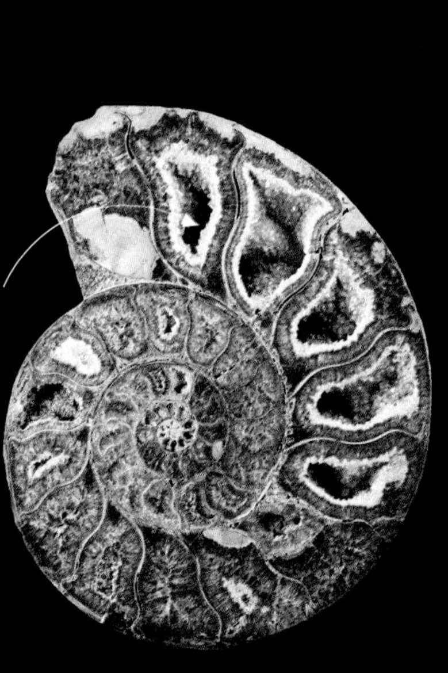

Crystals growing inside one of the chambers

Ammonite fossil
Sometimes crystals form inside the remains of animals preserved as fossils. For example, fossilized ammonites—extinct relatives of squid and cuttlefish—may contain cavities ("chambers") in which the organism once lived. When mineralizing fluids interact with the chambers, crystals can develop.

Roughly spherical shape with a bumpy, uneven exterior

Geode lining consists of banded agate and chalcedony

HOW GEODES FORM

Geodes begin as voids in rock or sediment. These can be the spaces left behind by gas bubbles in magma, cracks formed in rocks due to tectonic activity, or natural cavities within fossils. When fluids percolating through tiny fractures or pores in the rock or sediment enter the void, minerals in the fluid separate out, or precipitate, forming layers of crystals around the edges of the void. The crystals then continue to grow.

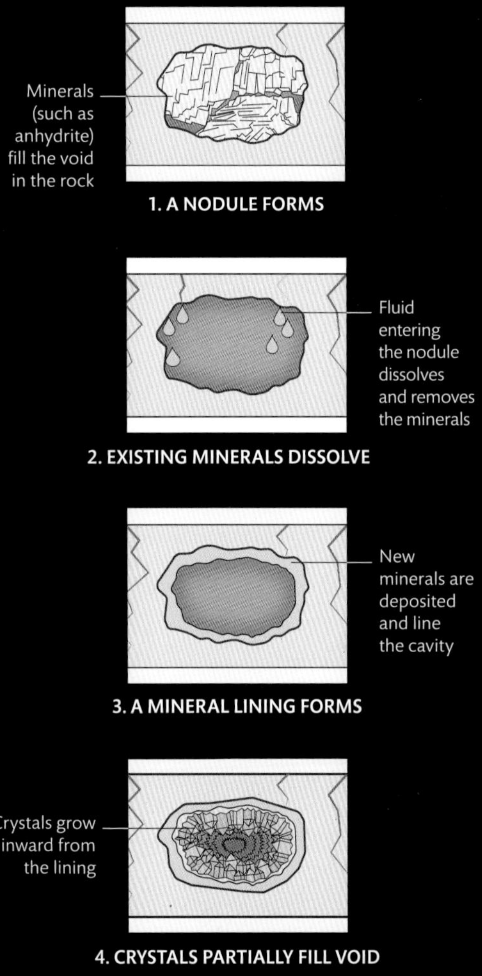

crystalline cavities

A geode is a hollow, usually rounded rock structure that from the outside resembles a nondescript nodule. However, when split open the geode reveals a cavity lined with minerals, which typically grow into beautiful crystals that project inward. Unlike crystals growing in confined conditions, crystals inside a geode usually have enough space to develop well-formed faces. As a result, they are highly sought after. Geodes range in size from an inch or so to several yards across.

Quartz geode
This geode, found in northern Bohemia in the Czech Republic, is filled with quartz varieties, including chalcedony, banded agate, and clear crystalline quartz. Other minerals that grow in geodes include the quartz varieties amethyst and jasper, calcite, dolomite, and celestite. The lining can be made of successive, concentric layers, giving the geode a banded appearance.

HARDNESS

A gemstone's hardness is a key factor when it comes to jewelery, since the cut stones must survive abrasion from repeated use. The Mohs scale (see p.42) is a hardness test that rates minerals in order of their "scratchability" on a scale of 1 (very soft) to 10 (very hard). Ideally, gemstones are at least as hard as quartz (7 on Mohs scale). Diamonds are by far the hardest gemstone.

Colorless, octahedral crystal

DIAMOND
10 (Mohs scale)

Cluster of orange-red crystals with a vitreous luster

GARNET
7–7.5 (Mohs scale)

Long, hexagonal, prismatic crystal is a greenish blue variety of beryl

AQUAMARINE
7.5 (Mohs scale)

COLOR

In diamonds, clarity and an absence of color make for the most desirable gemstone. However, in most other cases, specific colors make gemstones highly desirable and add to their value. When beryl is colorless, it is of only moderate value, but emeralds (see opposite)—the green variety of beryl—are among the world's most sought-after gemstones.

Purple crystals get their color from traces of iron

AMETHYST

Green variety of olivine is colored by iron

PERIDOT

Reddish, tablet-shaped crystal (variety of corundum)

RUBY

Blue, six-sided bipyramidal crystal (variety of corundum)

SAPPHIRE

Orange-brown crystal fragment, with perfect cleavage surface and vitreous luster

TOPAZ

Different-colored layers are produced by traces of different chemicals

TOURMALINE

Tough, interlocking, granular texture and a greasy to vitreous luster

JADEITE JADE
7 (Mohs scale)

Sky-blue nodular tourmaline, with dark veining and no visible crystals

TURQUOISE
5.5–6 (Mohs scale)

Multicolored blue and purple crystal fragment; although soft for a gemstone, fluorite is prized for its exceptional colors

FLUORITE
4 (Mohs scale)

Emerald
Together with diamond, sapphire, and ruby, emerald is one of the world's most valuable gemstones. Emeralds are found all over the world, but the three main producers are Colombia, Brazil, and Zambia. An alternative name for emerald is green beryl—the silica-rich mineral from which it derives.

Six-sided emerald crystal with typical rich green color

gemstones

The three key attributes that qualify a mineral to be classed as a gemstone (a jewel or stone used in jewelery) are durability, beauty, and rarity. The beauty of gemstones is enhanced by polishing and cutting the faces of the stones at specific angles to maximize their brilliance and color. Gem dealers assess gemstones based on their clarity, how they are cut, their color, and their weight (measured in carats). Just 130 minerals—or fewer than 4 percent of all minerals—are considered gemstones. In addition, some noncrystalline minerals of organic origin, such as pearl, coral, and amber (see pp.60–61), are classed as gemstones.

HOW A PEARL FORMS

When mollusks sense the presence of foreign bodies, such as sand particles, between their shell and mantle (the organ that secretes the shell), they form a defensive coating around the object. The mantle epithelium—the layer of tissue cells that surrounds mollusks' bodies—starts to coat the object with concentric layers of aragonite and conchiolin, called nacre (or mother-of-pearl). Over time, the layers enclose the object, forming a pearl.

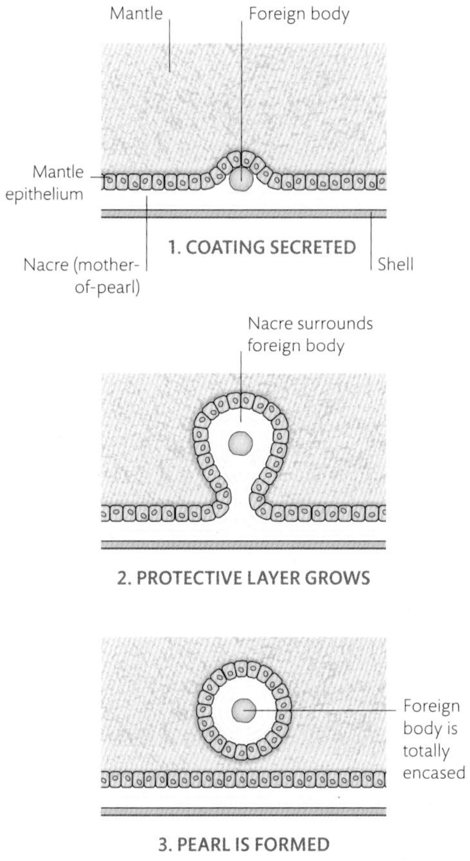

Minerals from living organisms
Organic minerals are abundant in nature. The soft-bodied parts of bivalves (such as mussels, clams, and oysters) and gastropods (which include snails, whelks, and conchs) build their own homes by secreting calcium carbonate—a chalky mineral—to make shells. Coral skeletons are also made of calcium carbonate, secreted by organisms called polyps. The remains of trees that become buried and heated may become organic materials such as coal, anthracite, and jet.

Layers of calcium carbonate secreted by gastropods form a spiraling shell

GASTROPOD SHELL

Dark specks trapped inside the amber could be plant fragments

organic minerals

Some minerals are formed by living organisms through biological processes. Although these are not minerals in the strictest sense, since they are not created inorganically, they can be referred to as organic minerals. The process by which living organisms produce minerals is called biomineralization, and it results in the formation of materials that include shell, pearl, coral, coal, and amber. Sometimes minerals form in the presence of microbes; for example, stromatolites—which are among the oldest known fossils—are made of sediment trapped by living cyanobacteria.

OYSTER SHELL WITH PEARL

CORAL SKELETON

ANTHRACITE

JET

Amber

Resin is a thick, sticky substance that is exuded from trees with damaged bark to seal gaps and tears and prevent pathogens and insects from entering. When hardened resin from ancient forests became buried in sedimentary deposits, it formed amber, or fossilized tree resin. Amber can form nodules, rods, and droplets.

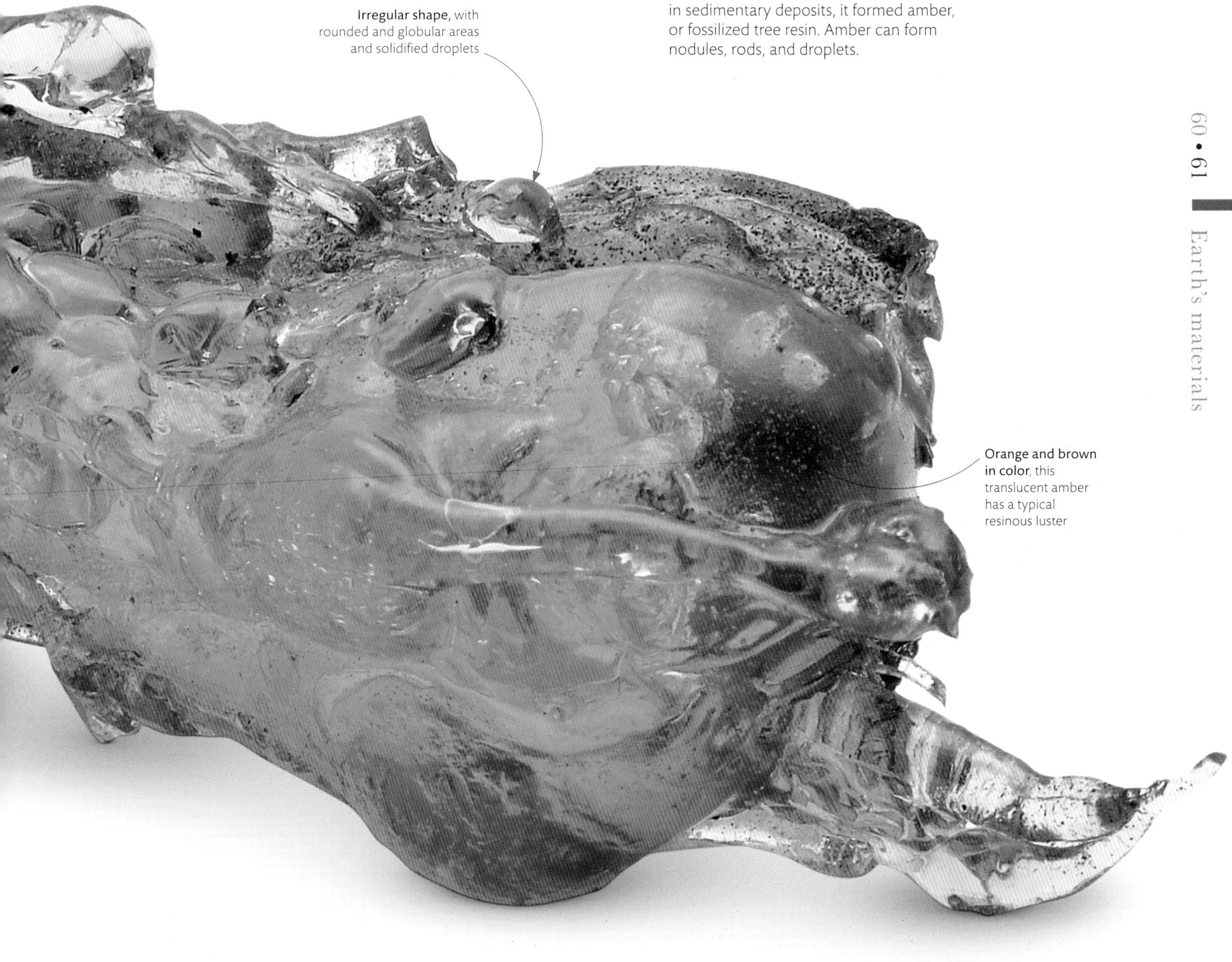

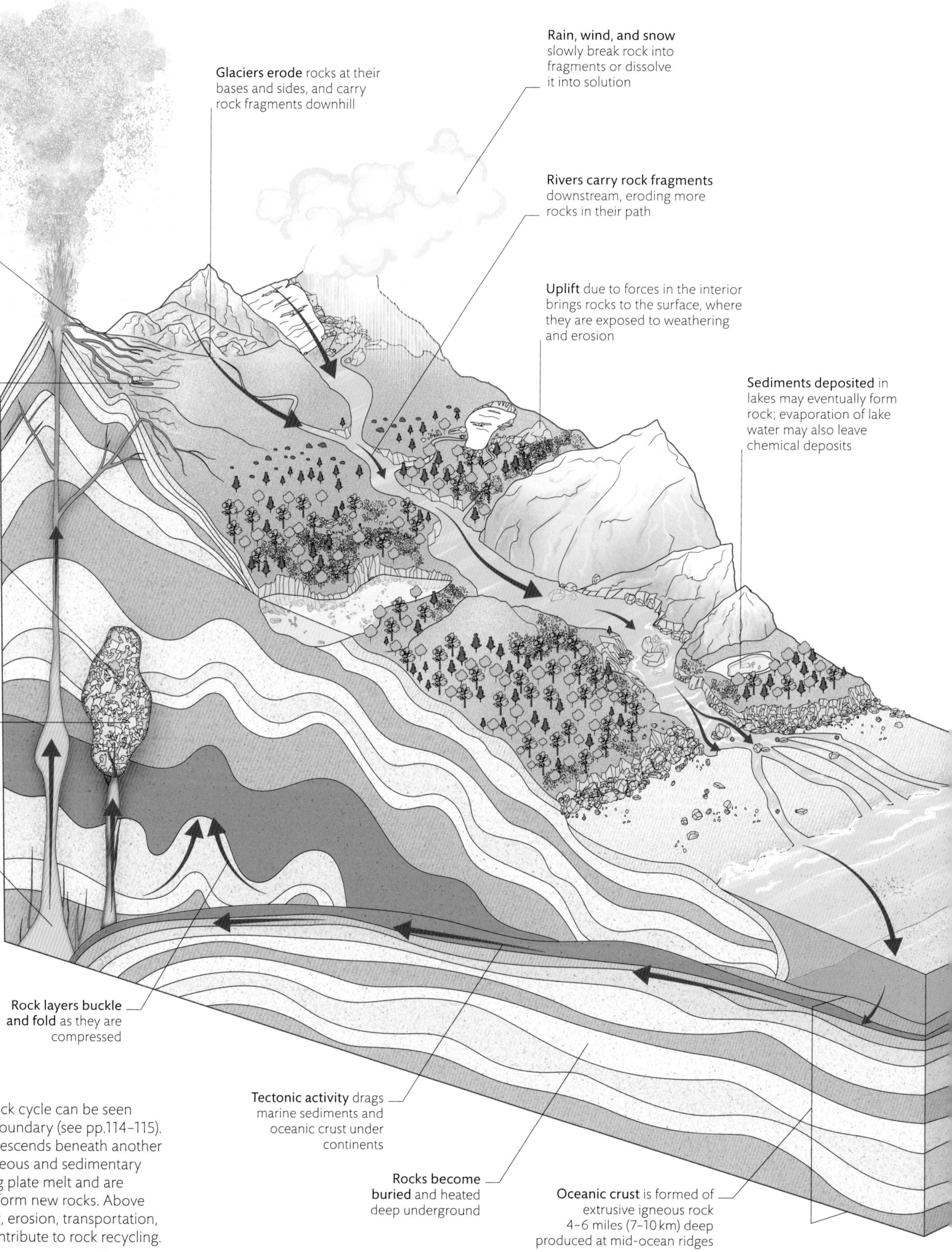

A cycle of change
Several stages of the rock cycle can be seen at a convergent plate boundary (see pp.114–115). As one tectonic plate descends beneath another into Earth's interior, igneous and sedimentary rocks in the descending plate melt and are eventually recycled to form new rocks. Above the surface, weathering, erosion, transportation, and deposition also contribute to rock recycling.

From igneous to sedimentary rock
The Jokulsa A Bru river in the Studlagil Canyon, Iceland, flows between cliffs made of columns of igneous rock that formed from slow-cooling magma. Fragments of the cliffs that are eroded and transported by the river will be deposited downstream or in the sea, where they may eventually become sedimentary rock.

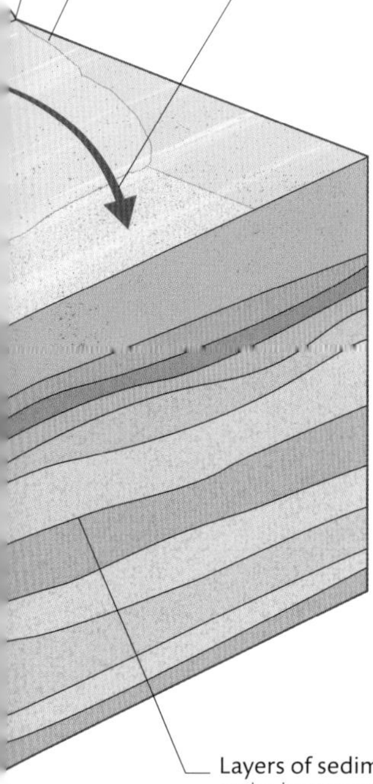

Waves break up rocks at the shoreline and transport the fragments along the coast or out to sea

Rock fragments carried downstream by rivers are deposited along the coast or offshore, with large particles being the first to drop out of suspension

Fine sediments are deposited on the seabed

Layers of sediment accumulate at the bottom of ocean basins, later becoming compacted and cemented to form new rock

the rock cycle

Over huge expanses of time, the three basic rock types—igneous, sedimentary, and metamorphic—are constantly changing in a process called the rock cycle. This is governed by the combined effects of heat below the surface, movements of Earth's crust, and surface erosion and deposition. Igneous rocks form when molten rock cools and solidifies below the surface or above ground after being erupted as lava. Sedimentary rocks form when rock fragments, or grains, are loosened or dissolved by the processes of weathering and erosion before being transported, deposited, and then compacted and cemented together. The raw material from which sedimentary rocks form may also be created by animals or plants or by chemical processes. Metamorphic rocks form when igneous or sedimentary rocks are subjected to higher temperatures and pressures that alter their structure or mineral content.

Moss and lichens are growing on the basaltic pillow lava
Rounded shapes of basaltic lava that cooled and formed basalt rock in ancient oceans
Basalt rock, with grooves and ridges

igneous rocks
of the oceans

Earth's thin outer shell is called the crust. There are two distinctly different types: oceanic and continental crust (see pp.66–67). The oceanic crust covers about 70 percent of Earth's surface and consists of several layers of mainly igneous rock, which combine to form a mass typically 4–6 miles (7–10km) thick. The oceanic crust is created at mid-oceanic ridges—elevated volcanic structures that form on the ocean floor, at the boundary between two diverging tectonic plates (see pp.110–111). As the plates spread apart, hot magma rises and pools in magma chambers. Some of the magma cools inside the chambers, forming igneous rocks such as gabbro. The rest erupts as lava from fissures on the seafloor and solidifies as basalt.

Pillow basalts
Igneous rocks known as pillow basalts, or pillow lavas, are formed when molten basaltic lava erupts into cold sea water on the ocean floor. The lava cools very quickly and forms a thin skin; with increasing eruptions and pressure from the lava, the skin expands and inflates into rounded, pillow-shaped mounds, typically up to 3 ft (1 m) in diameter. These pillow basalts are in Iceland.

Gabbro
The igneous rock gabbro has the same chemical composition as basalt, but since the magma cools and solidifies more slowly below Earth's surface, there is time for larger crystals to form.

OPHIOLITES

The ocean crust is consistent in structure and thickness, and includes a layered sequence of sediments overlying different rock types. The best place to see oceanic crust is where a section of ancient oceanic crust (called an ophiolite suite or sequence) and the underlying upper mantle have been pushed up onto continental crust, thus exposing it above sea level.

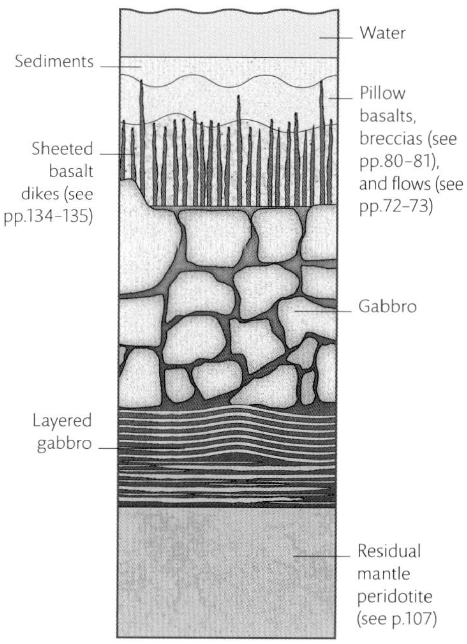

CROSS SECTION OF OCEANIC CRUST

igneous rocks
of the continents

The continental crust—the part of Earth's crust that makes up the continents and continental shelves—is composed mainly of granite, and is older, thicker, and less dense than the oceanic crust (see pp.64–65). Away from tectonic plate boundaries, the interior portions of continents have remained relatively stable over geological time, since the rocks of the continental crust are relatively buoyant and resist subduction (the process of one tectonic plate sliding under another, which can result in earthquakes and volcanoes). The oldest rocks in the world—about 4 billion years old—are found in the continental crust.

Pink potassium feldspar gives this specimen its color

Pink granite
An igneous rock that cools under Earth's surface in magma chambers, granite can be gray, white, or pink, depending on the chemical composition of its minerals. Its crystals are visible to the naked eye.

Mount Fitz Roy
Many of the world's mountain belts are made of granite from the continental crust. These rocks formed deep within Earth's mantle. Over time, the granite rocks were uplifted into mountains and were eroded, which is how the jagged peaks of Mount Fitz Roy, in Patagonia—on the border between Chile and Argentina—were formed.

CONTINENTAL CRUST

The continental crust can extend to about 44 miles (70 km) under mountain ranges, although where continents are stretched and thinned, the crust can be as shallow as 12 miles (20 km). It is more varied in composition than oceanic crust, mainly because it is lighter and has not been recycled within Earth's interior to the same extent as denser oceanic crust. As a result, it tends to remain near the surface, where it undergoes repeated cycles of erosion, formation into sedimentary rocks, and metamorphosis.

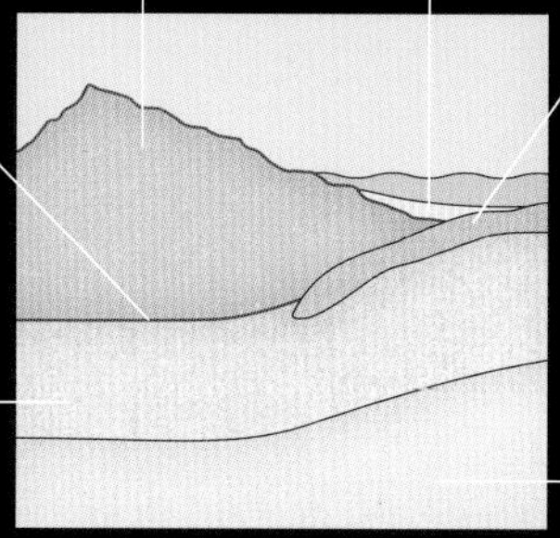

EARTH'S LAYERS

Smooth granite face carved out by glaciers

spotlight El Capitan and Half Dome

The steep cliffs of El Capitan tower 3,300 ft (1,000 m) above the floor of Yosemite Valley in California. At the valley's head, the sheer rock face of Half Dome presents an even more spectacular sight. The striking landscape of Yosemite is a result of heat, water, and ice. The granite of El Capitan and Half Dome was intruded as hot magma, deep beneath layers of sedimentary rock, more than 100 million years ago. Erosion of the sediments had exposed the resistant granite by about 65 million years ago.

Yosemite lies near the crest of the Sierra Nevada Mountains, which were uplifted and tilted from about 25 million years ago, increasing the erosive power of the area's streams, which cut deep canyons into the rock. By about 3 million years ago, the mountain range was high enough for ice fields to form along its

crest when the climate began to cool. Glaciers scoured back the valley's sides, carving out the steep cliff faces.

When the ice receded, a large lake formed on the valley floor, trapped behind the rock debris carried down by the glaciers. The lake eventually filled with sediment, giving the valley its flat floor, which is covered by a mix of forest and meadows. Yosemite Falls, North America's tallest waterfall at 2,430 ft (740 m), is one of several cascading down the valley's steep sides. Yosemite was the first wilderness area to be given legal protection by the US government, in 1864. It became a national park in 1890.

Half Dome

Half Dome rises 8,836 ft (2,693 m) above sea level

Yosemite valley
Seen here catching the evening sun, El Capitan guards the entrance to Yosemite Valley. On the opposite side, Bridalveil Fall plunges 620 ft (190 m) from a hanging valley. The snow-capped peak of Half Dome can be seen in the distance.

Pele's tears
These small pieces of volcanic glass, up to ¾ in (20 mm) long, are fragments of volcanic rock solidified from low-viscosity lava that cooled rapidly at Earth's surface. They are named after the Hawaiian goddess of fire, Pele.

Lava fountains
Streams of molten lava spew out of fissures in the Kilauea volcano, on Hawaii's Big Island. The lava cools and solidifies into gray-black igneous rock with a jagged, blocky, or ropelike texture.

rocks from lava

Hot molten rock (called magma) is formed in Earth's upper mantle. Because it is less dense than the surrounding rocks, it rises up into the crust, where much of it cools and solidifies inside magma chambers in the upper crust. Magma that reaches the surface is erupted from fissures and volcanoes as lava, which then cools and solidifies to form extrusive igneous rock. The explosivity of the eruptions and the rocks that they form depend on the composition and viscosity of the magma. An igneous rock's texture often contains clues to how the rock formed (see panel, below).

IGNEOUS ROCK TEXTURES

When magma cools slowly underground, it has time for large, coarse crystals to form. At Earth's surface, cooling is faster and the resulting crystals are microscopic. Cooling may be so rapid that crystals cannot form and a natural glass is produced. Slow cooling followed by rapid cooling as the magma moves to the surface may result in a porphyritic texture—a mix of large and small crystals. When gas bubbles are present in the lava, they form holes in the rock, giving it a porous appearance. Debris from an explosive eruption forms pyroclastic rocks.

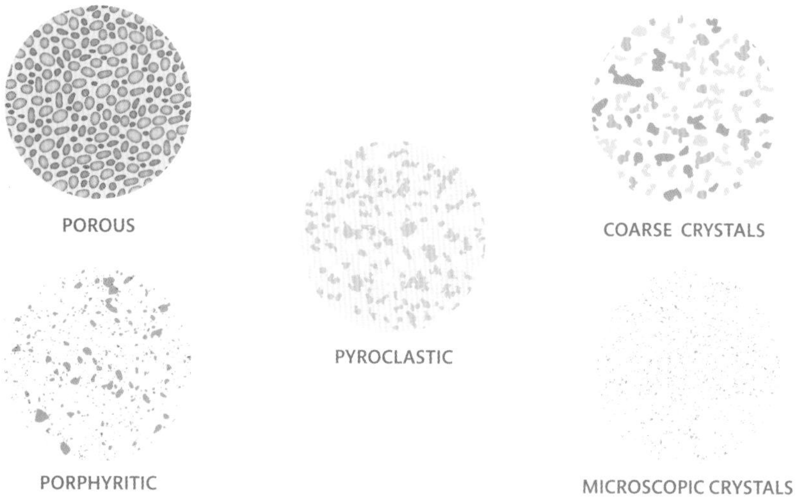

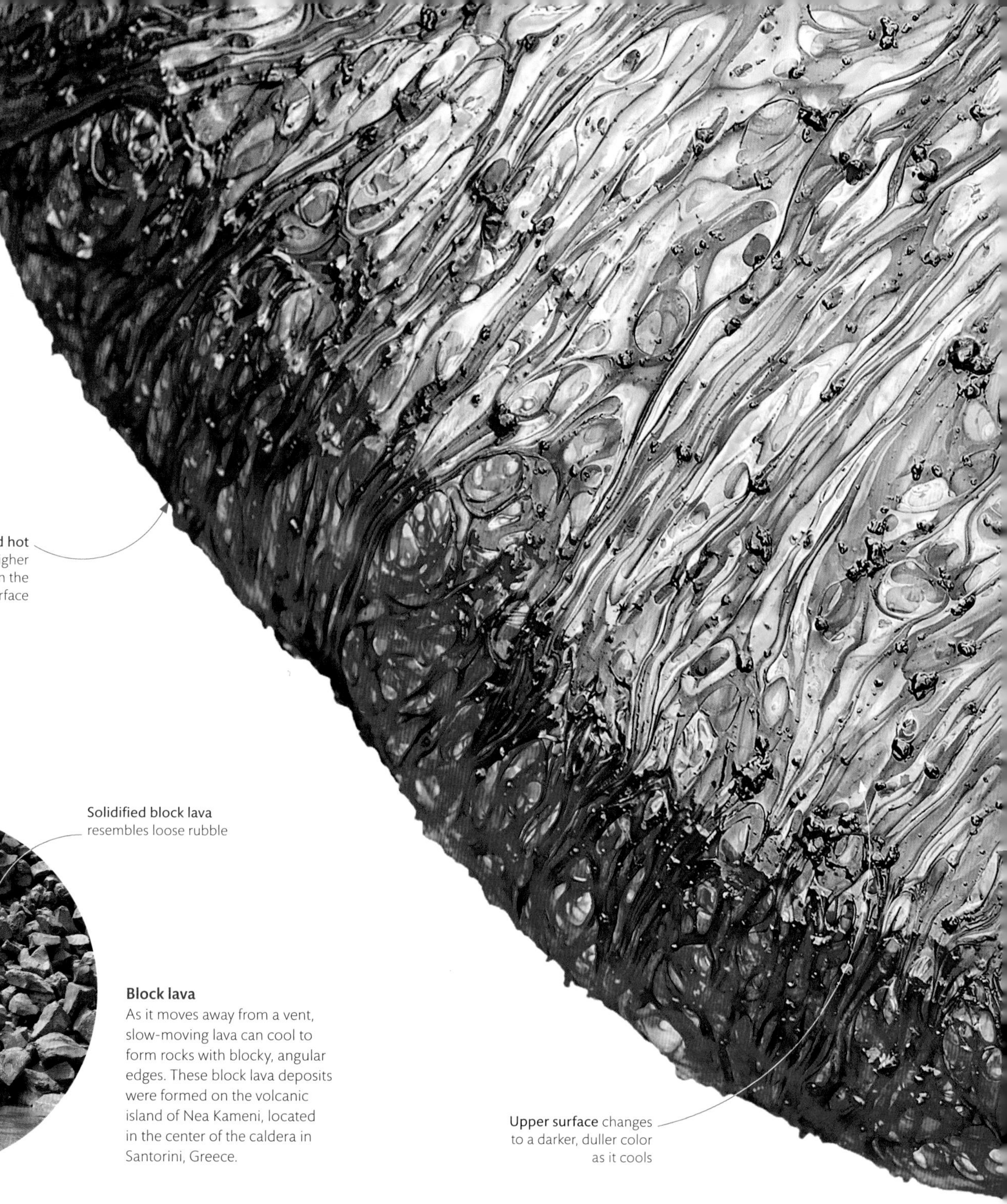

Block lava
As it moves away from a vent, slow-moving lava can cool to form rocks with blocky, angular edges. These block lava deposits were formed on the volcanic island of Nea Kameni, located in the center of the caldera in Santorini, Greece.

lava

There are many types of lava. A key factor is the amount of silica in the original magma. Magmas that are hot and have low silica content are more likely to rise to the surface and erupt as basaltic lava flows. Because gases escape these magmas easily, they tend to form quiet, or effusive, eruptions. Cooler lavas with a high silica content are thicker. If they reach the surface with a high gas content, they can erupt explosively.

Pahoehoe lava
The surface of a basaltic lava flow can form an elastic skin that is dragged into ropelike folds by the lava underneath. This results in a type of lava called pahoehoe, meaning "ropy" in Polynesian dialect. This close-up image of slow-moving lava in Kilauea, Hawaii, shows a lobe of pahoehoe, which is at more than 1,800°F (1,000°C) in places.

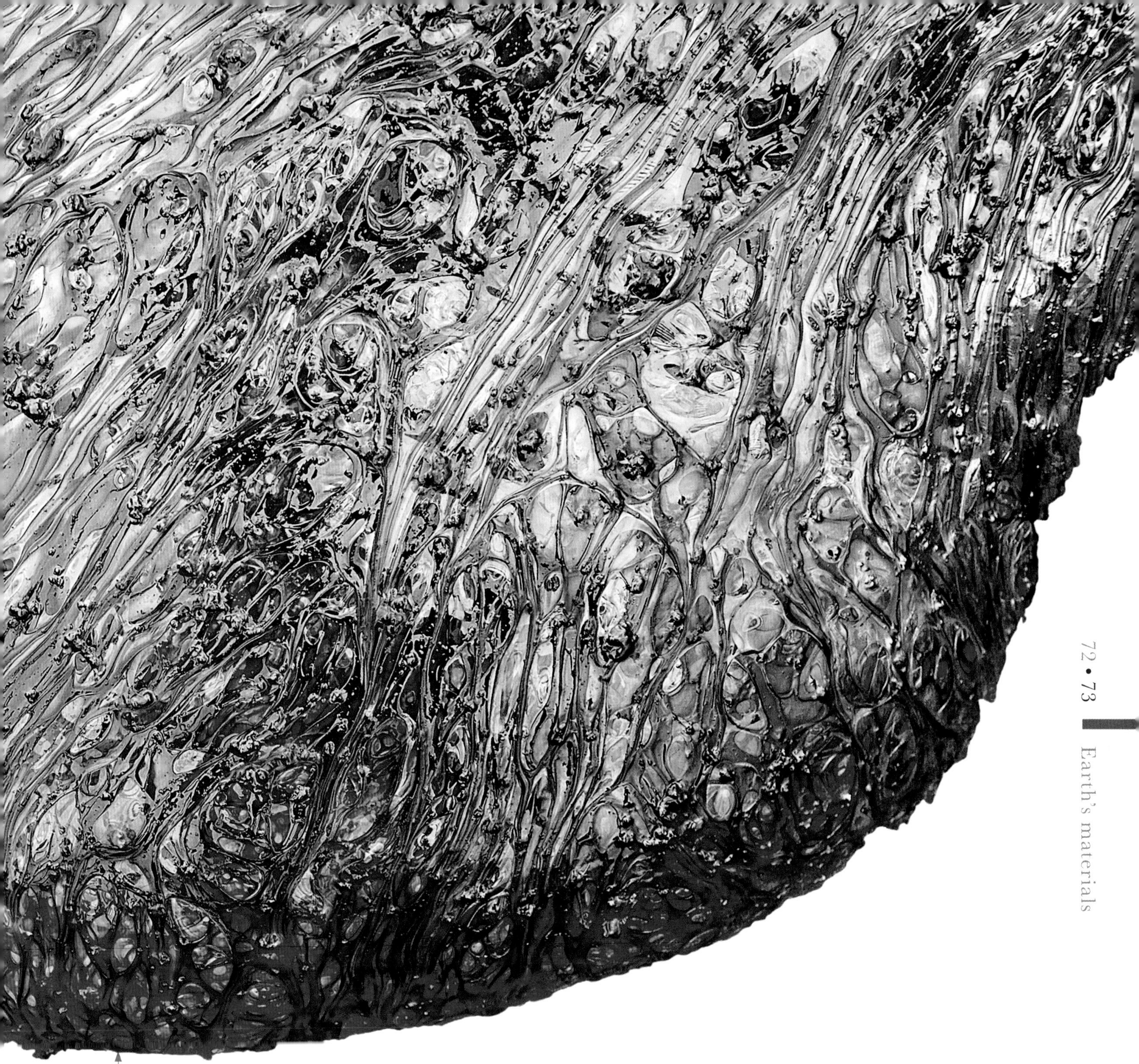

Lava has spread into a thin, lobe-like sheet at one edge of a large lava flow

LAVA VISCOSITY

Lava flows can be classified according to their viscosity, or resistance to flow, which increases with silica content. Lava that is rich in silica can flow only a short distance before solidifying, but low-silica lava can keep flowing for several miles. As viscosity increases, lava solidifies into different rocks, from basalt to andesite, dacite, and rhyolite.

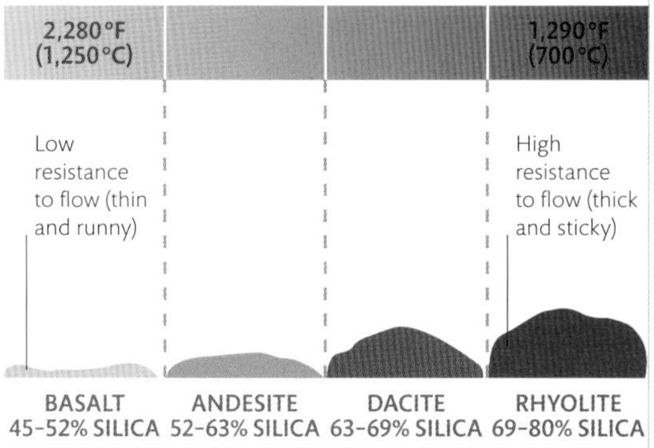

Shale
Sedimentary rocks make up about 5 percent of the rocks in Earth's crust. Almost 80 percent of these are fine-grained sedimentary rocks called shale.

Cathedral Gorge
Roughly 1 million years ago, Cathedral Gorge in Nevada was covered by a freshwater lake. The rock formations that make up the gorge are the layered remnants of the sediments that were deposited in the lake, namely silt, clay, and volcanic ash. Later, erosion and weathering of the rocks by wind and rain created gullies and canyons in the rocks.

fine-grained **rocks**

Siltstone, mudstone, and shale are fine-grained sedimentary rocks. They consist of tiny particles smaller than sand grains that are deposited in relatively still waters—including lakes, lagoons, swamps, deep ocean basins, and river floodplains—and which become buried and compacted to form parallel layers of rock. These fine-grained rocks vary in texture and hardness. Silt is deposited to form siltstone with a grain size 0.004–0.06 mm in diameter. Clay is deposited to form shale or mudstone made of finer grains (less than 0.004 mm in diameter). Shale splits into thin layers and mudstone breaks into clumps.

LOESS PLATEAUS

A yellowish-brown silt deposit called loess covers over 10 percent of Earth's land surface. It forms in various environments, including deserts near silt-rich floodplains. The wind carries silt from the floodplain to the desert; fine sediment, like desert dust, being light, is deposited a long way from its source, while coarser, heavier particles are deposited closer. When thick deposits build up, they are called loess plateaus. The largest loess plateaus are in China.

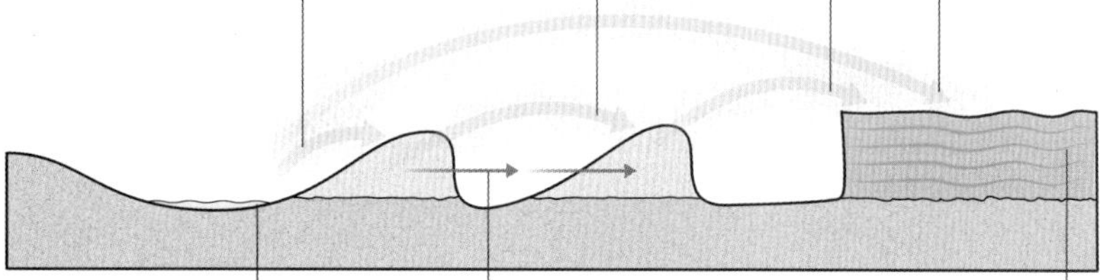

Bedding layers are visible among sedimentary rock and ash deposits
Pinnacles and gullies shaped by erosion

CROSS-BEDDING

Sedimentary rocks form in layers, called bedding. If the sand-size grains are deposited in still water, the sediment forms layers of rock parallel to the main bedding plane. However, many sandstones display cross-bedding, with grains deposited at an angle to the main bedding plane. Cross-bedding indicates that sediments were transported by a moving current such as the wind, a river, or a stream. The cross-beds in dunes form as sand is carried over the top of a dune and deposited on the other side. As more sand is deposited, sloping layers form.

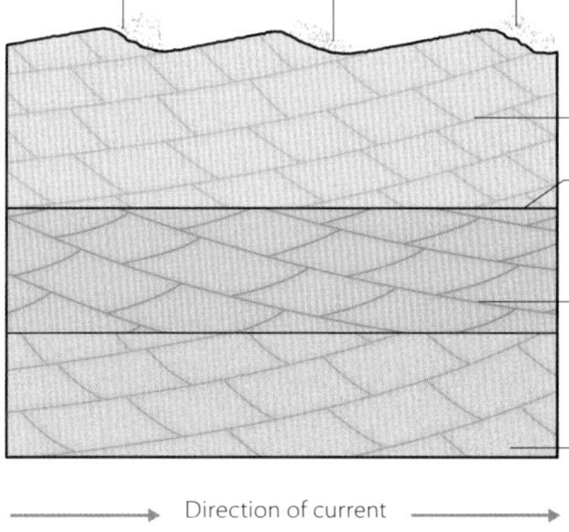

sandstone

Sedimentary rocks made up of mainly sand-size grains (0.004–0.08 in/0.125–2 mm across) are called sandstones. The grains—which consist of minerals, rock fragments, and organic materials such as shell debris and skeletal remains—are transported by wind, streams, rivers, and oceans, and are deposited when the current is no longer strong enough to carry the particles. Over time, the grains become compacted and cemented together to form layers of rock. There are different types of sandstone, depending on the composition of the grains. For example, graywacke, arkose, quartz sandstone, and calcarenite consist mainly of rock fragments, feldspar, quartz, and calcium carbonate respectively. Most sandstone contains high quantities of quartz.

Navajo sandstone
Antelope Canyon in Arizona is made from Navajo sandstone, which was formed after sand was deposited by desert winds about 190–170 million years ago. Iron oxides present in the sand give the rock its red color. Over thousands of years, fast-flowing flood waters eroded the sandstone to form a narrow, steep-sided gorge called a slot canyon. The wavelike shapes were caused by wind-blown sand eroding the rock.

Sea urchin spines, each made of a single crystal of calcite

Sand under a microscope
From skeletal remains of organisms to fragments of different rock types, the tiny grains that make up sand display many shapes, sizes, and colors when viewed under a microscope.

cemented fragments

Conglomerates and breccias are sedimentary rocks made of fragments called clasts. In increasing order of size, these clasts range from fine-grained (with the particles larger than sand grains) to pebble- and cobble-size, with the largest clasts being up to 10 in (25 cm) across. Once the fragments are deposited, they become buried under other rocks and lithify, or transform into solid rock. This process begins with compaction, when the weight of the overlying rocks forces the clasts closer together. When minerals—such as clay, iron oxide, silica, or calcium carbonate—precipitate in the spaces between the clasts, cementation occurs and binds the clasts together. The main difference between breccias and conglomerates is that breccias have clasts with angular edges, while in conglomerates the edges are more rounded. The composition of these rocks gives clues as to the environment and location in which the rock formed.

Conglomerate
This specimen contains jasper and agate, both varieties of microcrystalline quartz (see pp.48–49). The rounded shape and smooth texture of the clasts, typical of conglomerates, indicate that the fragments were deposited far from their point of origin. It is likely they were transported in a fast-flowing river or other high-energy environment capable of carrying the large fragments a long distance and wearing down their sharp edges in the process.

Breccia
The relatively angular edges of breccia clasts indicate that the fragments did not travel far from their source. Breccias are often formed at the bases of cliffs or steep hill slopes where weathering debris accumulates.

TYPES OF BRECCIA

There are two types of breccia, depending on the ratio of clasts to matrix (the groundmass of fine-grained particles in which the clasts are cemented). When the clasts touch each other and the matrix fills the remaining voids, the breccia is called clast-supported. When the clasts appear to be "floating" in the matrix and are not in contact with one another, the breccia is called matrix-supported.

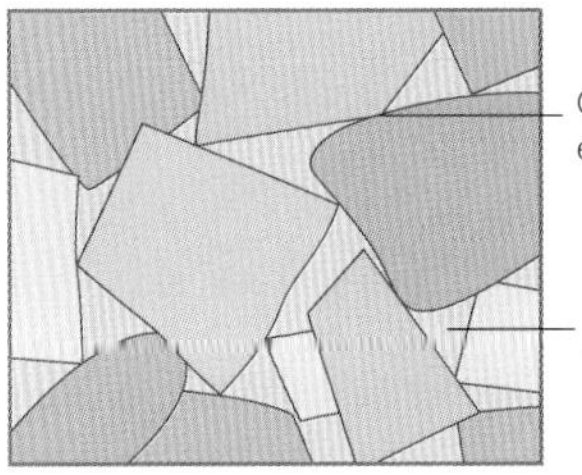

CLAST-SUPPORTED BRECCIA

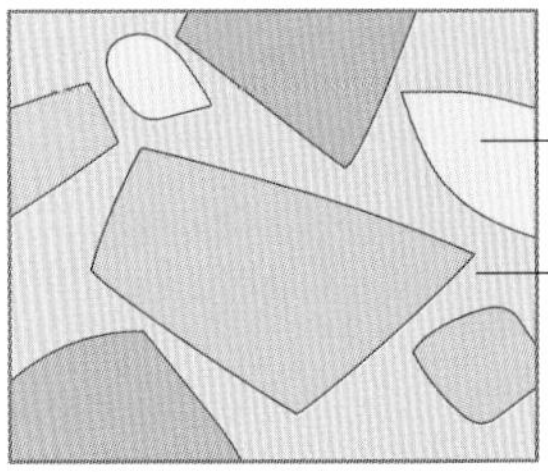

MATRIX-SUPPORTED BRECCIA

chemical deposition

Sedimentary rocks formed from fragments of preexisting rocks and minerals are called clastic sedimentary rocks. Others, called chemical sedimentary rocks, are made of chemicals precipitated out of solutions, often as crystals, and are inorganic or organic in origin. Inorganic rocks formed by chemical deposition include evaporites, dolomite, and inorganic limestone. Organic sedimentary rocks include chert, coal, and organic limestone, mainly formed by marine organisms (see pp.82–83).

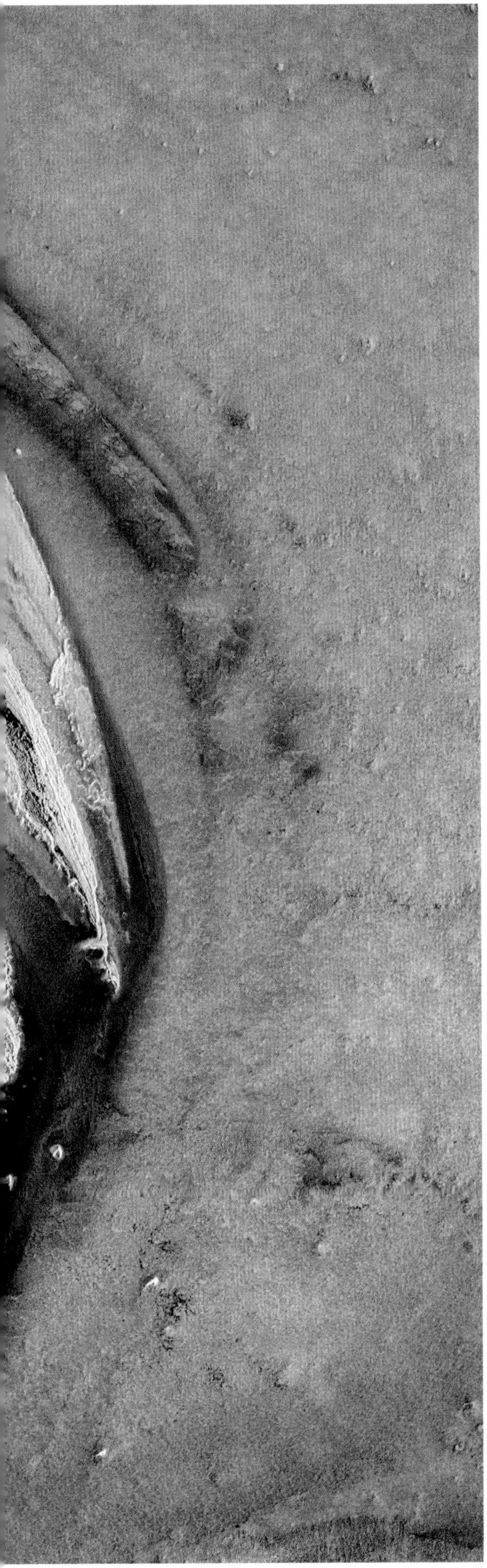

Alternating layers of dark, iron-rich minerals and silica-rich minerals colored red by iron

Banded-iron formation
About 3.7 to 1.8 billion years ago, oxygen produced by photosynthetic organisms in the ocean reacted with iron dissolved in sea water, creating chemical deposits with a banded appearance (see pp.252–53).

Kati Thanda–Lake Eyre
Spanning about 4,280 square miles (11,080 square km) of desert in South Australia, Kati Thanda–Lake Eyre consists of a thick succession of sediments deposited about 60 million years ago. This aerial view shows the lake bed, which floods periodically, covered in salt crusts left behind by repeated evaporation of a shallow body of mineral-rich water. The bright colors are believed to arise from the presence of bacteria.

HOW SALT CRUSTS FORM

In deserts, lakes containing salt water often evaporate faster than they can be replenished by rainfall (1). When this occurs, precipitation leads to a buildup of salt deposits, including halite and gypsum, in a dry basin called a salt pan (2). Repeated episodes of replenishment and evaporation lead to further layers of salt deposits (3). A hard, dry crust forms on the surface, which may break into polygons.

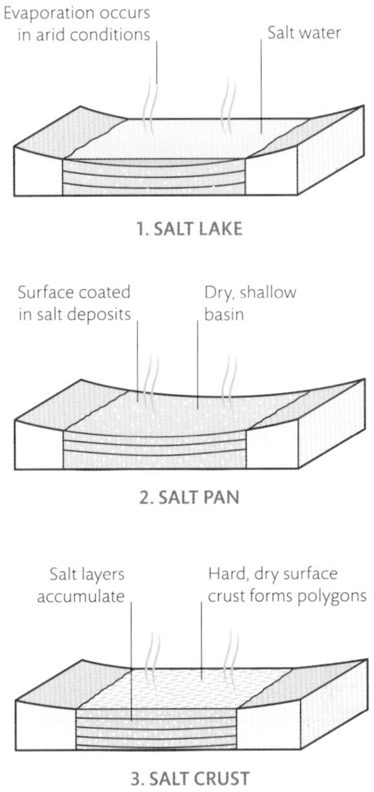

HOW FOSSILIFEROUS LIMESTONE FORMS

Most limestones are formed from the calcium carbonate in the shells and skeletons of marine organisms such as mollusks, sea snails, corals, and crinoids. When the marine organisms (1) die, their hard parts accumulate as calcareous sediment (2). Burial and compaction of these parts in the sediment lead to the formation of limestone (3).

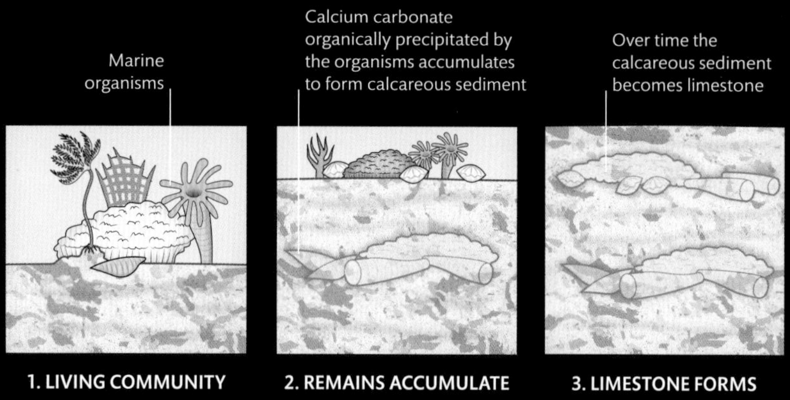

limestone

Accounting for 10–15 percent of all sedimentary rocks, limestone is composed mainly of calcite (a crystallized form of calcium carbonate). Most limestone is formed from the accumulation of the shells and skeletons of dead marine organisms (see panel, above). However, it is also formed by chemical processes, when calcite is precipitated from sea water or lakes. Caves develop in limestone when slightly acidic water seeps through the rocks and dissolves the carbonate deposits, creating a hole in the rock. When limestone undergoes metamorphism, the calcite recrystallizes to form marble.

Chalk
Made from calcite, chalk is a type of limestone formed from the remains of microscopic marine organisms that form a very fine calcareous mud when they die. This mud becomes chalk.

Fossiliferous limestone
Limestones are among the most fossiliferous rocks, with well-preserved fossils providing a useful window on past life. This specimen, which formed about 400 million years ago in a shallow tropical sea, contains abundant fossils of various marine invertebrates, including brachiopods, trilobites, crinoids, and bryozoans.

LOCH GLENDHU
LOCH GLENCOUL
QUINAG
LOCH ASSYNT
An Stuchd

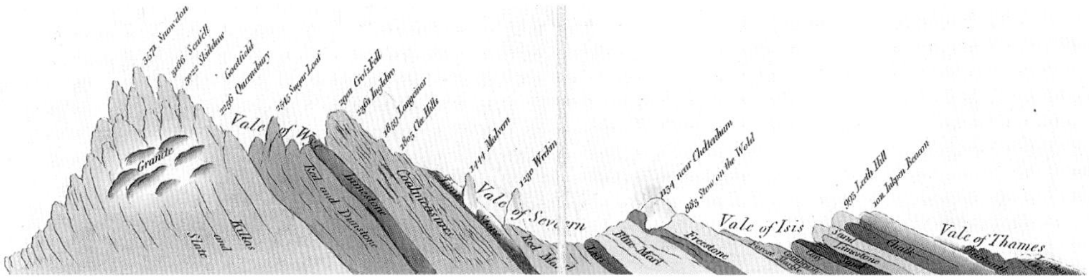

Groundbreaking cross section
This visualization of a cross section of the rocks of England and Wales (right) is part of William Smith's 1815 map *A Delineation of the Strata of England and Wales, with Part of Scotland*. It shows the dip of the rock layers toward the southeast, with the older rocks of the Welsh mountains in the west (left) overlaid by successively younger rocks to the east (right).

history of Earth science

making maps of rocks

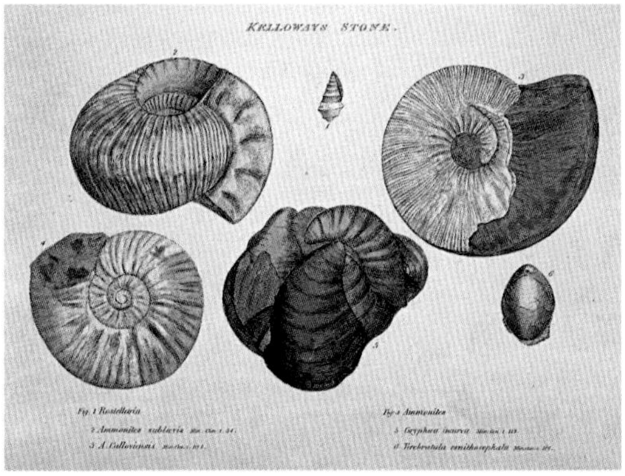

Indicator fossils
This illustration shows some of the characteristic Upper Jurassic fossils found in a pair of rock layers called the Kellaways Formation. William Smith used his understanding of fossils to link outcrops of the same rock layer in different locations.

Geological map of Assynt, Scotland
Compiled at the end of the 19th century, this was one of the first maps to show the complex arrangement of rocks in the heart of an ancient mountain range. As well as using color to show the rock type, lines represent geological faults, and arrows the direction of dip.

For several millennia, maps have been used to record the locations of economically valuable deposits, such as building stones and precious metals—the earliest preserved example is a papyrus map from Egypt, dating from 1150 BCE. But it was not until the 18th century that people started to fill in the gaps between such observations to create systematic and continuous maps of the different rock types beneath their feet.

French geologists Philippe Buache and Jean-Étienne Guettard heralded a new era of map-making when they published a map showing the extent of a layer of chalk across northern France and southern England in 1746. In 1809, Scots-born amateur geologist William Maclure went a step further and produced a geological map of the eastern United States, which classified the surface rocks into four types: primitive, transition, secondary, and alluvial.

Meanwhile, William Smith—a young surveyor in the west of England—had been gaining much more detailed insight into the arrangement of rock layers as he inspected coal mines and canal works. He noticed the same sequence of rocks as he descended mine shafts in different locations. He also found that these rock layers could be differentiated by the characteristic fossils of ancient flora and fauna that they contained, from which he devised his own principle of "faunal succession".

As Smith supervised canal excavations, he was able to test his ideas across a wide area and to extend his vertical sampling of rock layers into a horizontal representation of their surface outcrops. He started in 1799 with a map of the rock types in the close surroundings of the city of Bath, where he lived. He spent many more years traveling England, Wales, and southern Scotland, collecting fossils and mapping the rocks in which they were found. In 1815, he produced the first geological map of an entire country, identifying 23 rock layers, each hand-painted in a distinctive color.

Smith's pioneering work laid the foundations of modern geological mapping. Many of the names he gave to the rock layers he identified are still used by geologists today.

> "Organized Fossils are … the antiquities of the earth; and very distinctly show its gradual regular formation."
>
> WILLIAM SMITH 1817

Serpentinite

The metamorphic rock serpentinite is an example of a rock altered by burial. It is formed when igneous rocks of the oceanic crust are heated to temperatures up to about 400°F (200°C) in the presence of water, and their minerals undergo a metamorphic process called serpentinization.

Colors are evenly distributed in this specimen but they may also occurs in bands

Coarse grains can be seen easily with the naked eye

Mylonite
The compact, fine-grained metamorphic rock mylonite (right) is formed by the shearing of rocks along a fault. The parallel banding, or foliation, is typical of rocks produced by dynamic metamorphism.

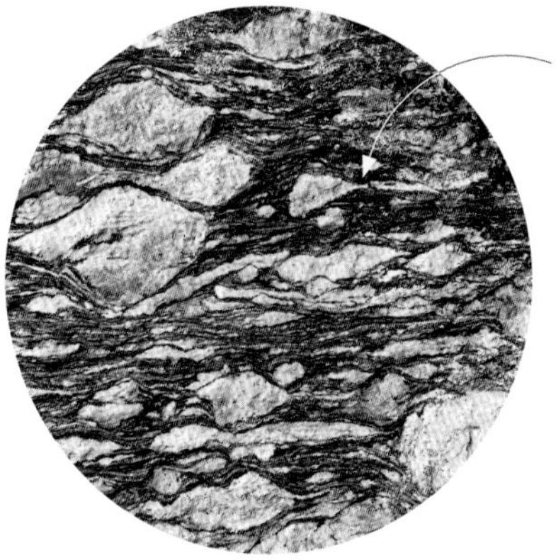

dynamic and burial metamorphism

Metamorphic rocks are formed in various ways. Dynamic metamorphism occurs when pressure is applied in a particular direction during large-scale movements in Earth's crust, especially along fault planes and along active plate margins. When one part of Earth's crust slides, or shears, past another, the surface along which they slide is called a fault. When faults occur at depth, minerals in the rock partially recrystallize, meaning the rocks become fragmented and re-form to create new rocks with a different texture. Rocks can also be transformed when the temperature and pressure increase due to burial, a process called burial metamorphism.

SHEAR ZONES

At shallow depths in Earth's crust, down to about 6 miles (10 km), rocks are relatively cool and brittle. When deformed, they tend to fracture. But in the middle and lower crust, where temperatures are higher, deformation causes rocks to behave like heated plastic and flow rather than break. Over long periods of time, the strain on rocks in these deep locations, called shear zones, causes their minerals to recrystallize and results in the banded textures seen in mylonite (see above).

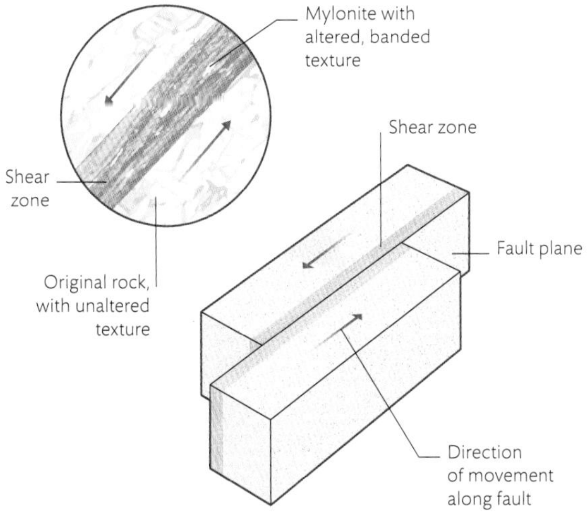

Gneiss
The metamorphic rock gneiss typically forms in high-pressure, high-temperature conditions (over 950 °F/500 °C). The light and dark banding is due to chemical reactions that cause the minerals in the rock to segregate into separate areas with a different composition. The light areas contain mainly quartz and feldspar; the dark banding contains minerals such as pyroxene and amphibole.

regional metamorphism

Rocks that have been altered by pressure or heat, or both, over a large area have undergone a process called regional metamorphism. This typically occurs when tectonic plates collide to form mountain belts (see pp.114–115 and pp.120–121). In this situation, rocks in the crust become compressed (see panel, below), and rocks that were once at Earth's surface can be dragged tens of miles down into the crust, where temperatures are much higher. The type of metamorphic rock that results depends on the original rock type, as well as the heat and pressure it is subjected to.

Welsh slate
A fine-grained metamorphic rock, slate is formed by the effects of heat and pressure on shale. Since slate can be split into thin layers along its natural cleavage planes (see panel, below), it is useful as a roofing and paving material.

FOLDING AND FOLIATION

When rock layers are compressed due to movements in Earth's crust, they fold, or bend. Also, minerals in the rock partially recrystallize without melting. The new crystals are aligned perpendicular to the direction of compressional stress and parallel to the axial plane of the fold. This alignment is called foliation. In the case of slate, the parallel planes (called slaty cleavage) are closely spaced and are areas of relative weakness, which is why the rock can be easily split into thin sheets to use as tiles.

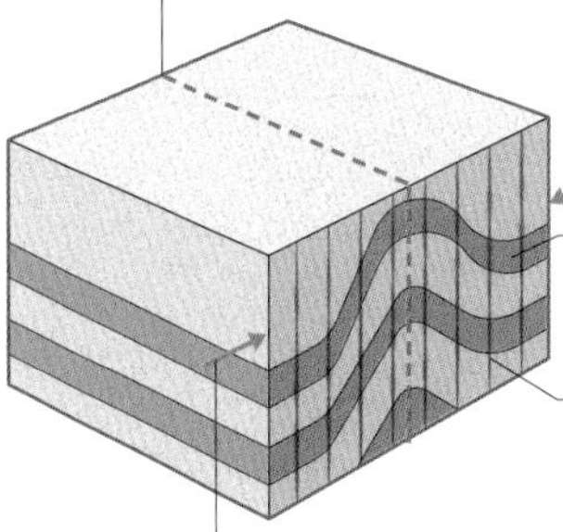

SECTION OF COMPRESSED ROCK

contact metamorphism

As intensely hot migrating magma starts to cool and solidify deep below Earth's surface, it forms a body of igneous rock called an igneous intrusion (see pp.134–135). This intrusion releases heat into the surrounding rocks, which undergo a process of chemical change called contact metamorphism. Several types of metamorphic rock are formed in this way, including hornfels—which forms when heat from magma comes into contact with shale, slate, or basalt—and skarn (see right). Since contact metamorphism occurs as a result of temperature increases, rather than pressure, folding and foliation do not occur during the process—unlike in dynamic and regional metamorphism (see pp.86–89).

Skarn
When the intense heat from an igneous intrusion reaches carbonate rocks such as impure limestones, the rocks are altered by contact metamorphism. The resulting metamorphic rock is called skarn.

METAMORPHIC AUREOLES

The area around an igneous intrusion that is affected by contact metamorphism is called a metamorphic aureole. The degree of alteration decreases with distance from the source of the heat, producing a series of distinct concentric zones. In the example shown here, molten magma has solidified to form a granite intrusion. The rocks closest to the magma (hornfels) have undergone greater metamorphosis than those situated farther away (spotted rock). The sedimentary rock shale is beyond the aureole and has not been affected by metamorphism.

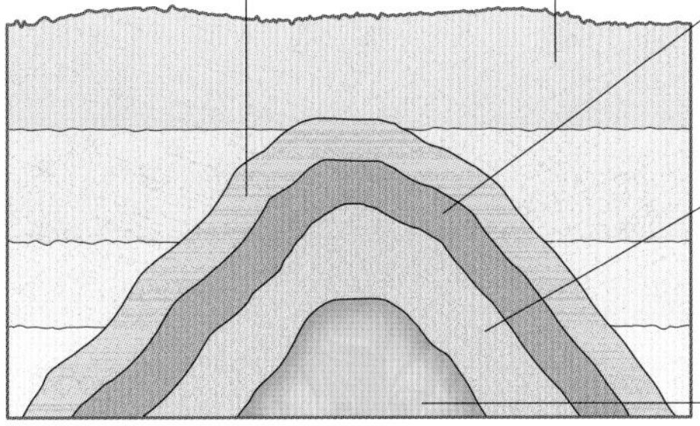

Altered by heat

Among the layers of sedimentary rocks that make up the mountains surrounding the Grinnell Glacier in Montana, a sill has formed. This occurs when molten magma is thrust between layers of sedimentary rock and solidifies to form a band of igneous rock that follows the sedimentary layers. The intense heat from the sill has altered the rocks immediately above and below through contact metamorphism.

Dark band is a sill made of the igneous rock dolerite—the source of heat that resulted in the alteration of the surrounding rocks

Sedimentary rocks have been deposited in distinct layers over time

Tilted layers
The colorful Rainbow Mountains in north-eastern China are made from sandstone and siltstone that were deposited in the Cretaceous Period, 145–66 million years ago. The layers, which started off as horizontal, were compressed and folded about 50 million years ago during the collision between the Indian and Eurasian continental plates, which resulted in the formation of the Himalayas.

ANGULAR UNCONFORMITIES

The relative positions of rock strata help reveal their geological history. Sediments are deposited in horizontal layers (1), but tectonic activity can result in layers folding and tilting (2). Erosion of those layers can then wear away the top part of the fold to a flat plane (3). If new rock layers are deposited on top of the tilted, eroded layers, an angular unconformity is created (4)—a feature that represents a gap in the rock record.

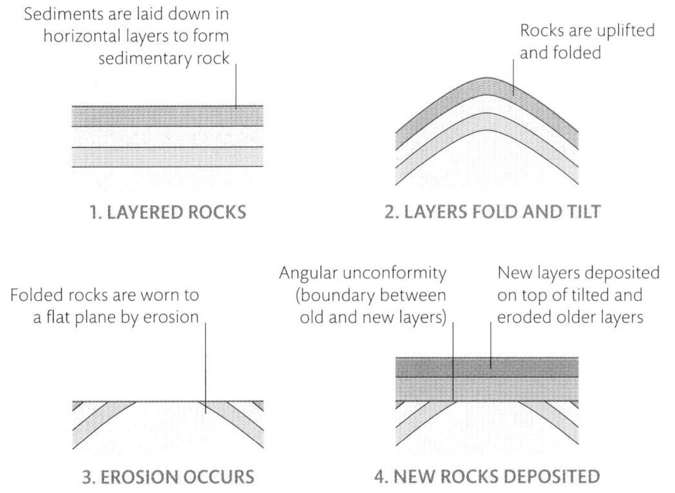

layers of rock

Studying layers of rock allows Earth scientists to understand how and when rocks were formed. Sedimentary rocks are deposited in horizontal layers, with the oldest rocks at the bottom of the sequence and the youngest ones at the top, from which the relative ages of the rocks can be determined. However, the layers can be tilted or overturned by tectonic activity, and erosion can result in gaps in the geological record (see panel, above). By studying fossils embedded in the rock, scientists can assess the ages of the layers, and whether layers that are geographically separated were deposited at similar times.

Horizontal layers
This cliff (right) on the west coast of Wales, in the UK, forms part of the group of stratified sedimentary rocks known as the Aberystwyth Grits. The base layer is thought to have been laid down about 488–443 million years ago.

spotlight the Grand Canyon

The Grand Canyon winds across the Colorado Plateau in the southwestern US. Here, the Colorado River has cut 1 mile (1.6 km) through the crust, creating a breathtaking landscape and opening a unique window on 1.8 billion years of Earth's history, revealed in the layers of rocks exposed on its walls. At its widest, the canyon is 18 miles (29 km) across.

The rocks of the Colorado Plateau were deposited from about 575 to 270 million years ago, mostly in shallow seas, beaches, and swamps. The area was then lifted 10,000 ft (3,000 m) above sea level between about 70 and 30 million years ago, probably as a result of a collision between tectonic plates. The canyon itself started to form about 6–5 million years ago when the

Colorado River found an outlet to the sea in the Gulf of California. Falling about 2,000ft (600m) over its journey of 277 miles (446km) through the canyon, the river's course is steep, giving its water tremendous speed and down-cutting power. The water volume and rock load can increase dramatically during times of flood, and would have been much higher at the ends of glacial periods, when upstream glaciers were melting.

The main river has now reached basement rocks that are much harder than the overlying sediments, so the canyon is now growing wider faster than it is becoming deeper, as tributary streams erode its sides.

The canyon at sunrise
The Grand Canyon's cliffs display colorful layers of limestone, sandstone, shale, and lava that overlie a basement of ancient igneous and metamorphic rocks. The uplift of the Colorado Plateau happened without much deformation, leaving the rock strata mainly horizontal and the stratigraphic column easy to read.

Moss is growing on the organic layer on top of the soil

Dark, fertile topsoil, with a high content of organic matter and minerals

Light gray layer is paler than layers above due to leaching of organic matter and iron

Subsoil layer, where leached organic matter and minerals (especially iron oxides, giving the reddish color) have accumulated; roots also extend down to subsoil

Dark humus layer, which contains live and decomposing plant and animal matter, stores large amounts of carbon

SOIL PROFILES

Most soils are divided into layers, or soil horizons, and their arrangement is called a soil profile—a vertical section through the soil from the surface to bedrock. Not all horizons are present in every soil, but most soils consist of (from top to bottom): a humus layer with plant matter; a dark, fertile topsoil layer; a lighter-colored, mineral-rich subsoil layer; an infertile, stony layer; and a solid layer called bedrock.

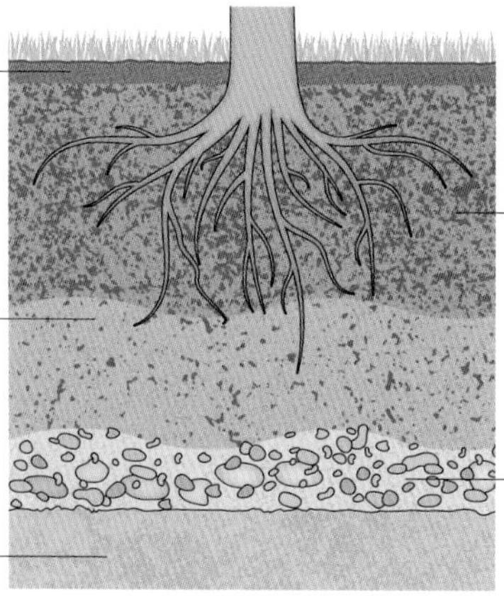

the soil layer

Comprising a mix of organic matter, minerals, water, air, and living organisms, soils play a vital role in Earth's ecosystems. They absorb and store carbon, recycle dead organisms into nutrients, and provide a key growing medium for plants and a home to a quarter of all known species. Soils also soak up and hold water for plants and other organisms, and control the rate at which excess water drains through. In addition, soils rich in microorganisms break down harmful chemicals, reducing contamination of groundwater.

Parched soil
When drought and dry conditions persist in an area, like this barren plain in Victoria, Australia (right), soils release water and shrink, leading to the formation of cracks. This typically happens in clay-rich soils.

Dry soil has cracked, forming deep fissures

Layers of a podzol
A type of acidic soil called a podzol is distinguished by a light gray layer below the humus or topsoil. It is formed by leaching, when iron and organic matter are washed down into the lower subsoil, usually due to rainfall and rapid drainage. Podzols are found in coniferous forests.

Water droplets
The formation of water droplets, and their roughly spherical shape (left), are due to surface tension, which has a cohesive effect on water molecules interacting at the surface layer (see panel, below).

Water droplet lingering on leaf-tip is held together by surface tension

properties of **water**

Water is the most abundant chemical in living things. It also has many unusual or unique properties, largely due to its molecules' ability to form hydrogen bonds (a chemical bond that attaches one molecule to another). Few other substances naturally occur at Earth's surface in all three states of matter: as a liquid (water), a solid (ice, see pp.100–101), and a gas (steam or vapor). In addition, water can absorb large amounts of heat without a significant rise in temperature, and releases stored heat when temperatures fall. The relatively constant temperature of water helps sustain life.

Three states of matter
As the sun sets over this landscape in Loch Shiel, in the Scottish Highlands, ice is forming a solid surface over the liquid waters of the lake, while water vapor is condensing to form water droplets hanging in the air as mist.

SURFACE TENSION

Water has a high surface tension, which has advantages for some animals and plants—for example, it enables some insects to walk on water and helps transport water from the roots to the leaves of plants. Surface tension is created by the cohesive forces between water's molecules. Within a body of water, the water molecules below the surface are pulled in all directions by surrounding molecules. In contrast, the water molecules at the surface are pulled only inward or sideways, as there are no water molecules to bond with above the surface. As a result, the molecules at the surface are pulled into the bulk of the liquid, which forces liquid surfaces to contract.

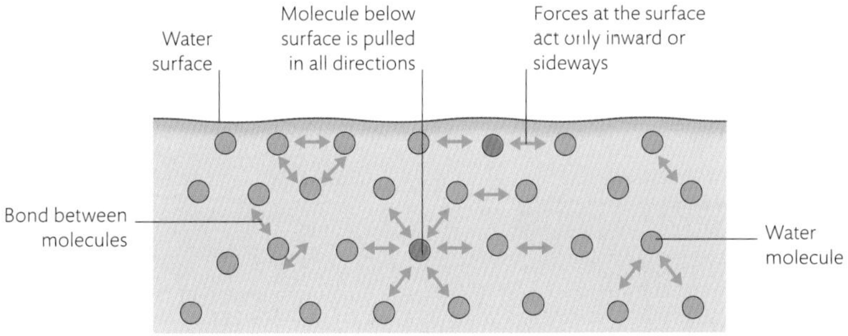

LATENT HEAT

In order for a substance to change state (for example, when a solid such as ice turns into liquid water, or water turns into vapor), heat energy called latent heat is required. The temperature of the substance remains constant during this process. The amount of energy required depends on the state of the substance. Water has high latent heat, since it needs a lot of heat energy to alter its atomic bonds. This means as snow and ice warm, they melt relatively slowly, staying the same temperature until they are liquid.

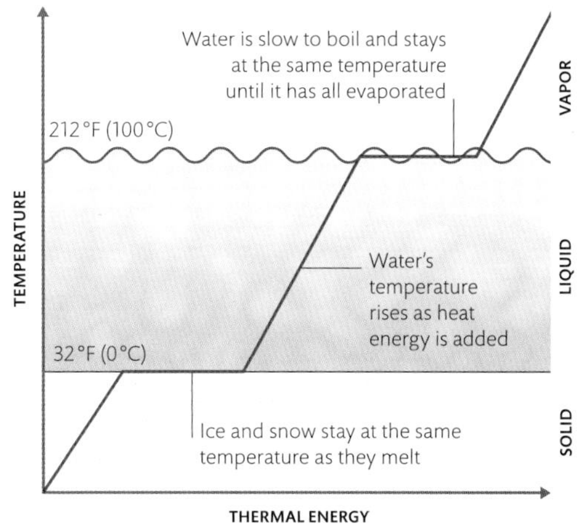

frozen water

When liquid water cools below its freezing point, 32 °F (0 °C), it starts to turn into solid ice. The molecules in ice are more loosely packed than in water. This means that, unlike most other substances, water expands as it freezes, growing in volume by about 9 percent. Ice is therefore less dense than water, which is why it floats. Ice is essentially a mineral, being a naturally occurring compound with a defined chemical formula and crystal structure. Snowflakes are ice crystals that have fused together and naturally arranged themselves into intricate hexagonal structures (see pp.230–231).

Lake Baikal
The ice that sits on the surface of the world's deepest lake, Lake Baikal—in Siberia, Russia—is crisscrossed with cracks (left). The natural cracking of ice is caused by daily fluctuations in the temperature of the air, which causes ice to expand during the day and contract at night.

Dendritic, or branching, patterns are typical of frost flowers

Frost flowers
These flower-shaped ice crystals grow on young sea ice (mainly in polar regions) and thin lake ice in calm conditions. They form when water cools rapidly when exposed to much colder air.

Descending ice cloud
Snow and ice hurtle down a steep face of Baltistan Peak (or K6), Pakistan. Avalanches such as this one can descend at speeds of hundreds of miles per hour.

avalanches

The sudden movement downslope of a mass of ice or snow, often incorporating earth and rock, is known as an avalanche. Avalanches start when a layer of snow or other material becomes unstable. This can be triggered by snowfall, temperature change, earthquakes, or disturbance caused by human activities such as hiking and skiing. There are several types of avalanche. Loose snow avalanches, also called sluffs, occur when fresh snowfall moves down steep slopes. Slab avalanches occur when a large block of solid snow breaks loose and slides down a slope. Powder snow avalanches comprise components of both loose snow and slab avalanches. Wet snow avalanches occur when water and snow move downslope together.

SLAB AVALANCHE FORMATION

These highly destructive avalanches form when heavy, new snowfall accumulates on a weak, crusty snow layer that is lying on top of older, consolidated snow. The new snow layer can fracture into blocks or slabs that on a steep slope may slide downhill, often at great speed. As the slab descends, the front edge disintegrates and forms a cloud of icy particles, which is sometimes preceded by a powerful blast of air.

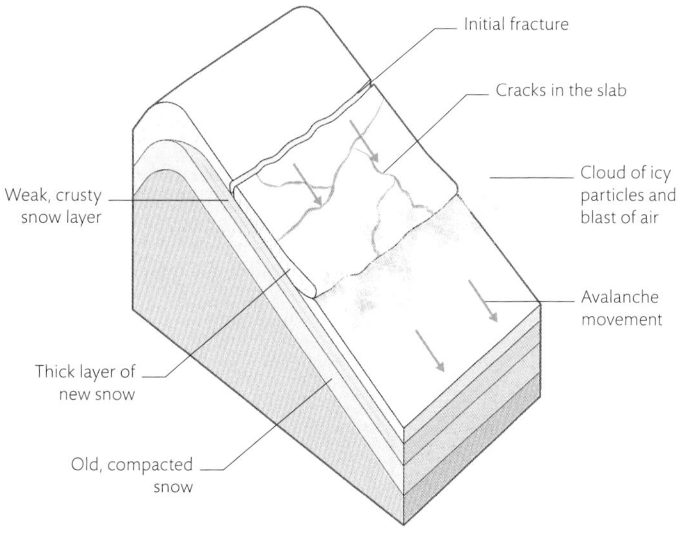

dynamic Earth

Our planet is a complex structure made up of three primary layers. The outermost of these layers is in a state of constant change, driven both by forces in the interior and processes of weathering, erosion, transportation, and deposition at the surface. Many of these changes happen too slowly to be perceived in a human lifetime, but occasionally they happen with sudden speed and violence.

Green crystals are omphacite, a sodium-rich variety of the mineral pyroxene

Red crystals are garnet, which is a mineral used as a gemstone (see pp.58–59)

Coarse grains are evenly distributed in this specimen, although they may also form bands

EARTH'S INTERIOR

Earth's outermost rocky layer, called the crust, is typically up to 6 miles (10 km) thick under the oceans (oceanic crust) and up to 44 miles (70 km) thick beneath the continents and continental shelves (continental crust). The thick layer of denser rocks directly below the crust is called the mantle. The solid, brittle top layer of the mantle, which extends down to about 60 miles (100 km) deep, is called the lithospheric mantle. Below this is the asthenosphere, which is made of soft, partially molten rocks and extends to about 220 miles (350 km) beneath the surface. Here, high temperatures and pressure cause the rocks to flow like wax and change shape. Lower down within the mantle, pressure from overlying rocks keeps materials solid. The core, which is composed mainly of iron and nickel, has a liquid outer part and a solid inner part.

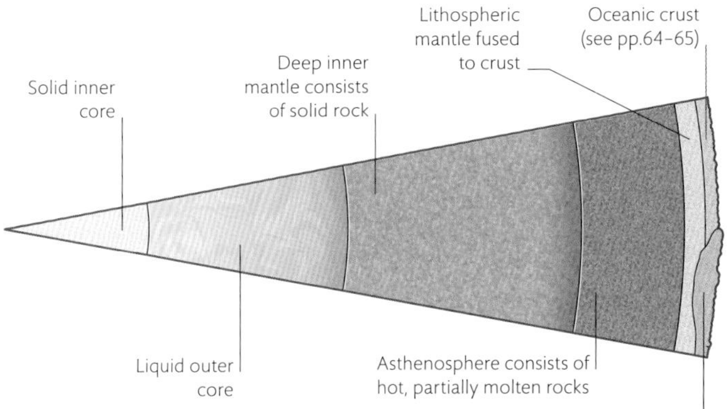

CROSS SECTION THROUGH EARTH

Earth's structure

Earth is made up of concentric layers, each of which has different properties and is made of different materials. The three main layers are: a relatively thin outer crust, a large mantle mostly made of the igneous rock peridotite, and a central core; within each layer, there are further subdivisions (see panel, above). Much of what we know about Earth's structure and composition comes from analyzing seismic waves—vibrations that are generated by an earthquake, a volcanic explosion, or a landslide and move through different materials in Earth's interior at different speeds. For example, scientists assume that the outer core is liquid because certain types of seismic waves cannot move through liquids. Earth's crust, being closer to the surface, can be sampled directly through drilling.

Large rock fragments from Earth's interior eroded from the mountains towering in the background

Rocks from Earth's interior
The Gros Morne National Park in Newfoundland, Canada (above), provides a rare opportunity to see rocks from Earth's mantle, including peridotite, and oceanic crust. Plate tectonics caused the rocks to thrust through the continental crust around 470 million years ago.

Garnets are globular, giving eclogite its distinctive spotted appearance

Eclogite
A rare, coarse-grained metamorphic rock, eclogite forms when igneous rocks low in silica, mainly basalt and gabbro, are dragged deep down into Earth's lowermost crust and mantle—sometimes to depths of 90 miles (150 km)—following the convergence of two tectonic plates (see pp.114–115). The intense heat and high pressure cause the igneous rocks to recrystallize and form new rock. Eclogite consists of greenish omphacite and red garnets, but it also contains smaller quantities of other minerals such as quartz and feldspars.

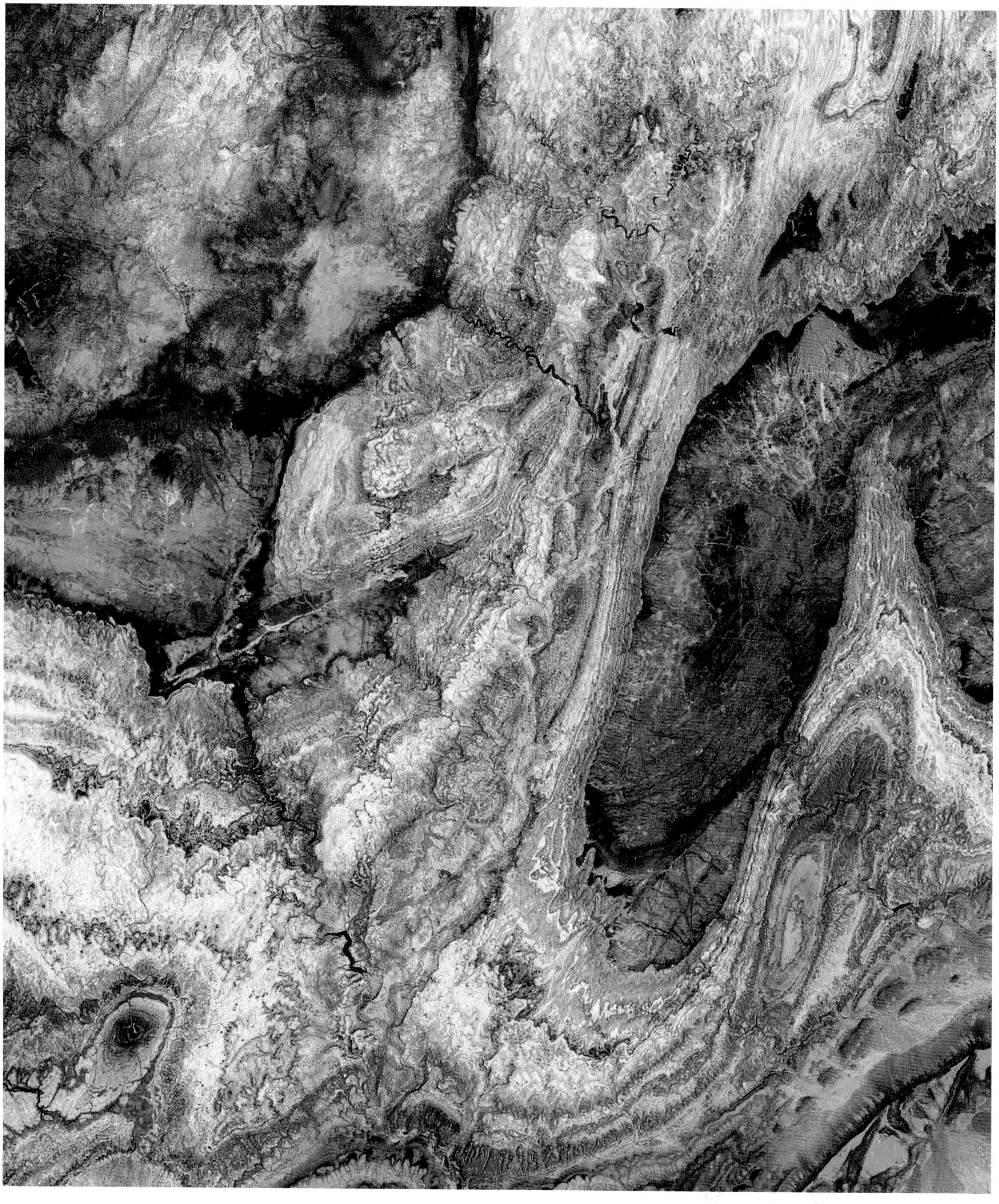

Landsat 9
Orbiting from pole to pole once every 99 minutes at a height of 443 miles (705 km), Landsat 9, launched in 2021, takes more than 700 images per day and revisits each part of the land surface every 16 days.

history of Earth science

satellites and Earth science

The use of satellites to study Earth and its atmosphere began in 1958 when a research satellite detected Earth's radiation belts. The first successful weather satellite was launched in 1960, and since 1972 detailed images of the land surface have been taken by a series of Earth observation satellites.

Colors represent differences in height from a sphere

Gravity "potato"
Low-orbiting satellites are sensitive to small variations in the local gravity field. They have been used to record the shape of the Earth's geoid—the reference surface that the ocean would have without winds and tides. These readings have been used to create visualizations such as the "gravity potato" (above).

Like the early weather satellites, Landsat 1, the first Earth observation satellite, carried a TV camera. It was also equipped with a new multispectral scanner with an infrared channel that enabled it to monitor vegetation growth. Later Landsats added more channels to view more of the spectrum, making it easier to identify different rock types.

During the 1980s and 1990s, other satellites explored different parts of the electromagnetic spectrum: ultraviolet light to track atmospheric ozone; thermal infrared for the temperature of land, ocean, and clouds; and microwave emissions to show water in all its forms—as liquid in the soil, frozen as snow and ice, and as vapor and droplets in the atmosphere. Radar satellites can see through clouds and at night. They are used to measure the height of land and sea, and changes to the thickness and flow of ice sheets.

Satellites allow us to quickly survey remote and inhospitable areas such as the oceans, tropical rainforests, and the polar regions. In the 21st century, geophysical satellites have even opened up a window on Earth's interior, detecting minuscule variations in the planet's gravity and magnetic fields to probe rock density variations in the mantle and the flow of material in the outer core.

Geology from space
This image of the Anti-Atlas Mountains, Morocco, was taken in 2001 by the Advanced Spaceborne Thermal Emission and Reflection Radiometer (ASTER), designed to distinguish different rock types. Layers of limestone, sandstone, and gypsum appear here in yellow, red, and light green, while underlying granites are shown in dark blue and green.

> "Man must rise above the Earth … for only thus will he … understand the world in which he lives."
>
> ATTRIBUTED TO SOCRATES, 470–399 BCE

Continental crust
Rocks on continental plates are generally old and relatively stable. The oldest continental crust is concentrated in areas known as shields, such as the Australian Shield (see left), in southwest Australia. The rocks in shields are 570 million years old or more.

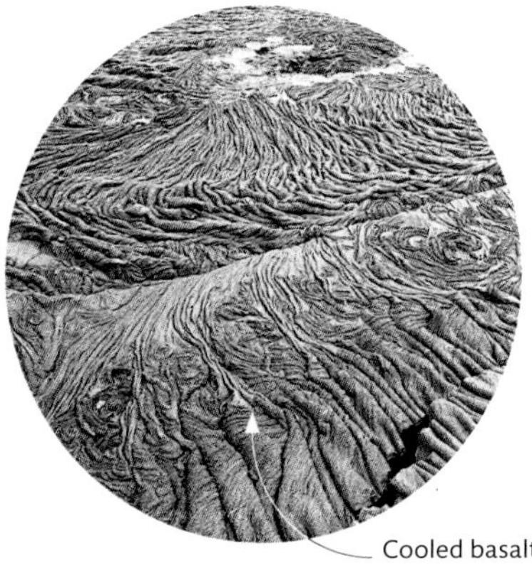

Oceanic crust
These basalts in the Galápagos Islands formed when the Pacific Ocean's Nazca Plate moved over a plume of hot mantle material. The rising magma welled up to Earth's surface and erupted as lava flow.

Cooled basaltic lava has a twisted, ropelike texture

tectonic plates

Plate tectonics is a widely accepted theory that Earth's solid outer shell, called the lithosphere (made up of the crust and the top layer of the mantle), is divided into large, rigid blocks (plates) that float on the hot, partially molten rocks of the asthenosphere. This causes the plates to move (see panel, below), typically at a rate of ¾–8 in (2–20 cm) per year. The movement of the plates—which may collide, move away from each other, or move past each other—results in the formation and modification of Earth's major surface features, with most activity taking place at the boundaries (see pp.112–113). There are two main types of plates: continental and oceanic, which are about 90 miles (150 km) and 40 miles (70 km) thick respectively.

CAUSES OF PLATE MOTION

Plate motion is thought to be driven mainly by heat rising from the interior of the planet. Convection currents in the mantle lift Earth's lithospheric mantle, and hot, molten magma rises at mid-ocean ridges (see pp.116–117), causing plates to move. The pull of gravity at subduction zones (where a tectonic plate descends into the mantle) and at mid-ocean ridges is also thought to be significant in the movement of plates.

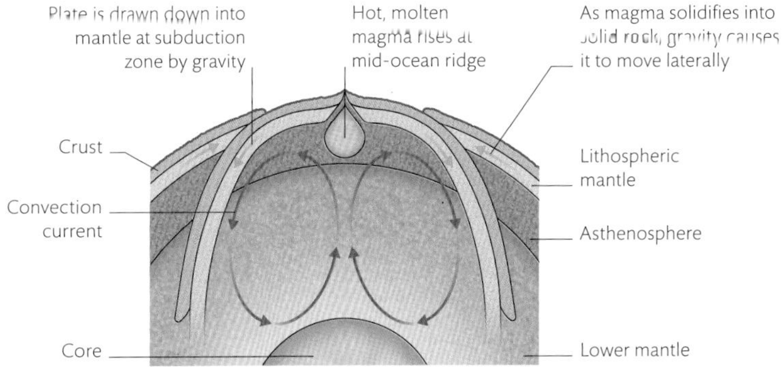

CROSS SECTION THROUGH EARTH

TRANSFORM BOUNDARIES

When tectonic plates slide past each other horizontally, in opposite directions, the boundary between them is called a transform boundary or transform fault. This usually occurs on the ocean floor or at the edges of continents. In continental transform boundaries, the continental crust and the lithospheric mantle—the solid, brittle part of the upper mantle fused to the crust—move over the underlying asthenosphere (see below and panel, pp.106–107). The motion between the plates causes the rocks on either side to strain and deform, sometimes resulting in earthquakes.

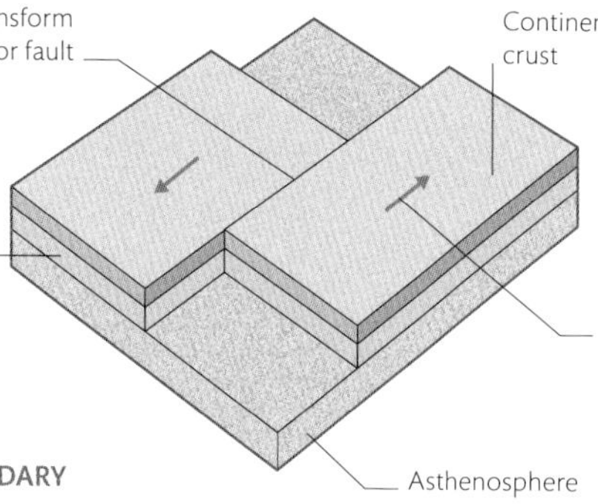

CONTINENTAL TRANSFORM BOUNDARY

plate boundaries

Some of the most intense tectonic activity on Earth takes place at the boundaries of tectonic plates (see pp.110–111). There are three main types of plate boundary: convergent boundaries, where two tectonic plates meet and collide (see pp.114–115); divergent boundaries, where two plates move away from each other (see pp.116–117); and transform boundaries (see panel, above), where the plates slide past one another horizontally. Each type of plate boundary is associated with different geological processes, including mountain-building, volcanic activity, and earthquakes. A single plate can have multiple types of boundary.

Speckled appearance due to pale feldspar mixed with black hornblende and biotite crystals

San Andreas Fault
On the west coast of North America, two adjacent tectonic plates are moving in opposite directions, offsetting hills and landforms. Their transform boundary, which is 750 miles (1,200 km) long, is associated with powerful earthquakes in the region.

Diorite
The igneous rock diorite is associated with convergent plate boundaries. It is an example of an intrusive rock, with visible crystals that solidified slowly in a magma chamber.

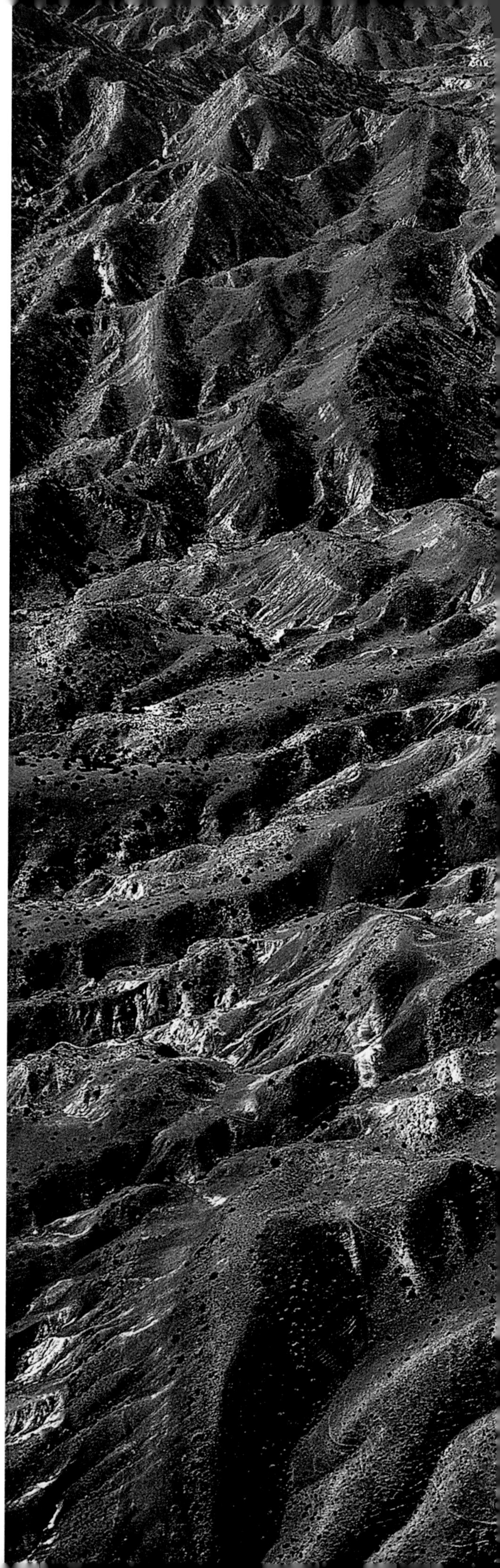

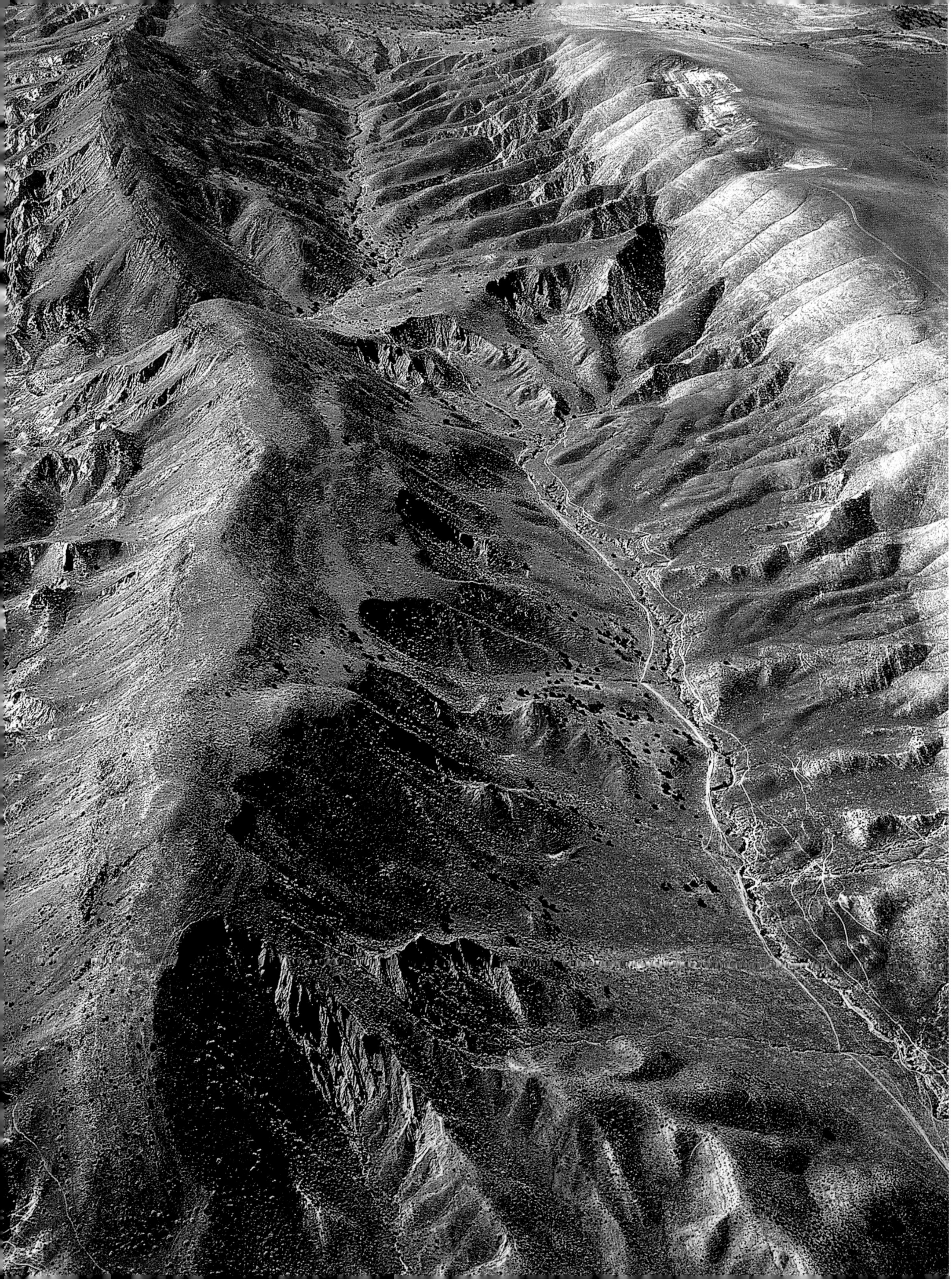

Continent-continent collision
This aerial view shows part of the Ladakh Range in the western Himalayas. Like the rest of the range, they are the result of a collision between sections of continental crust in the Indian and Eurasian plates.

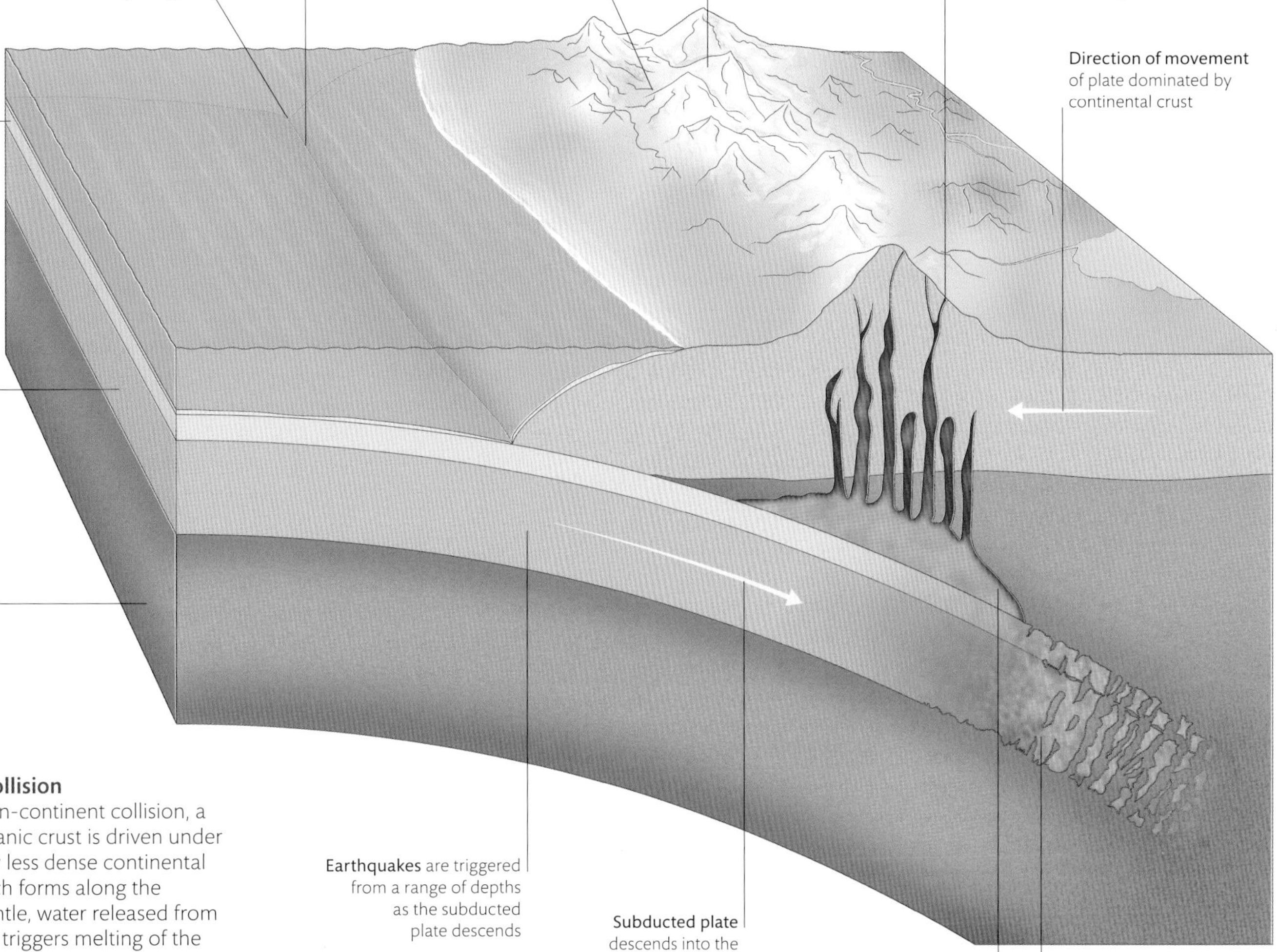

Ocean-continent collision
At the site of an ocean-continent collision, a plate with dense oceanic crust is driven under a plate dominated by less dense continental crust. An ocean trench forms along the boundary. In the mantle, water released from the descending plate triggers melting of the mantle, which gives rise to a chain of volcanoes.

colliding plates

Earth's tectonic plates are constantly moving. When two plates move toward one another and collide at a plate boundary, that boundary is termed convergent. There are three main types of convergent boundary: ocean-continent, where oceanic crust collides with and is forced down (subducted) under continental crust; ocean-ocean, where the denser of two sections of oceanic crust is subducted under the other; and continent-continent, where two continents collide, merging into a single landmass. Because continental crust is of relatively low density, it does not become subducted, but crumples to form mountain ranges. The boundary between two plates in which one is subducted is known as a subduction zone. Where continents collide, the boundary is called a suture zone.

diverging plates

When two tectonic plates stretch, thin, and break apart, the zone where they meet is called a divergent boundary. Where this occurs beneath oceanic crust, magma rising from the mantle forms a deep fissure. New magma then flows into the fissure, spreads outward, and solidifies to form new crust. If the process (seafloor spreading) is repeated, it can lead to the formation of a submarine mountain range called a ridge (see panel, opposite). Where a divergent boundary occurs under thicker continental plate, the plate fractures to form a rift (trough-shaped valley) with faults on either side. Water then flows into the valley from rivers and streams, forming a lake; where the rift drops below sea level, sea water may flow in. If the rift deepens and widens, a new ocean basin can form.

Great Rift Valley
In East Africa (including Tanzania, above), valleys and gorges have been created by the thinning of continental crust close to a divergent boundary under the Red Sea and Gulf of Aden.

MID-OCEAN RIDGES

Spreading ridges have different topographies depending on the speed at which the tectonic plates move. In fast-spreading ridges, such as the East Pacific Rise, the ocean floor moves away from the ridge before cooling and subsiding, or sinking, forming broad, gentle landscapes. In slow-spreading ridges, such as the Mid-Atlantic Ridge, the ocean floor does not move far before cooling and subsiding, forming narrow, steep landscapes with faults.

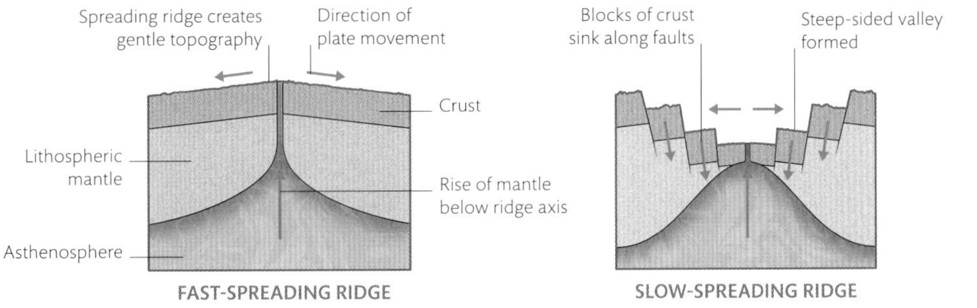

Oceanic divide
The Mid-Atlantic Ridge extends about 10,000 miles (16,000 km) along the center of the Atlantic Ocean. The Silfra Fissure in Iceland (shown here), which straddles the Mid-Atlantic Ridge, is one of the few places on Earth where people can dive and swim between two tectonic plates—the Eurasian and North American plates.

Rocks on the limb of the fold, originally part of a syncline, have been compressed and overturned
Smaller folds, called parasitic folds, have formed on the limb of the overturned fold

TYPES OF FOLD

When horizontal layers of rock become compressed laterally, they crinkle into a series of troughs called synclines and arches called anticlines. The innermost rocks of synclines are the youngest, and the innermost rocks of anticlines are the oldest. If the compression persists, one side, or limb, of the fold can move more than the other, forming an overturned fold, with the sequence of rock layers on the lower limb reversed (oldest layer on top).

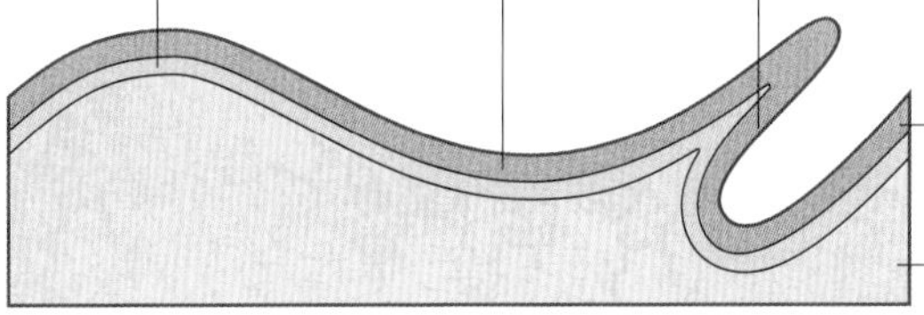

Earth's folds

When hot rocks deep within Earth's crust become compressed by the movement of tectonic plates (see pp. 110–111), rather than fracturing or breaking, they can become folded or bent. Folds often form at the boundaries of colliding plates (see pp. 114–115) and can occur on a variety of scales, from a small bend affecting a hand-size rock specimen to the folds that make up entire mountains, such as in the Swiss Alps and Canadian Rockies. By studying how rocks are folded, geologists can better understand the geological history of an area.

Mountain-size fold
These layers on the flank of the Dent de Morcles in Switzerland (left), were deposited in the Mesozoic Era (about 245 to 65 million years ago). They were folded during a collision between Europe and Africa about 65 million years ago.

V-shaped folds
These limestone and chert layers in Crete, Greece, have been folded into V-shapes called chevron folds. These folds have straight limbs that bend abruptly on one point, called the hinge zone.

Snowy spires
The Dolomite mountains of northeastern Italy are still gradually growing, buckling upward as the Eurasian and African continental plates collide. They are composed of dolomite, a hard carbonate rock, and their dramatic jagged peaks are the result of erosion by rain and ice.

mountain-building

Some mountains are created by the accumulation of basalt that forms from lava erupted from volcanoes on land or on the ocean floor, while others are formed when rivers erode steep gorges into preexisting plateaus. But one of the most common settings for mountain formation is at the boundary of two continental plates. When two continental plates collide, the crust buckles, thickens, and folds to create high mountains. About 60–40 million years ago a collision between the continents of India and Asia gave rise to the Himalayas (see pp.124–125). Other mountains ranges—for example, the Andes—formed at the boundary between an oceanic plate and a continental plate.

MOUNTAIN ROOTS

Earth's crust is thicker below high mountains than below the surfaces of low elevations. When mountains are built, the crust thickens both above and below the surface. Much like icebergs, mountains have deep roots underneath, which mirror and exaggerate the surface topography. The roots, which are made of a lighter, more buoyant, crustal material, float on the denser mantle underneath. Earth's crust is in a state of buoyant equilibrium called isostasy. According to this principle, the depth to which a floating object sinks depends upon its thickness and density. Over time, the upper surface of a mountain is lowered by the action of water, wind, and ice, and the buoyant root rebounds to balance the loss of mass on the surface.

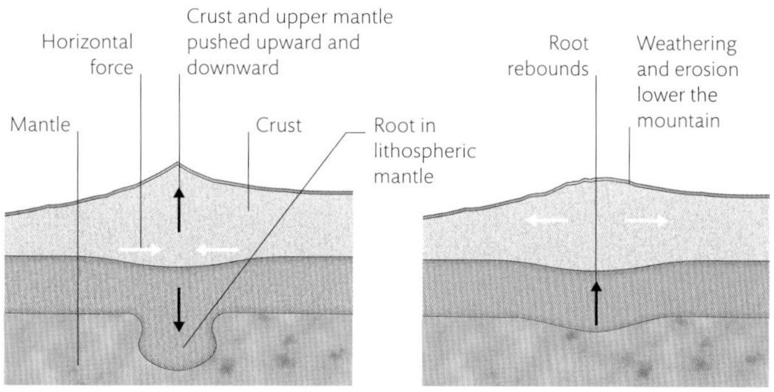

MOUNTAIN BUILDING

EROSION AND ROOT REBOUND

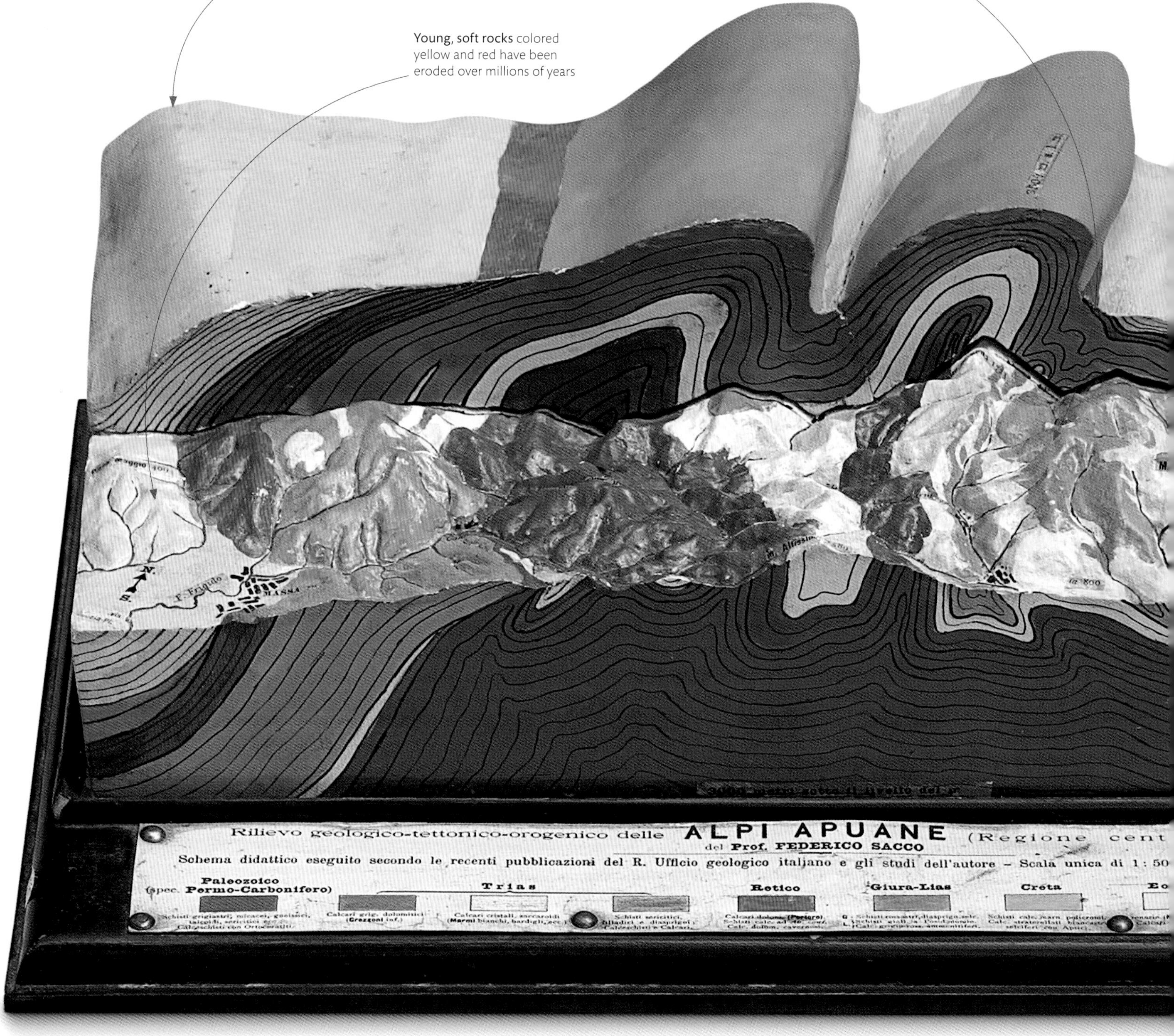

The upper part of the model includes possible sections of folds that have been removed by erosion

Older, harder rocks indicated in purple and gray have greater resistance to the forces of erosion and form peaks

Young, soft rocks colored yellow and red have been eroded over millions of years

> “ … the work of Suess marks the end of the first day, when there was light. ”
>
> MARCEL-ALEXANDRE BERTRAND, FRENCH GEOLOGIST, 1897

Albert Heim
Swiss geologist Albert Heim (1849–1937), pictured in later life with local mountain guides in the Maderaner Valley, Switzerland, made pioneering studies of the Alps from the 1880s until the 1900s.

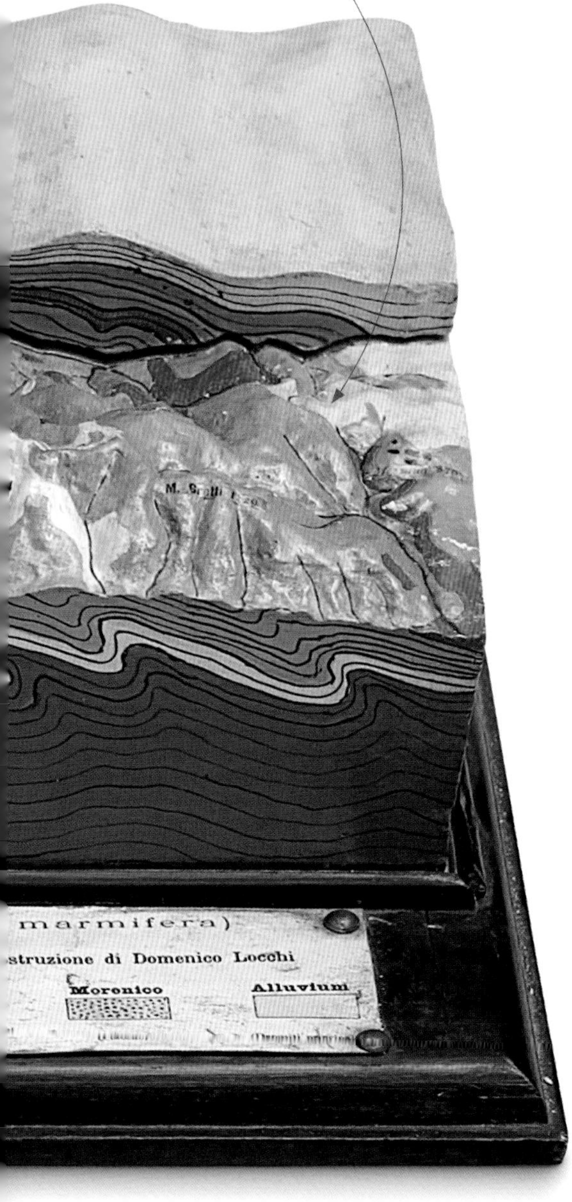

Relief model of the Apuan Alps
Many early geologists created cross-sectional relief models to explain their ideas. This example was made by the Italian geologist and paleontologist Federico Sacco (1864–1948).

history of Earth science

understanding mountain-building

Much of our understanding of how mountains form is based on the study of the Alps. From the late 1700s, geologists from Switzerland and France began to explore and survey the Alps to produce maps and later three-dimensional models that helped them understand the structure of the mountains and the processes that gave rise to them.

Swiss scientist and Alpine explorer Horace Bénédict de Saussure (1740–1799) made detailed studies of the rocks of the Alps. In the 19th century, further systematic surveys of the rock types of the region were carried out, and in 1853 Arnold Escher von der Linth, a Swiss geologist, completed the first geological map of the whole of Switzerland. But these new maps often raised more questions than they answered. Across the Alps, older rocks were often found overlying younger ones, reversing the expected order of strata.

One of the first to recognize that horizontal movement could be greater than vertical movement in mountain building, Austrian geologist Eduard Suess (1831–1914) concluded that the much-studied Glarus Alps were the result of a 21-mile (35-km) northward thrust fault that left older rocks on top of younger rocks. He mistakenly believed that the huge horizontal forces were due to the gradual contraction of Earth's crust.

French geologist Marcel-Alexandre Bertrand (1847–1907) took these ideas further, identifying immense thrust sheets across the Alps. He found similar features elsewhere, leading him to identify distinct mountain-building episodes—the Caledonian, Hercynian, and Alpine.

As the Alps became better understood, they became the testing ground for new geological theories. During the 20th century, continental drift and plate tectonics became accepted by geologists partly because they offered better explanations for features that had been studied in the Alps.

spotlight the Himalayas

There is no greater expression of the power of tectonic forces than the world's highest mountain chain, the Himalayas, which stretch 1,800 miles (2,900 km) across Central Asia. There are at least ten peaks in the Himalayas over 26,000 ft (8,000 km). No other range in the world has a single summit above this elevation. Here, the fossils of ancient sea creatures can be found on the roof of the world. The summit of Mount Everest consists of marine limestone. These sea floor sediments have not only been lifted 29,032 ft (8,848 m) above sea level but also thrust 1,250 miles (2,000 km) north into the heart of Asia.

About 200 million years ago, the ancient supercontinent of Pangaea began to break up. The Indian subcontinent moved rapidly north, colliding with Asia about 60–40 million years ago. Coastal sediments from the shores of both continents were crumpled and piled high to form the Himalayas as the Indian Plate drove under the Eurasian Plate, compressing and

thickening the crust to the north and lifting up the Tibetan Plateau. The plateau is bounded to the west by an extension of the Himalayas, the Karakoram Mountains. The formation of the Himalayas has transformed the region's climate, leading to the establishment of the annual Southeast Asian monsoon. Several major rivers, including the Ganges and the Brahmaputra, also have their sources in the range.

The Himalayas are a young mountain range that is still rising. GPS measurements show that the peak of Mount Everest is rising 0.4 in (1 cm) every ten years, or about 1 ft (30 cm) every 300 years.

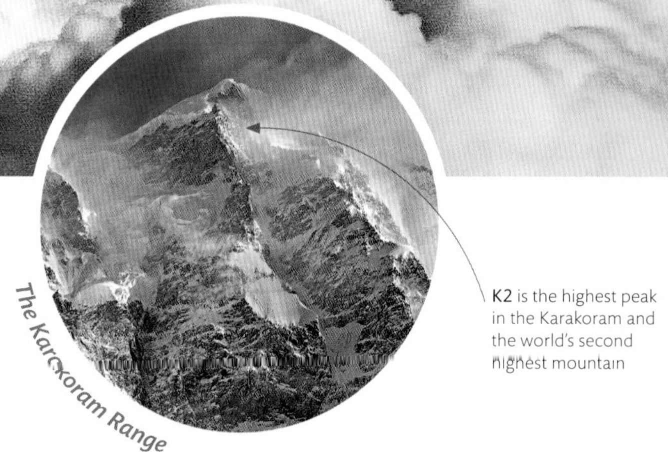

The Annapurna Massif
Four peaks carry the name Annapurna. They are part of a massif that stretches 34 miles (55 km) through the central Himalayas. Annapurna I is the highest, at 26,545 ft (8,091 m) above sea level. Its steep sides, unpredictable weather, and the risk of avalanche make it one of the world's deadliest mountains for climbers.

TYPES OF FAULT

There are three key types of fault. In a normal fault, the hanging wall block (the block above the fault plane) moves downward relative to the footwall block (the block below the fault plane). In a reverse fault, the hanging wall block moves upward relative to the footwall block. In a strike-slip fault, the two blocks slide past each other in opposite directions along a horizontal plane, with little or no vertical movement.

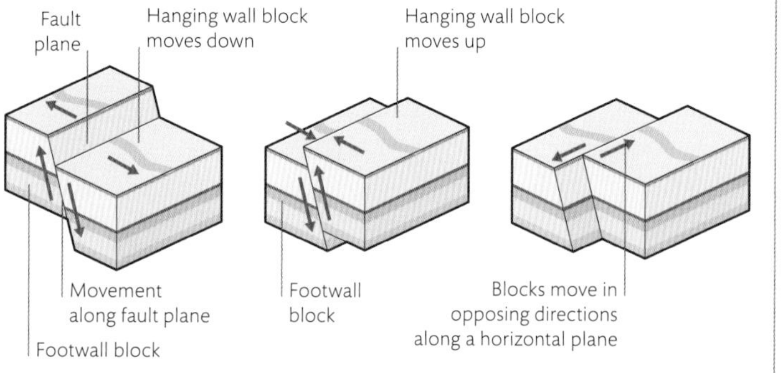

lines of fracture

When blocks of rocks in the brittle outer layers of Earth's crust and upper mantle fracture and move relative to one another, the fracture is called a fault. When the movement is sudden, it generates shock waves—vibrations that travel outward through the surrounding rocks—causing earthquakes on Earth's surface. The relative motion between adjacent blocks of rock is called fault movement, and it ranges from a few millimeters to thousands of miles. Movement, or slip, takes place along a roughly flat surface called a fault plane (see panel, above). The rock masses on either side of a fault plane are called fault blocks.

Fault lines in sandstone
These colorful sandstone beds, which were deposited in parallel layers separated by bedding planes, have been displaced by a normal fault.

Thrust faults
These red and green sandstones and creamy limestones in the Tien Shan Mountains in China (right) have fractured along thrust faults, sas older layers of rock have been pushed up on top of younger layers. Thrust faults are a type of reverse fault with movement at relatively low angles of 45 degrees or less.

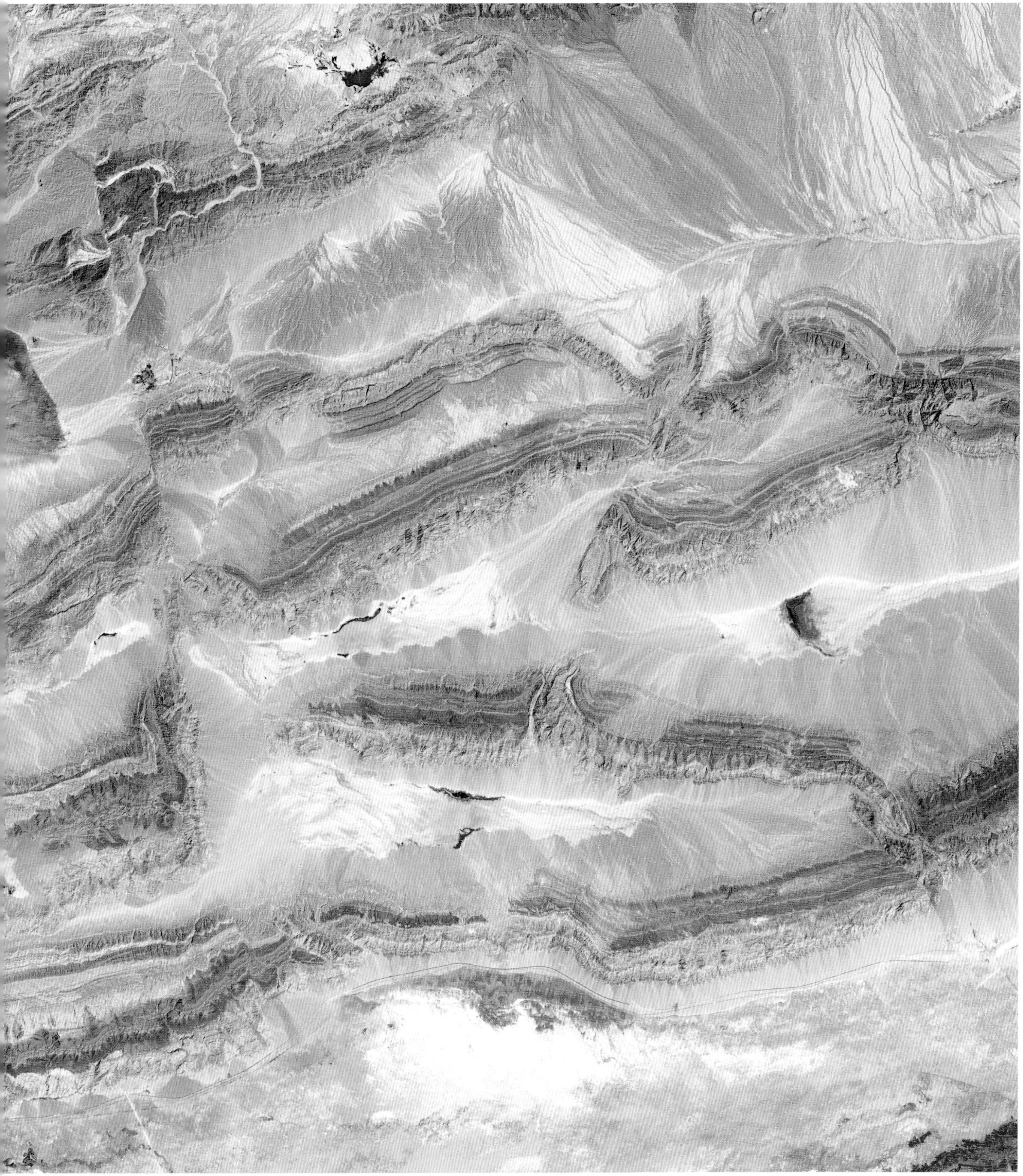

Aftermath of a landslide
A powerful earthquake that struck the island of Hokkaido, Japan, on September 6, 2018, triggered a series of devastating landslides as a secondary effect, engulfing homes and bringing down power lines.

earthquakes and tsunamis

Earthquakes are ground tremors caused by a sudden release of energy deep underground. Motion at the surface occurs along fault lines, which are usually located close to plate boundaries. An earthquake's strength and destructive power depends on the amount of energy it releases. The magnitude of an earthquake, expressed on the Richter or moment-magnitude scales, is a measure of this power. Underground movement continues after the initial earthquake, creating smaller quakes called aftershocks. Earthquakes on the seafloor, barely detectable on the surface, can cause giant sea waves that grow into huge walls of water, known as tsunamis, as they reach coastlines.

WHAT CAUSES EARTHQUAKES

Underground movement of rocks along a fault causes energy to be released as seismic waves. Rocks begin to rupture or shift at a point underground called the focus. The point on Earth's surface directly above the focus is called the epicenter. Seismic waves, called surface waves, radiate out from the epicenter along Earth's surface, often triggering landslides, tsunamis, and damage to built structures.

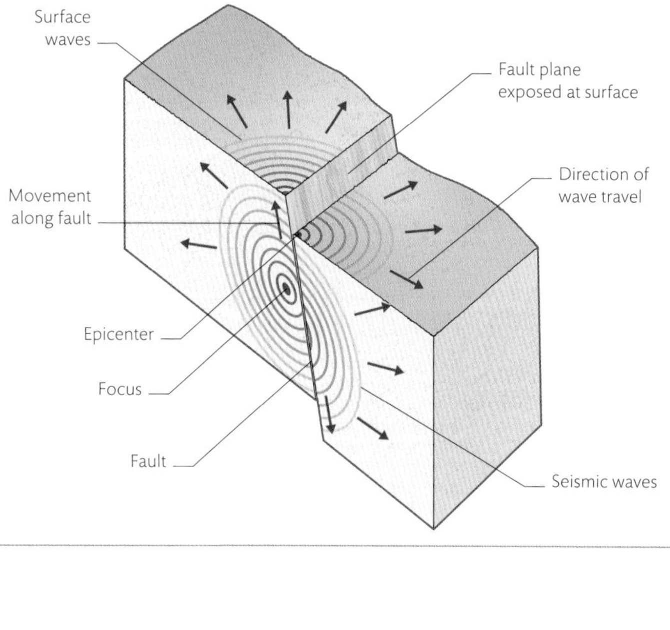

Palmieri's seismograph
Luigi Palmieri's 1856 device used mechanical sensors to detect large ground movements and electrical sensors to detect smaller movements. Four U-shaped glass tubes aligned to the compass directions were filled with mercury. When disturbed sufficiently, they completed an electrical connection triggering another part of the instrument to record the time and duration of the seismic event to paper tape.

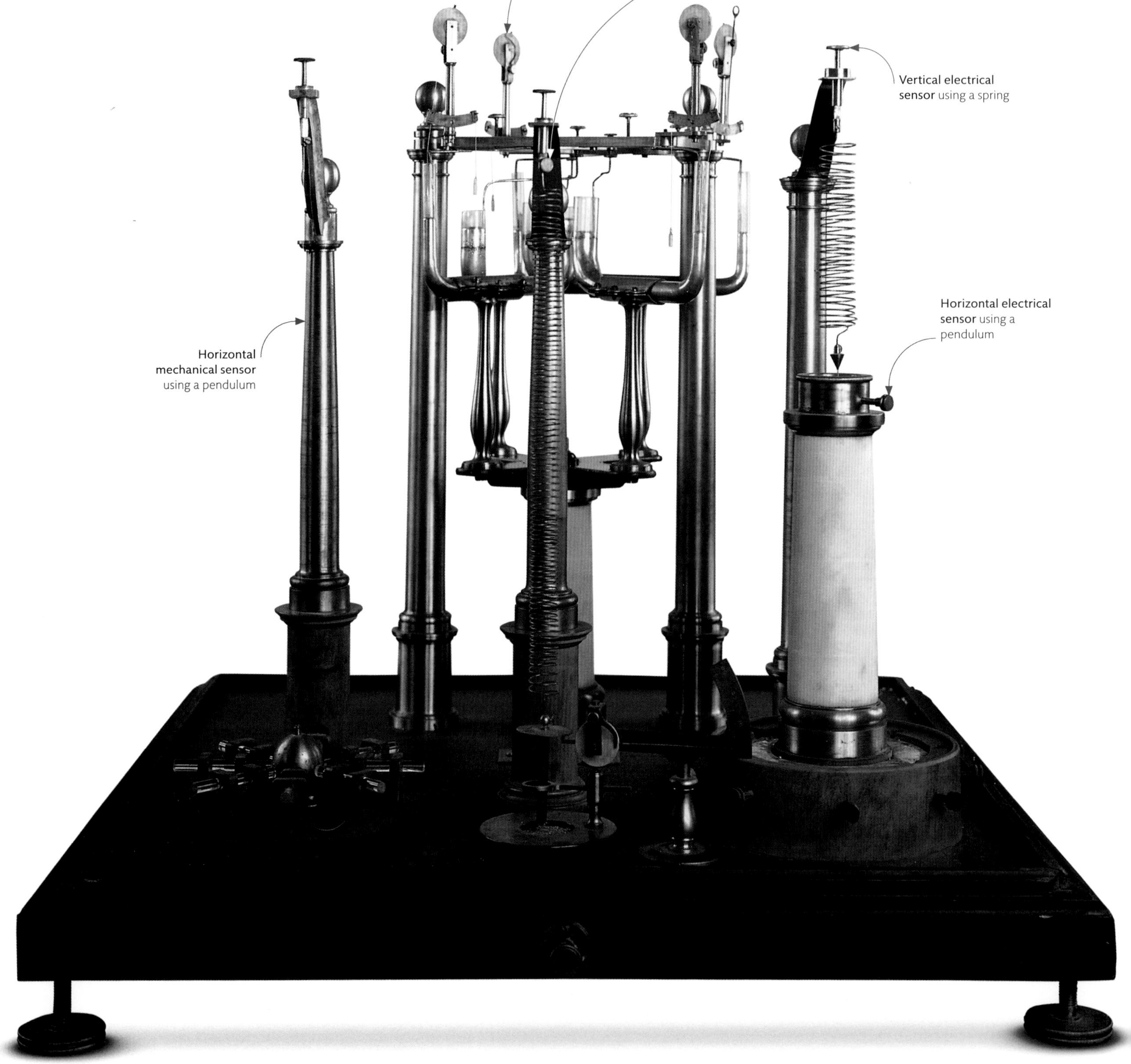

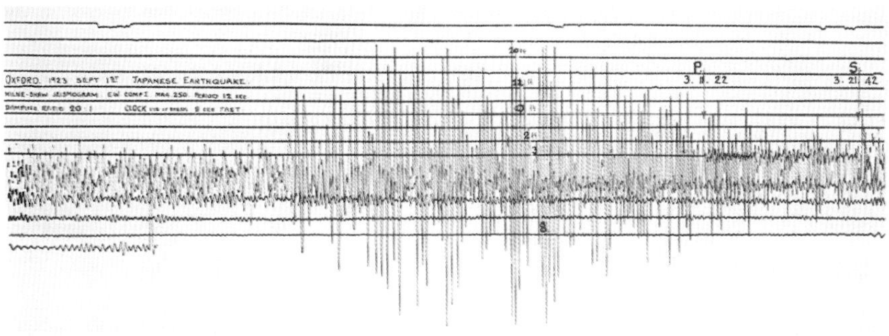

Tokyo earthquake on paper
A record produced by a seismometer is called a seismogram. This example, produced in Oxford, England, records the first disturbance from a major earthquake near Tokyo, Japan, shortly after 3 a.m. on September 1, 1923.

history of Earth science

measuring earthquakes

Seismometers on the moon
The most accurate seismometers are almost too sensitive for use on Earth due to the planet's constant seismic noise. But they were deployed on the moon in 1969 by the Apollo 11 astronauts, where they detected tiny moonquakes and the impact of micrometeors from 1969 until 1977.

Ancient civilizations had a wide variety of explanations for the causes of earthquakes, from the activities of subterranean creatures to the presence of openings to the Underworld and the effects of winds below the surface. An accurate measurement of an earthquake's strength and duration was not possible until the invention of the seismometer.

The first known instrument for measuring the strength of earthquakes was invented by the Chinese mathematician Chang Heng in 132 CE. It is likely, but not certain, that Heng's device used a pendulum to detect the ground motion, as did European scientists investigating earthquakes in the 18th century.

Most of these early seismometers used the inertia of a weight, suspended on a pendulum or a spring, to provide a fixed reference point against which ground movement could be measured. While the ground and the instrument's framework shake, the weight remains relatively still. The changing distance between the frame and the weight measures the strength of the earthquake.

> “... the seismograph ... enables us to see into the Earth and determine its nature.”
>
> RICHARD DIXON OLDHAM, IRISH GEOLOGIST, 1906

One of the first devices that could create a record of seismic events, known as a seismograph, was built by Luigi Palmieri at the Vesuvius Observatory in 1856. It recorded earthquakes that preceded the volcanic eruptions of Mount Vesuvius in 1861, 1868, and 1872 on a roll of paper.

Palmieri's seismograph was also the first to use electromagnetism to detect the motion. Instead of a spring or pendulum, modern seismometers use electronics to apply a magnetic or electrostatic force to keep the weight still. The most modern seismometers are sensitive enough to detect small, distant earthquakes, underground nuclear explosions, and even the tidal pull of the moon on Earth's rocks.

Widely separated seismometers can be used to triangulate the location of seismic events. Since the 1980s, computer analysis of data from a global network of seismometers has been used to give a three-dimensional picture of Earth's deep interior, similar to the images provided by computerized tomography (CT) scans in medicine.

MEASURING AND RECORDING EARTHQUAKES

A seismogram is a record of an earthquake's seismic waves, which are measured by an instrument called a seismograph. There are two types of seismic waves: body waves, which move from an earthquake's focus (the point underground where rocks start to rupture or shift) through Earth's interior; and surface waves, which travel from the epicenter (the point on Earth's surface located above the focus) along Earth's surface. The squiggly lines (amplitude) reflect the amount of energy released by the waves.

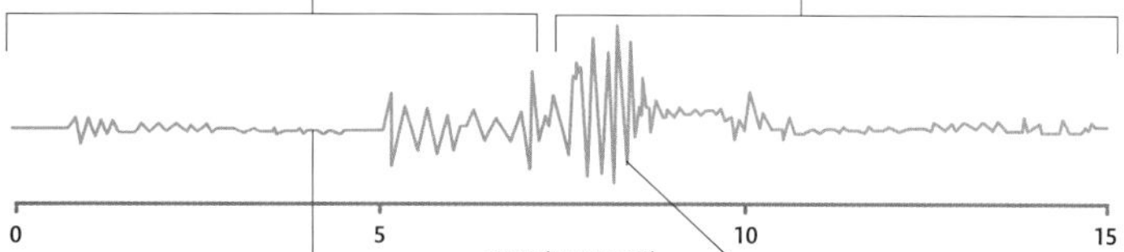

A TYPICAL SEISMOGRAM

shaking ground

Earthquakes are ground tremors caused by a sudden release of energy in underground rocks (see pp.128–129). Most last from a few seconds to a few minutes, but the vast amount of energy released can leave an imprint on a landscape for thousands of years. An earthquake's intensity at a given location is related to the distance from the fault, its depth, and the soil type. During all earthquakes, the ground starts to shake, but damage can come in many forms: large areas of rocks can move hundreds or thousands of miles along faults, landslides and mudflows can be triggered by seismic waves, and soil can liquefy into flowing mud.

Rainbow patterns show deformation and ground movement

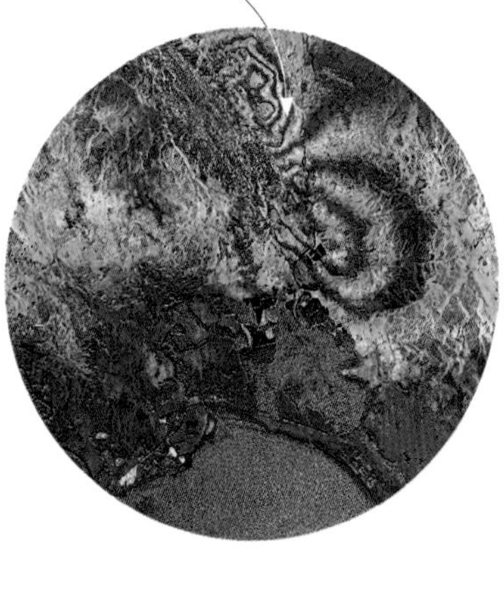

Ground-breaking damage
Earthquakes can cause immense damage to humans, landscapes, and infrastructure. The seismic waves that travel along Earth's surface are the ones that cause the most damage, often causing bridges and buildings to collapse, and roads (here, in Alaska, right) to fracture and break.

Earthquake by satellite
In 2014, the region of Napa Valley in California experienced its largest earthquake for 25 years. The colors on this image depict radar signals acquired by satellite during the event.

Dome-like structure flattened and lowered by erosion and weathering
Vast tower made up of individual, irregular-shaped columns, about 100ft (30m) high and 10ft (3m) wide

Magmatic dike
This dike was formed when hot, liquid magma pushed up through cracks in sedimentary rocks in Nahal Ardon, Makhtesh Ramon, Israel. The magma then cooled and solidified.

igneous intrusions

As molten magma rises from Earth's mantle through the crust, it squeezes through cracks in preexisting rocks, widening them and sometimes forcing overlying rocks to bulge upward. When the magma cools and solidifies beneath Earth's surface (rather than erupting through the surface as lava via fissures and volcanoes), it forms bodies of igneous rock called intrusions. If the overlying rocks become eroded, the intrusions can become exposed above the surface. Many of the world's mountain chains—such as the White Mountains in New Hampshire and the Yosemite Valley (see pp.68–69) in the Sierra Nevada in California—are made up of igneous intrusions that were exposed after the overlying rocks were eroded.

Devil's Tower
This dramatic geological feature (also known by its Lakota name Mato Tipila, meaning "bear lodge") rises above ponderosa pines in Wyoming. The tower is an intrusion formed from magma that cooled inside a magma chamber. It was once more than 1 mile (1.5 km) below Earth's surface but was later exposed by erosion and lowering of the surrounding landscape. Made of an igneous rock called phonolite porphyry, it is the largest structure of its kind in the world.

TYPES OF IGNEOUS INTRUSION

There are several types of igneous intrusion. Dikes and sills are both slab-shaped intrusions that form in layers of preexisting sedimentary rock called country rock. Dikes cut across the rock layers, while sills form parallel to them. Large, irregular-shaped intrusions are called plutons. The largest of these—batholiths—extend at least 40 square miles (100 square km) into the rock layers. The heat released by igneous intrusions can cause the surrounding rocks to undergo a process of chemical change called contact metamorphism (see pp.90–91).

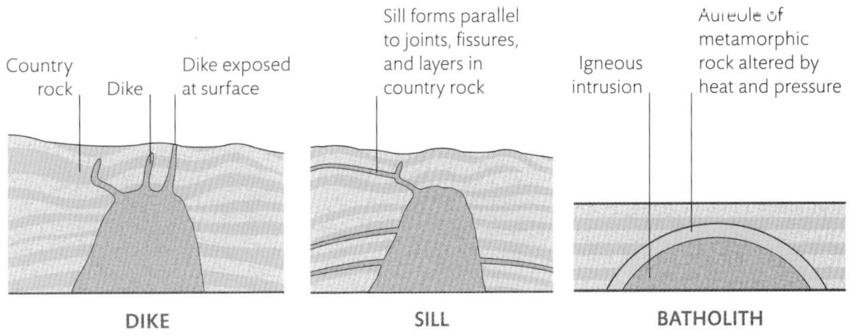

Like a square-rigger ship sailing across a dry sea, Shiprock rises above the high desert of northwestern New Mexico, trailing a series of low ridges in its wake. The ridges are volcanic dikes—thin walls of resistant rock radiating away from the central peak, which is the throat of a long-dead volcano. The rock is fragmented, but nevertheless hard, volcanic breccia—fine debris from a series of explosions triggered when magma

Shiprock spotlight

pushed up through fractures in Earth's crust and interacted with groundwater beneath the surface. Dating studies suggest that the rocks of the formation solidified about 30–25 million years ago. Since then, erosion has taken away as much as 3,000 ft (900 m) of the overlying sandstones and shales, exposing the volcano's internal plumbing.

The composition of the dike rock suggests it was formed from magma that melted at great depth. Similar rocks have been found in igneous intrusions and surface flows elsewhere in New Mexico and Arizona, including in the distinctive rock formations of Monument Valley. These intrusions are the result of the tectonic forces that drove the uplift of the Colorado Plateau, which extends across much of Utah, New Mexico, Arizona, and Colorado.

Standing 1,583 ft (482 m) above the surrounding plain, Shiprock is a tempting climb. But the rock formation is sacred to the Navajo people, and climbing it has been banned since the 1970s. Where European settlers saw a sailing ship, the Navajo (who refer to the rock as Tsé Bit'a'í), had always seen a "rock with wings," representative of the legendary great bird that brought the Navajo people to their land.

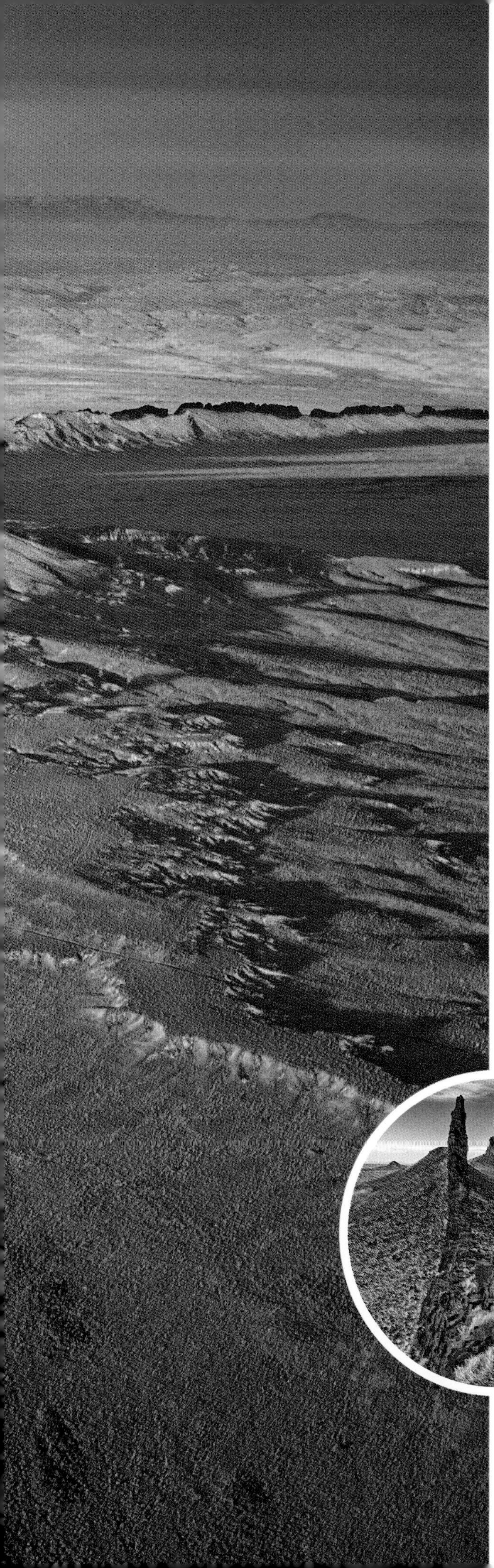

Pieces of breccia are scattered along one of the dikes

Aerial view
Two prominent dikes connect to the jagged central peak of Shiprock, which is a volcanic plug—a mass of rock that once filled the vent of an active volcano. Five smaller dikes can also be traced across the immediate area. Together, they reveal the internal workings of the now extinct volcano.

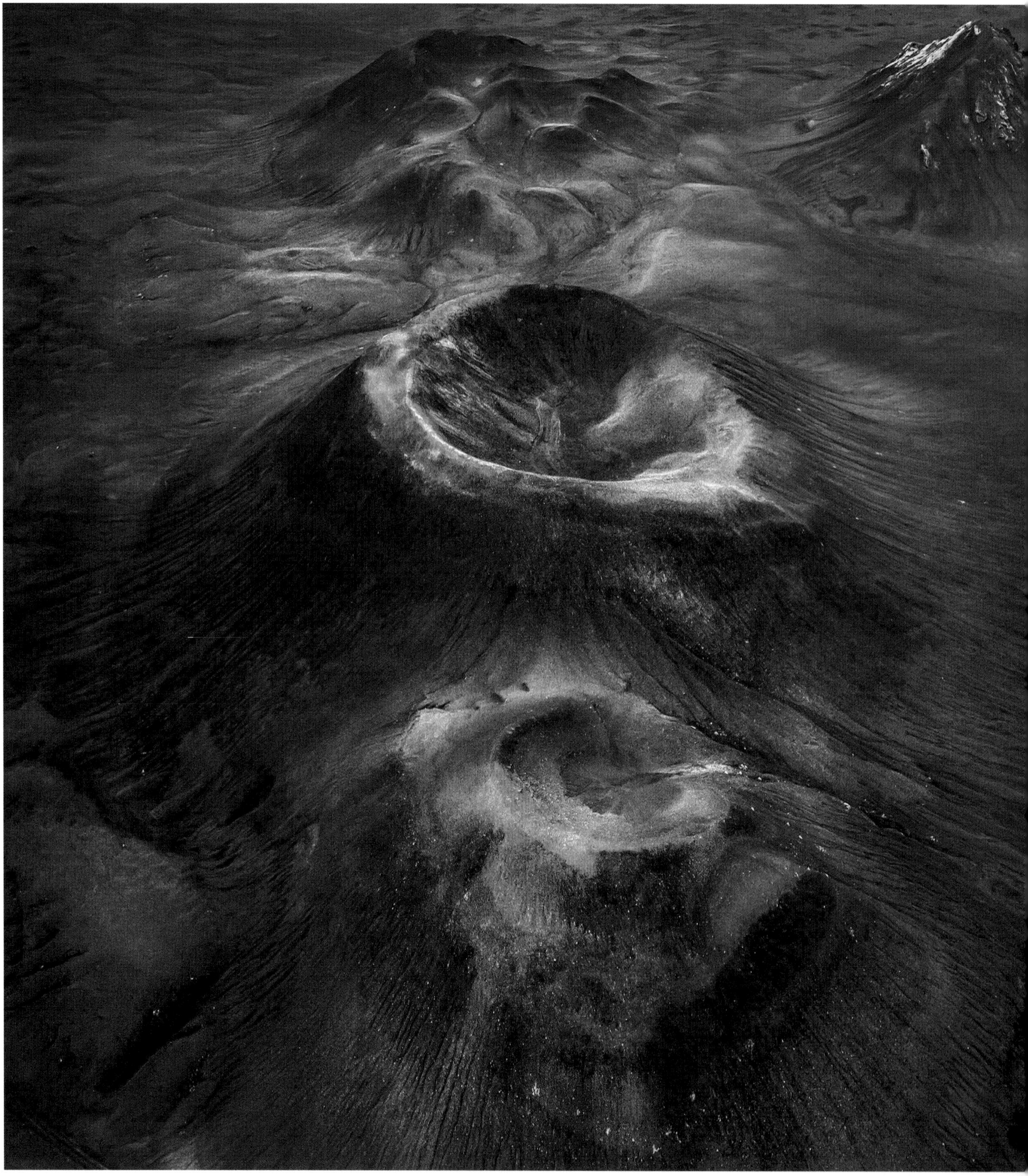

Mount Pinatubo
This stratovolcano (see panel, right) is situated on the eastern Kamchatka Peninsula in Russia. A lake of hot, acidic water has formed in the volcano's Troitsky Crater, which formed during an explosive eruption about 400 years ago.

Volcanic cones and craters
Dark gray basalt and red-rimmed craters give these low-lying volcanic cones and the surrounding landscape in Iceland an otherworldly appearance. The red color of the deposits on the rims of the craters is produced by oxidation of iron minerals.

VOLCANO SHAPES

There are several types of volcano. Lava domes (also called volcanic domes) form when viscous lava accumulates over and around the vent and solidifies, forming a dome that can explode violently. Cinder cones (also known as pyroclastic cones) are formed mainly from fragments of gas-charged magma that is ejected from the vent as lava, then solidifies and falls as cinders. Shield volcanoes have gently sloping sides produced by thin, fluid lava that flows without an explosive eruption. Stratovolcanoes (also called composite cone volcanoes) have steep sides made from many layers of lava and other rocks fragments ejected in explosive eruptions.

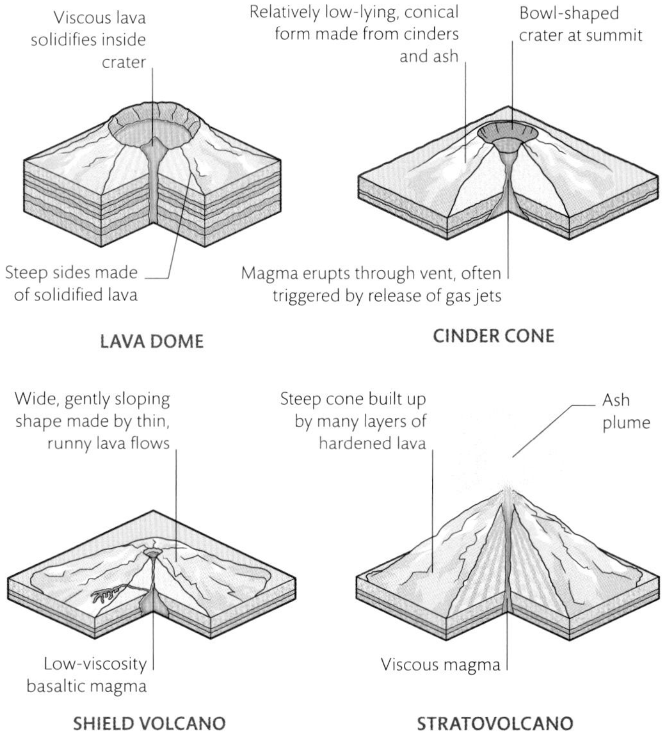

volcanoes

Volcanoes are landforms created when magma escapes from Earth's interior and reaches the surface through openings, or vents. They generally form at the boundaries of tectonic plates (see pp.110–111) or over mantle plumes—columns of extremely hot rock rising through the mantle and crust—and are found not only exposed on Earth's surface but also on the ocean floor and under icecaps. Once the magma erupts as lava, it cools and solidifies around the vents. Volcanoes consist of two main components: a volcanic cone, which is a mountain created by the accumulation of magma around the vent from successive eruptions; and a crater—a steep-walled, bowl-shaped depression surrounding the vent, from which magma erupts. If eruptions empty out all the magma held inside the reservoir (magma chamber) under the volcano, the cone may collapse, forming a circular depression called a caldera. There are various types of volcano; their shapes are determined largely by the properties of the magma and the way in which the volcano erupts (see p.141 and panel, above).

Tungurahua Volcano
Situated in Ecuador, Tungurahua (right) is a stratovolcano (see p.139) that erupts periodically in a Vulcanian or Strombolian style (see panel, below).

Vulcanian eruption
This dramatic Vulcanian eruption took place at the Fuego-Acatenango Massif—a series of volcanic vents in Guatemala. A large, dense plume of ash is ballooning over the volcano, while a pyroclastic flow hurtles down its flanks.

volcanic eruptions

Magma rising from Earth's mantle escapes to Earth's surface as lava through volcanoes during events called volcanic eruptions. These events can be very destructive or relatively calm, depending on the magma's chemistry and gas content. Magmas that contain large amounts of gas and are viscous (resistant to flow) tend to erupt explosively, with plumes of ash propelled into the atmosphere and pyroclastic flows—avalanches of ash, hot gases, and rock fragments that roll down the flanks of the volcano. Runny magmas with a low gas content tend to erupt as quiet, oozing lava flows. Explosive eruptions tend to be associated with steep-sided volcanoes, while quiet eruptions are associated with flat-lying, broad landforms (see pp.138–139).

TYPES OF VOLCANIC ERUPTION

There are several types of volcanic eruption, although a volcano can erupt in a number of ways and a single eruption can include elements from each type. Strombolian eruptions are associated with lava bombs and small to no ash clouds. In a fissure eruption, runny lava flows from a linear fissure, solidifying into low-lying landscapes. In Plinian eruptions, a tall gas and ash plume rises and falls on the flanks, accompanied by loud explosions. Vulcanian eruptions are short-lived, violent, intermittent events, accompanied by dense ash and gas clouds and, often, lava bombs.

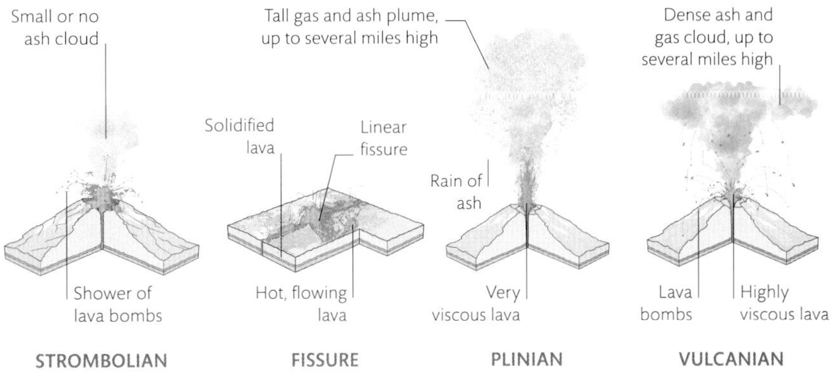

Hot spot volcano
La Palma, one of the Canary islands, is a volcanic island formed over a hot spot. In 2021 the Cumbre Vieja volcano erupted, causing lava to flow down the mountainside.

hot spots

When a column of hot mantle material, called a plume, rises to the base of the lithosphere, it creates a zone called a hot spot. In these locations, the magma may be hot enough to melt the overlying lithosphere and force its way to the surface to form a volcano. About 100 so-called "hot spot volcanoes" are known. Most of these are located within a tectonic plate rather than at the boundaries between adjacent plates, but a few appear on mid-ocean ridges, as in Iceland. Some hot spots lie within continents—for example, under Yellowstone National Park. Others underlie ocean basins, giving rise to volcanic island groups such as the Canary Islands and Hawaii.

FORMATION OF A VOLCANIC ISLAND CHAIN

When hot spots underlie oceanic plates, they form volcanoes on the seafloor above. As the tectonic plate moves over a hot spot, a chain of extinct volcanoes, or hot spot track, forms. A good example can be seen in Hawaii today. Volcanic eruptions occur only on the Big Island of Hawaii, the youngest island in the chain, which has three active volcanoes. The islands to the northwest are the remains of ancient volcanoes.

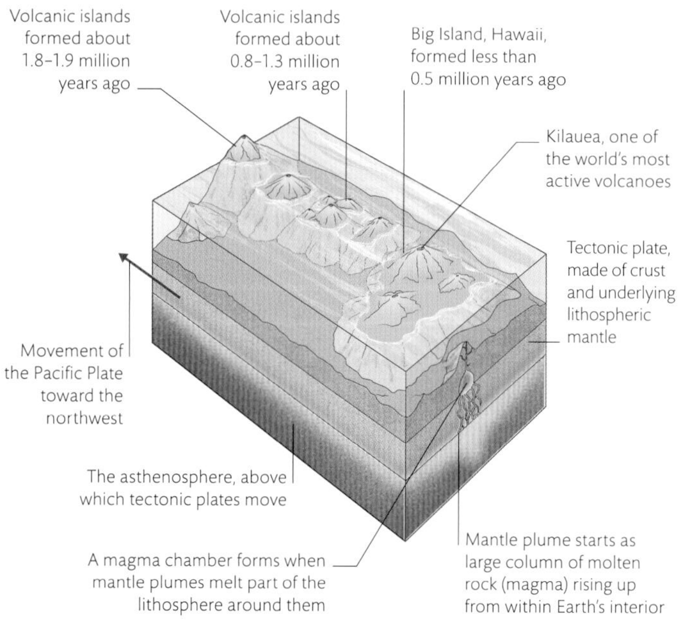

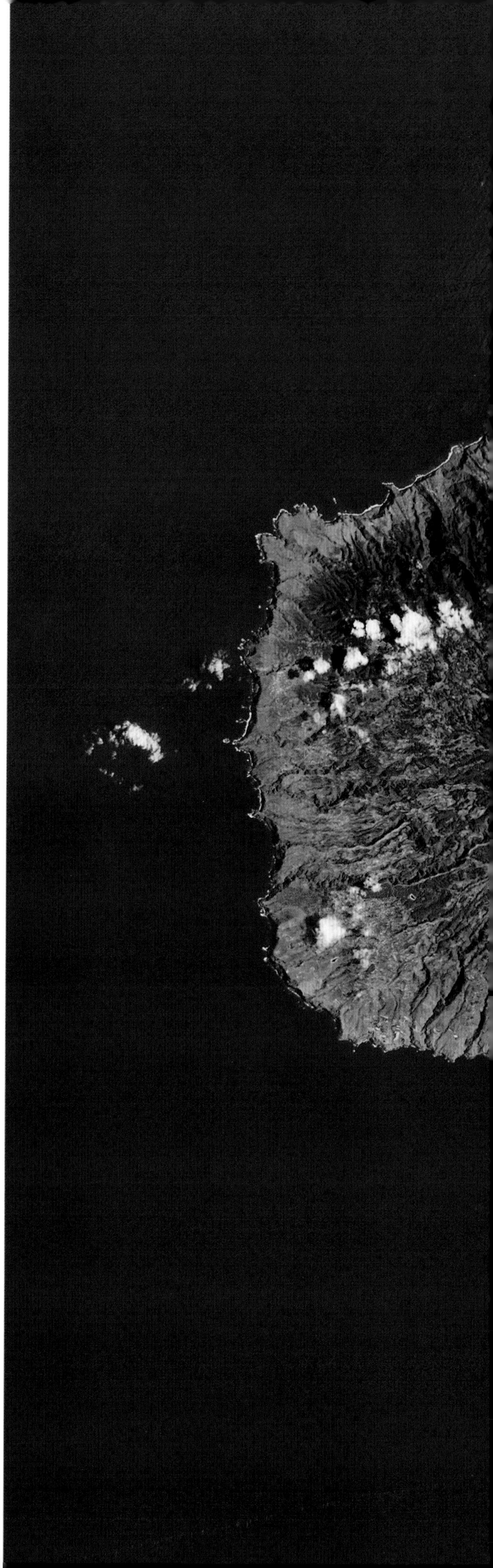

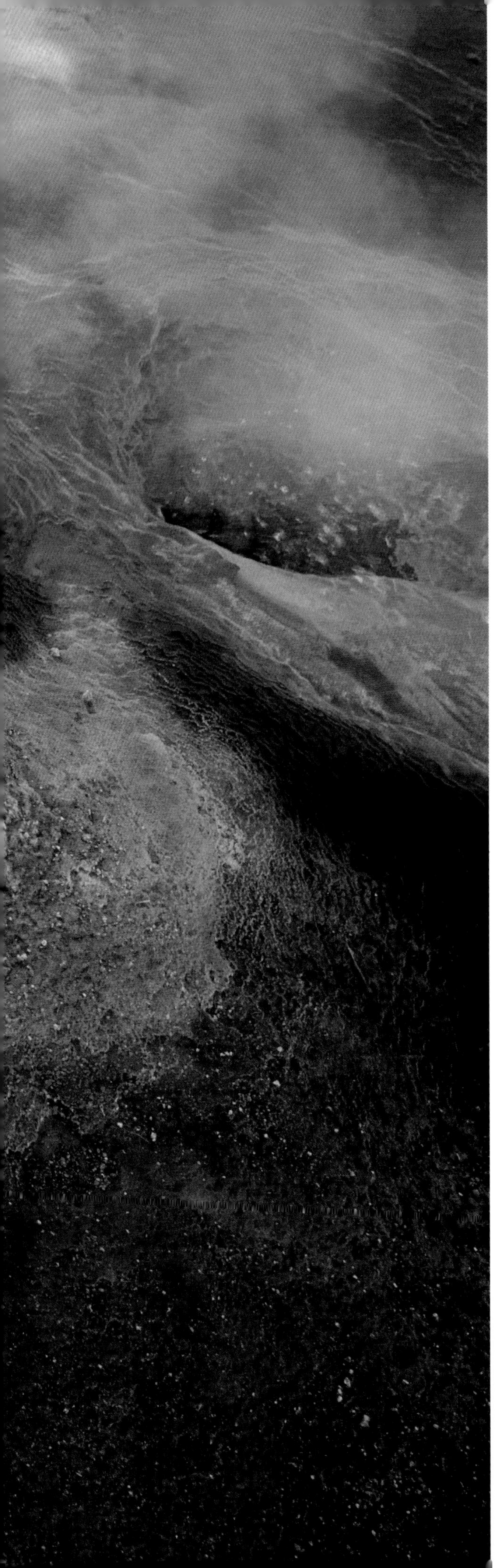

Colorful deposits
Some hot springs—such as this one in El Tatio, Atacama, Chile—form near active volcanoes. At the spring's surface, the hot water evaporates, leaving behind mineral deposits called sinters.

geothermal features

Geothermal activity provides some of the most striking evidence of Earth's processes, many of which are driven by its internal heat. The temperature of Earth rises with increasing depth, from the cooler crust toward the hotter mantle and core. A steep rise in temperature within a relatively shallow depth often leads to activity at the surface that can take dramatic forms. This occurs when groundwater seeps down into the subsurface through fractures and is heated by warm rocks or nearby magma. A combination of hot water and steam rises back to Earth's surface through geothermal features such as geysers (intermittent hot water spouts), hot springs (pools of hot water), and fumaroles (columns of hot gas and steam).

HOW GEOTHERMAL FEATURES FORM

Groundwater enters the subsurface as part of the water cycle. Nearby volcanoes or hot rocks can superheat the water to more than 360°F (180°C). Hot springs form when the water rises back to the surface and forms a pool. If water and steam become trapped in hollows underground, pressure builds up until they are periodically released in eruptions called geysers. If hot water turns into steam before reaching the surface, steam and other gases are ejected through vents called fumaroles.

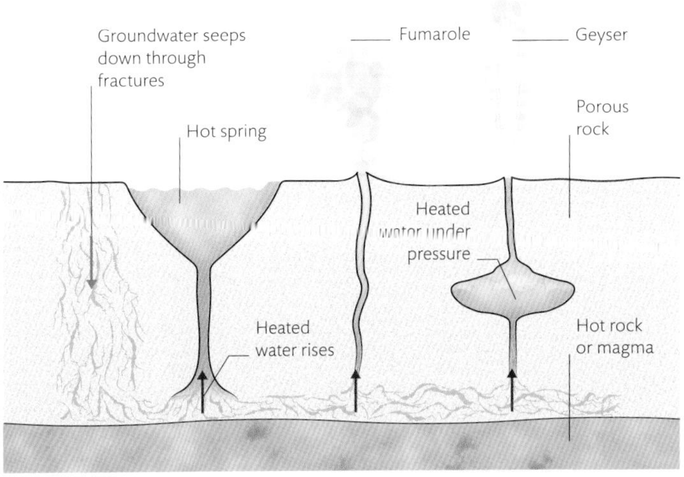

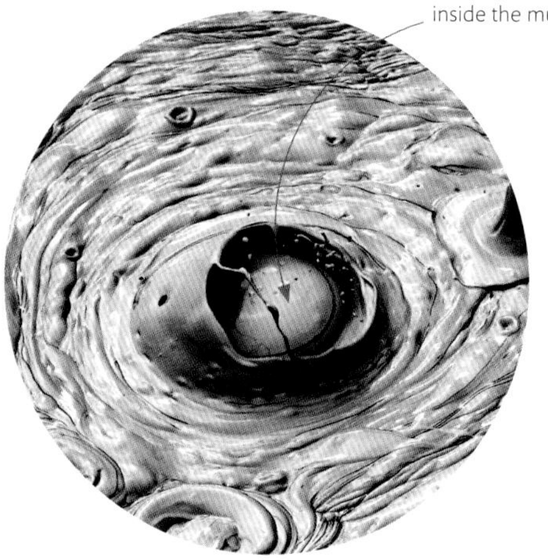

Bubbling mud pot
Groundwater rising through Earth absorbs gases that make it acidic. The acid in the water then breaks down rock, which turns into a pool of mud that bubbles and boils at Earth's surface. This image (left) was taken in Wai-o-tapu, New Zealand.

Fountain geyser
Iceland's Strokkur Geyser erupts every few minutes, releasing hot groundwater and steam in fountains about 100 ft (30 m) high. Silica-rich mineral deposits known as geyserites have precipitated around the edge of the water pool.

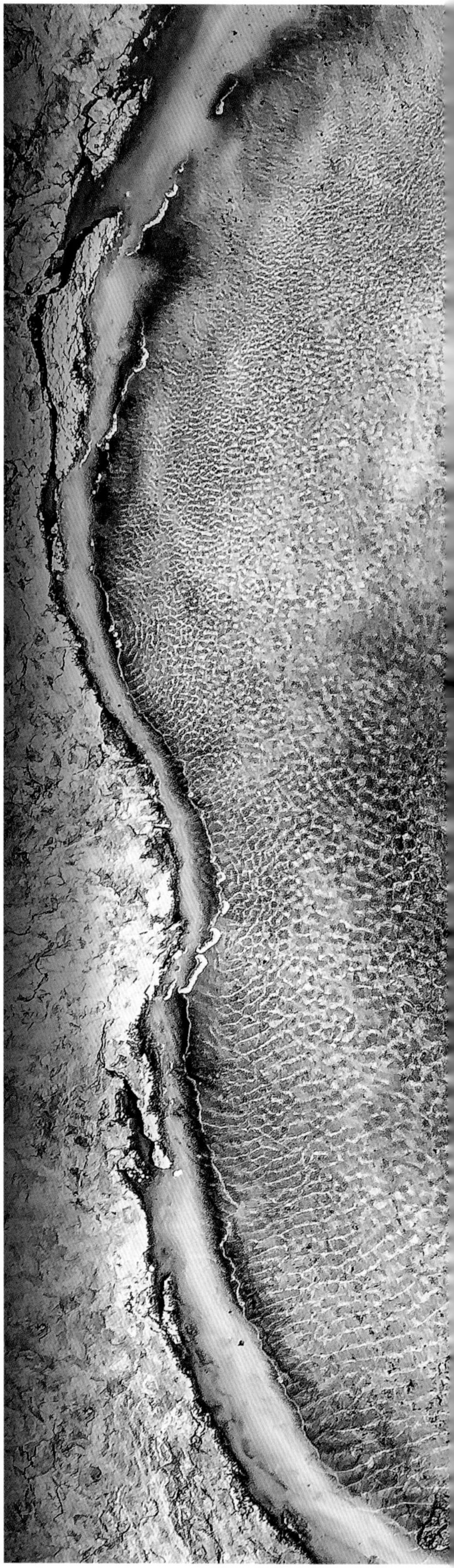

hot springs

Water constantly moves between Earth's surface and underground as part of the global water cycle. Groundwater seeping down through cracks in Earth's crust can become superheated (see pp.144–145) and then return to the surface in the form of hot springs. These range from slow seeps to flows of more than 40 gallons (150 liters) per second. As the heated water rises, it may dissolve minerals in the surrounding rock, which then precipitate out as solid, often vibrantly colored deposits when the water cools. Hot water can also erupt as intermittent steam and hot-water spouts called geysers (see panel, below), of which there are about 1,000 worldwide. Hot springs and geysers typically occur at the boundaries of tectonic plates (see pp.110–113) or near volcanoes.

HOW A GEYSER ERUPTS

A geyser occurs where superheated water flows through an underground system of tubes and chambers. A constriction in one of the tubes leads to an increase in pressure, which eventually results in a blast of hot water erupting at Earth's surface. As the water pressure drops, the water instantaneously turns to steam at the surface. After an eruption, groundwater percolates through cracks into the chamber, recharging it, and the process starts again. Deposits called geyserites left behind by mineral-rich waters line the chamber.

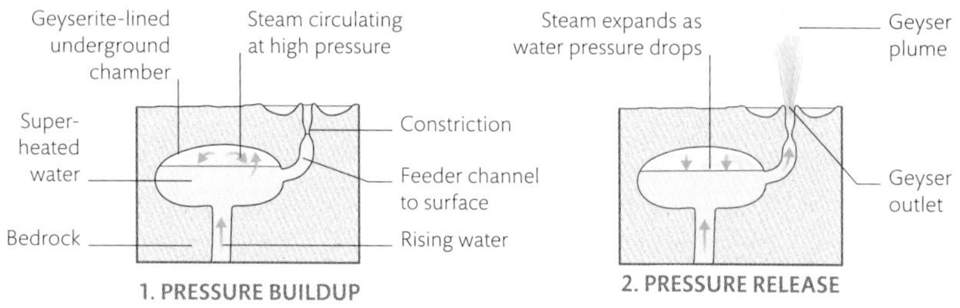

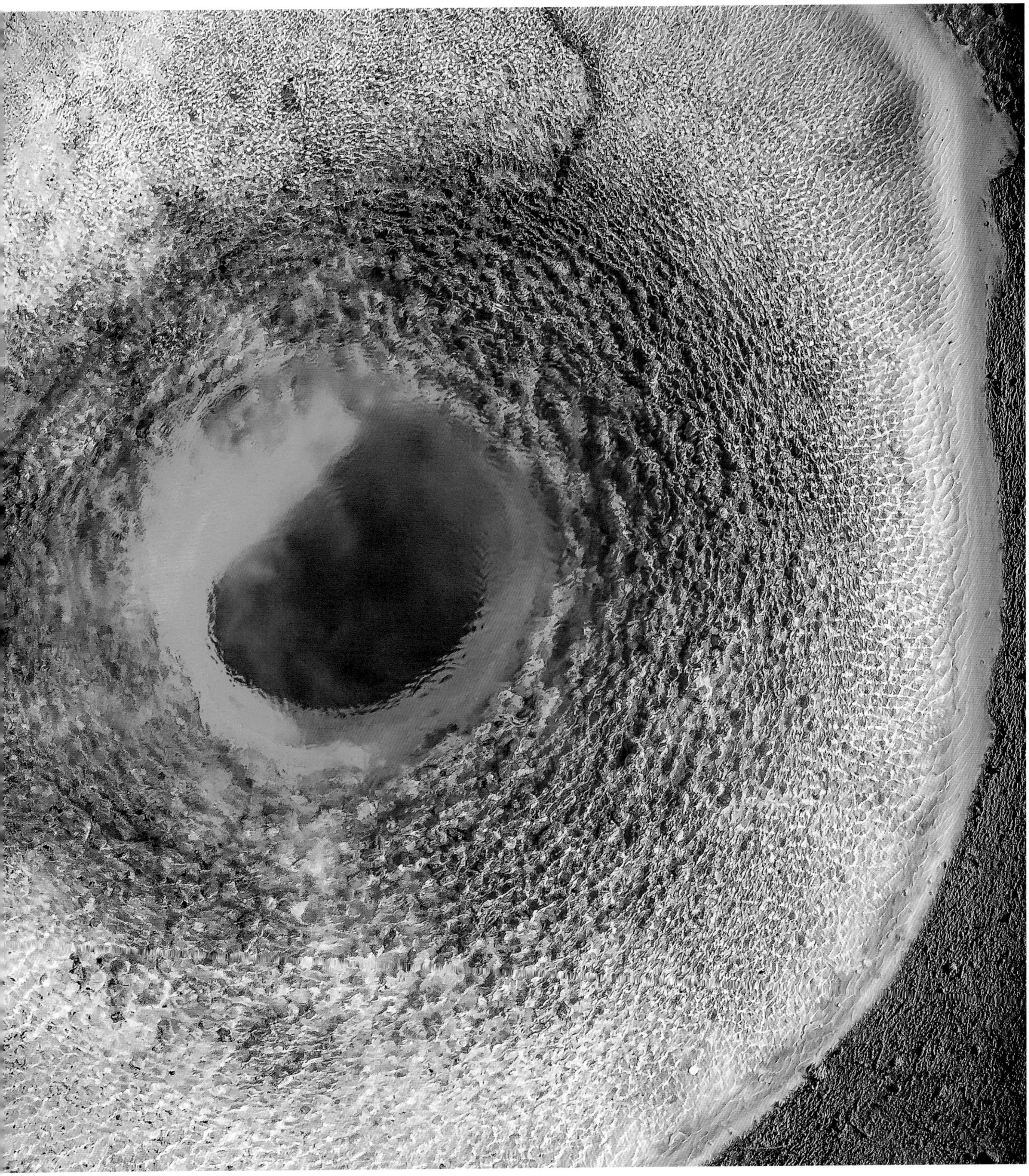

Potholes in rivers
Circular or cylindrical holes in the riverbed, called potholes, form when rock fragments carried by the river (here, Blyde River in South Africa, right) reach small depressions in the bedrock. The fragments whirl around the depression, eroding and scouring it.

Erosion from wave action
Coastlines are shaped by the impact of waves. Erosion cuts indentations, or notches, in coastal cliffs, usually at the level of high tide, as seen in these sandstone cliffs in Tasmania, Australia.

erosion by water

The water in oceans and rivers shapes the landscape. The repeated action of waves wears away coastlines, making them progressively straighter, and wave-based erosion of headlands—cliffs that project out into the sea—can carve out caves, arches, and sea stacks. As water from a river moves along its course, it erodes the rocks on which it flows, carrying sediments and rocks downstream and depositing them when the current slows. When the water is fast flowing—such as near the river mouth or in times of flood—it transports a greater volume of sediment.

RIVER TRANSPORTATION

Material eroded by water can be transported through several mechanisms, depending on the size of its particles. Fine-grained particles, such as silt and clay, are light enough to remain suspended in the water column while being carried along, while larger items such as pebbles are heavier and bounce along the riverbed—a process known as saltation. If the river is fast flowing, larger rocks and boulders can be dragged along the riverbed in a process called traction.

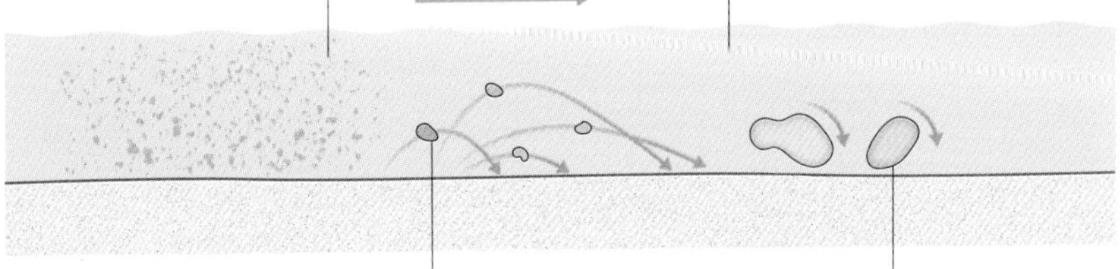

DESERT PAVEMENTS

Wind erosion in dry conditions can lead to the formation of a hard, stony surface called a desert pavement. This feature forms when fine-grained particles such as sand are removed from the desert's surface by wind in a process called deflation. As wind removes the finer particles, the coarse particles left behind become increasingly closely packed and concentrated. The land surface also lowers due to the particles' removal.

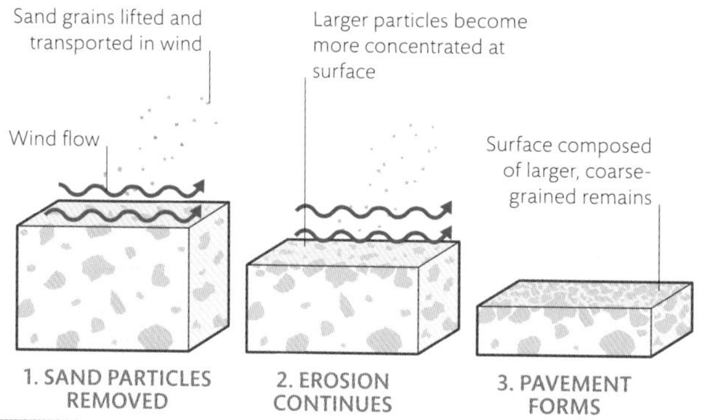

Carved by wind
In desert landscapes, eroded sand particles carried by winds collide with rock formations, sand-blasting them in a process called wind abrasion. Over time, this carves the rocks into smooth shapes, as seen in this image from the United Arab Emirates.

erosion by wind

Wind is a powerful sculptor. It can remove and transport large quantities of sand, soil, and dust particles. In the process, it shapes Earth's surface—particularly in desert environments, which are covered with loose material, and in areas with little vegetation, since plants stabilize the soil and act as windbreaks. In deserts, wind gusts remove sand- and silt-size particles, transporting them through the air and depositing them to form sand dunes, loess plateaus, and—over time—sedimentary rocks (see pp.74–77). The heavier particles, such as pebbles and cobbles, are left behind. Wind (aeolian) erosion can be caused by a light wind that rolls particles along the ground or a strong wind that lifts large numbers of small particles into the air to create dust storms.

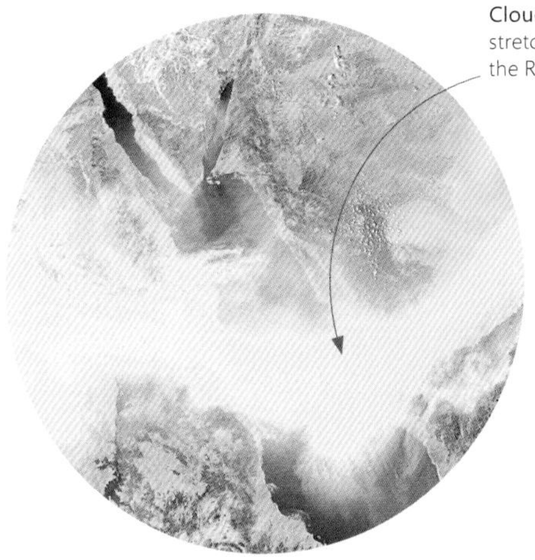

Satellite image of a sandstorm
When thunderstorms or cyclones occur, strong winds can lift large amounts of sand and dust from bare, dry soils and transport them hundreds to thousands of miles.

U-shaped glaciated valley
This valley (left) in Mount Rainier National Park in Washington was carved into a U-shape by glaciers—which are still present on the peaks—and later modified in its lower sections by rivers.

Elongated, parallel grooves indicate direction in which glacier was moving

Glacial scratches
These long scratches, called striations, were carved out by debris carried by glacial ice as it moved over the rocks.

erosion by ice

Glaciers are large, slow-moving bodies of ice. As they move—usually down a slope or mountain valley—they carve and modify the landscape. There are two processes involved in glacial erosion: quarrying, in which ice dislodges and transports large rock fragments; and abrasion, when rock fragments embedded in ice scratch and polish underlying rocks. Several landforms are created during erosion, including cirques (bowl-shaped, steep-sided hollows in a mountain), horns (angular peaks formed from several glaciers that have eroded a mountain in different directions), arêtes (narrow ridges that form when two glaciers move down opposite sides of a mountain), and roches moutonnées (see panel, below).

SCRATCHING AND ROUNDING

Ice sheets and glaciers leave evidence of their movement in the forms of scratched and rounded rocks. Outcrops in the path of moving ice are often rounded on the upstream side through abrasion, but on the downstream side cracks and joints indicate rocks have been ripped away (or quarried). The result is an asymmetrical feature called a roche moutonnée.

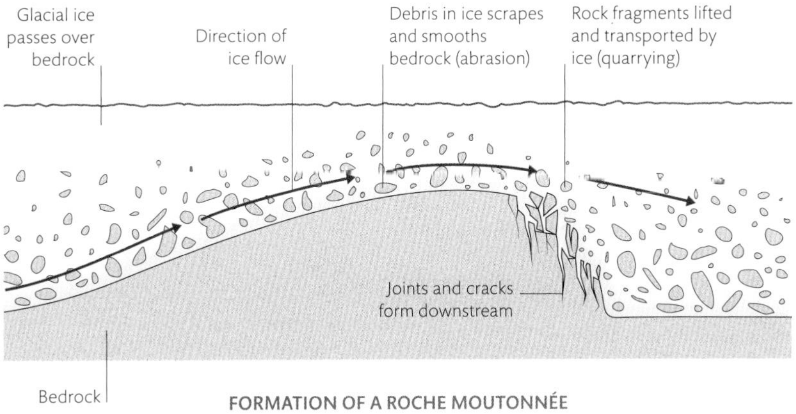

FORMATION OF A ROCHE MOUTONNÉE

weathering

The actions of water, wind, and temperature extremes can cause deterioration of rocks and minerals by breaking them down into smaller pieces or dissolving them. This process, called weathering, often acts in conjunction with erosion—weathering disintegrates or alters the rocks in situ, then erosion transports the disintegrated fragments away from the site. There are two main types of weathering: mechanical weathering (also called physical weathering), which involves the breaking down of rocks and minerals without altering their chemical composition; and chemical weathering, which changes their molecular structure. Typically, both types affect rock formations at the same time.

The Stone Forest
Groundwater and rain have chemically weathered this landscape in China's Yunnan Province to both sharp and rounded stone pillars. This type of weathering is typical of karst landscapes (see pp.166–167).

FROST-SHATTERING

Also called freeze-thaw weathering, frost-shattering is a form of mechanical weathering that occurs when water seeps into small cracks in a rock. When the water freezes, it expands by almost one-tenth of its volume, exerting pressure on the rock and enlarging the cracks, which allows more water to enter the rock. If a cycle of freezing and thawing is established, the rock will eventually break up. This type of weathering is typical of environments where surface water is abundant and temperatures repeatedly fluctuate around 32°F (0°C).

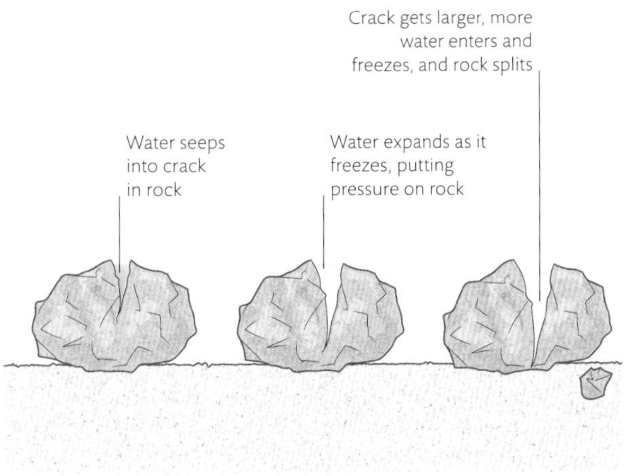

Castle of the Winds (Castell y Gwynt)
This rock formation, situated near the summit of Glyder Fach, in Snowdonia National Park, Wales, consists of volcanic rocks from the Ordovician period (about 450 million years ago). The landscape and rocks were shaped and eroded by glaciers (see pp.152–153) in the last ice age, and broken down by frost-shattering (see panel, opposite), leaving behind tors (prominent rock outcrops) and scree (piles of stones).

depositing sediment

Moving wind, water, and ice can all pick up and move particles of sediment. When the energy of movement drops—for example, when a river reaches a flatter section of its course and slows down—the effect of gravity means that the particles of sediment start to drop out of the flow, with the densest particles falling out first. Sediments often accumulate in layers in natural traps such as lake basins. If these sediments remain undisturbed, they may become compacted and cemented to form sedimentary rock, often with the original layers intact. Water, particularly rivers, play a major role in sediment transportation and deposition (see pp.158–159). The sediments they deposit form features ranging from sand bars and gravel bars to alluvial fans and deltas (see pp.162–163).

Pebbles and cobbles deposited by muddy glacial meltwater

Alluvial fan
This alluvial fan formed when a river flowing from the Kunlun and Altun mountains flooded an open plain in the Taklamakan Desert, China. During this rare flood, the river deposited its sediment in the desert. The area soon dried out again, leaving this intricate pattern of river channels and sand bars behind. The area shown in this satellite image is about 38 miles (60 km) wide. The white, green, and blue areas indicate differing levels of moisture.

Deposition by ice
The rocks and sediment carried by a glacier and deposited when the ice melts are known as moraine. This material may later be re-transported and reshaped by streams formed from meltwater (see pp.174–175).

RIVER CURRENTS AND SEDIMENT DEPOSITION

There are several reasons why a river deposits its sediment. Usually, it is because the water loses energy, for example as the river widens and the flow of water spreads out and slows down. Larger, heavier particles, such as pebbles and sand, fall out of suspension and are deposited first, while the lighter, finer particles, such as silt and clay, fall when the water current has slowed further. After the fine sediment settles, it can accumulate to form a sand bar or an alluvial fan.

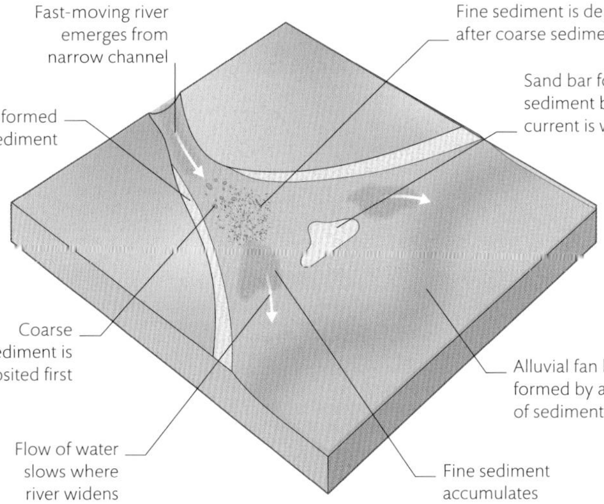

Water becomes shallower, with some rocks exposed above the surface

Fast-flowing water
Sudden drops in gradient along a river's course can result in the formation of rapids—very fast-flowing, turbulent parts of a river.

Meandering river
Streams called tributaries are joining this river (right) as it flows downstream in Iceland. Where the river bends, erosion has taken place on one side, resulting in the formation of river cliffs, while gravel is deposited on the other side (see panel, below).

rivers

A river is a large, natural stream of fresh water that starts on high ground, such as a mountain or a hill, then flows downstream along a defined course, or channel, to end its journey in a larger body of water—typically the ocean or sea, but sometimes a lake. A river's starting point is called its source, or headwater. The water can come from accumulated meltwater from glaciers and snow, from rainfall, or where groundwater flows up to the surface in the form of a spring. The river's end is called its mouth. A river shapes a landscape by eroding its bed and by transporting and then depositing sediment either downstream or across a wide area of its valley during a flood (see p.149 and panel, below).

HOW A MEANDER DEVELOPS

As rivers flow over relatively flat landscapes, they develop loops, or bends, called meanders. Because the velocity of the water is greater on the outer edge of a bend, the riverbed is more eroded there. Sediment is removed from the outside of the bend, sometimes forming a river cliff on the bank, and it is deposited downstream, in the inner bank of the next bend—an area called the slip-off slope. Over time, as increasing amounts of sediment are eroded and deposited, meanders become more pronounced. These features are called incised meanders.

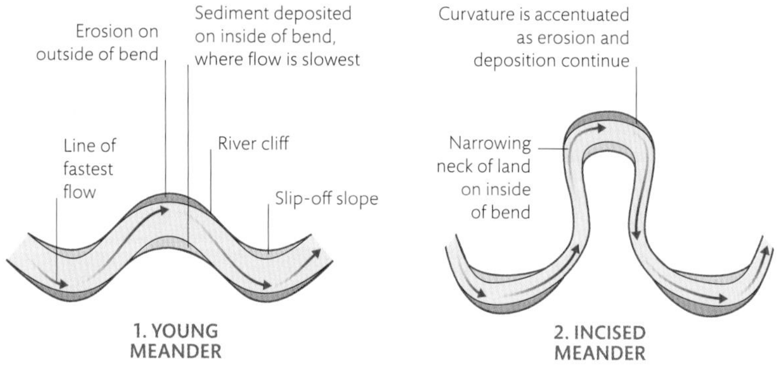

Partially frozen waterfall
In winter, some of the water cascading in waterfalls can freeze solid. This waterfall in Iceland (right) is partially frozen and partially liquid. In the plunge pool below, ice and liquid water coexist and mix.

Snow and icicles accumulate alongside cascading liquid water

Segmented waterfall
This broad waterfall in eastern Java, Indonesia, is cascading down a steep, amphitheater-shaped ridge of rocks. Several frothy streams of different sizes are plummeting down the ridge and converging into a plunge pool below. When a waterfall consists of several, distinct streams it is called segmented.

waterfalls

Also called cascades, waterfalls form when a river or other body of water flows over a steep rock ledge and falls into a pool (called a plunge pool) below. Waterfalls can form in several ways but are usually caused by erosion in the upper stages of a river, where fast-flowing water carrying sediment flows from hard rock to softer rock (see panel, below). Waterfalls that form in this way usually retreat upstream, sometimes forming a gorge. Dramatic changes in a landscape—for example, caused by an earthquake, a volcanic eruption, or a landslide—can also lead to the formation of a waterfall. The shape of a waterfall depends on the underlying geology as well as the size and shape of the river.

HOW A WATERFALL FORMS

Waterfalls form where a river flows over a layer of hard rock (such as granite) onto an underlying layer of softer rock (such as limestone or sandstone). Since the water erodes the softer rock faster than the harder rock, the harder rock creates an overhanging ledge, called caprock. Eventually, the caprock collapses since it is no longer supported, and rocks fall into the plunge pool below, where they swirl around and contribute to further erosion of the softer rock.

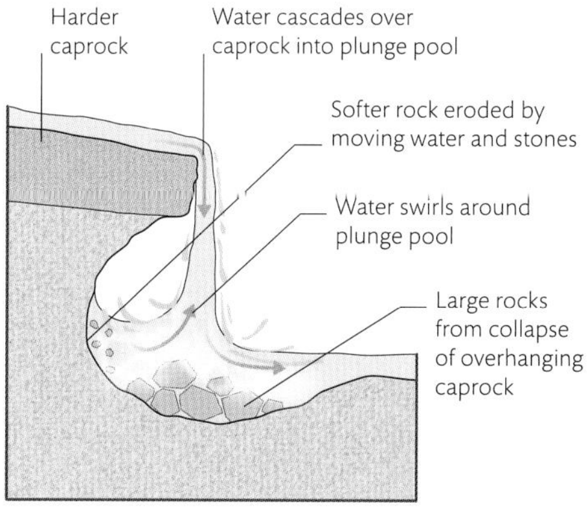

ESTUARY ZONES

An estuary can be divided into zones, with each zone having different degrees of tidal influence and proportions of sea water to fresh water. At its farthest end, where the river meets the sea, the estuary is dominated by sea water. The middle of the estuary contains an almost equal mixture of sea water and fresh water. At the upper zone of an estuary, fresh water begins to encounter salt water driven by tides.

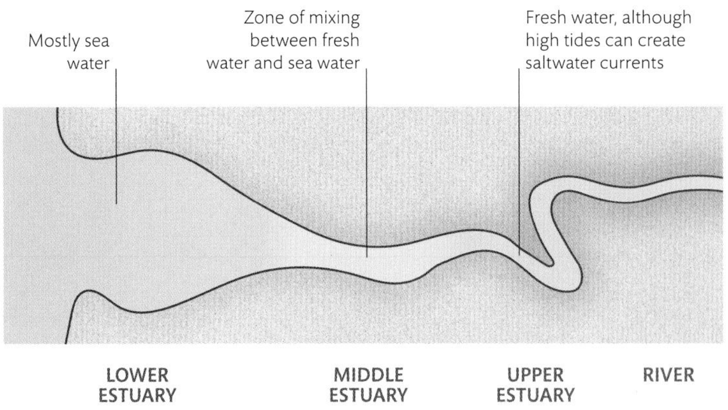

Inland river delta
The Okavango River flows almost 600 miles (1,000 km) from Angola through Namibia and into a depression in Botswana. Unlike most of the world's rivers, the Okavango does not empty into a body of water; instead, it empties into a plain in the Kalahari Desert, where it ultimately evaporates or is transpired.

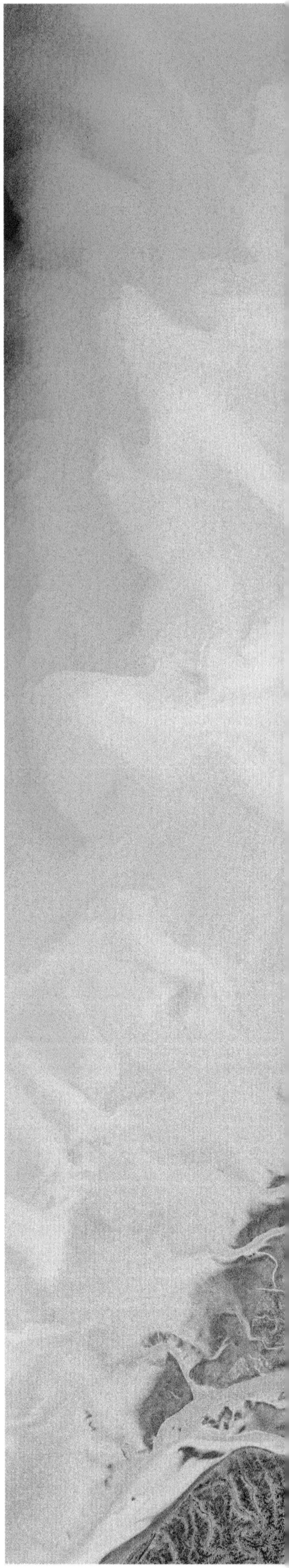

Yukon Delta
The Yukon River starts its life in British Columbia, Canada, and flows through the Yukon Territory before entering Alaska and forming a delta in the Bering Sea. This satellite image shows the many meandering channels that make up the fan-shaped delta. Scientists believe that sediment flow into the delta is increasing due to climate-related thawing of permafrost in the basin, which is releasing more sediment held in ice.

deltas and estuaries

Deltas form where rivers empty their water and sediment into a larger body of water—typically the sea, but sometimes a lake or another river—or, very rarely, into land (see above). Since a river loses energy and velocity as it nears its mouth, sediment often falls and settles on the riverbed. If large amounts of sediment (mainly mud, silt, and sand) build up, the river divides into a number of smaller, shallower channels, and new land (the delta) is formed. If a river flows more slowly and carries insufficient sediment to form a delta, sea water may enter the river mouth, creating a channel or inlet of brackish (slightly salty) water called an estuary. Within an estuary, the water level and salinity are affected by the tides (see panel, above).

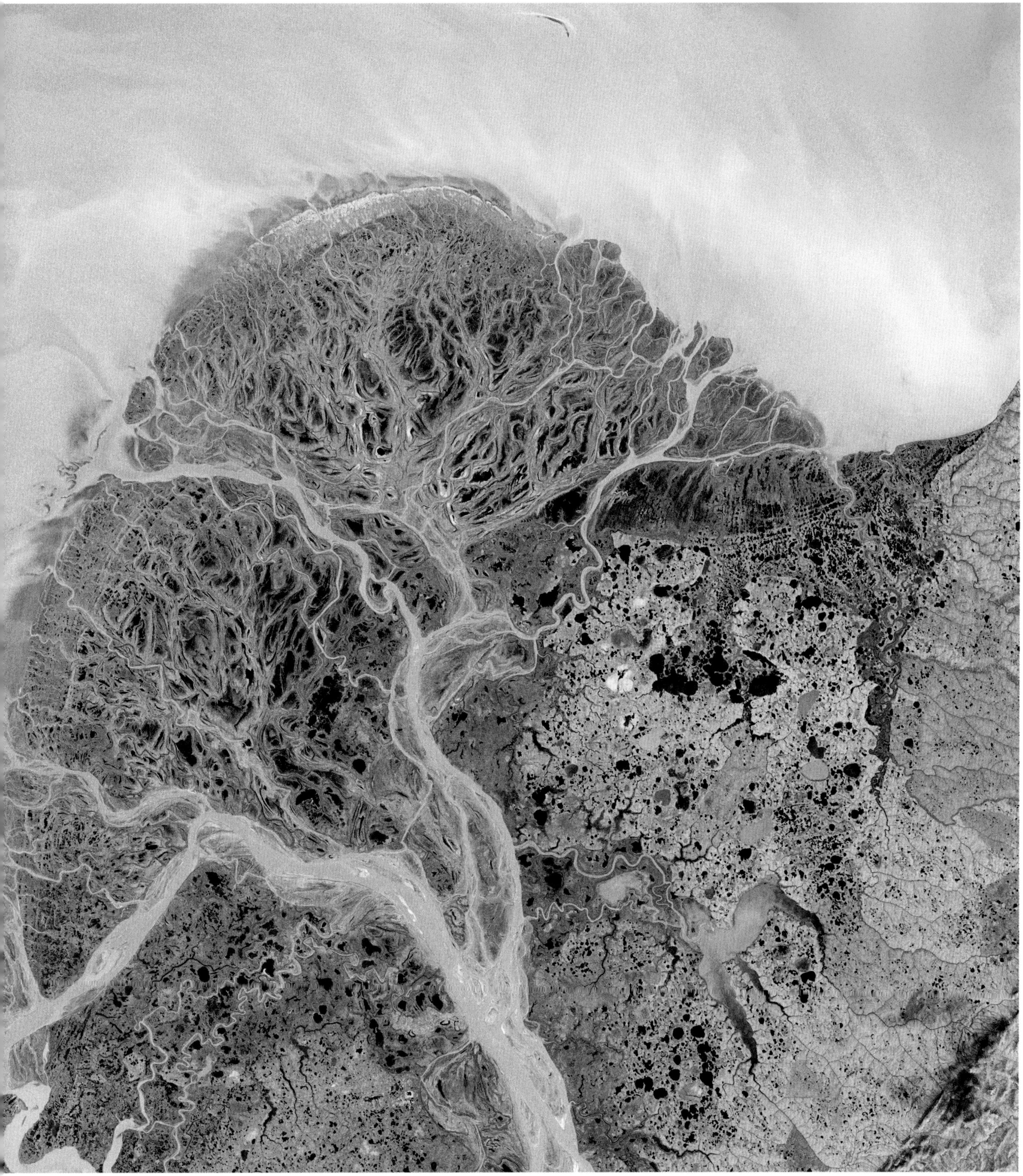

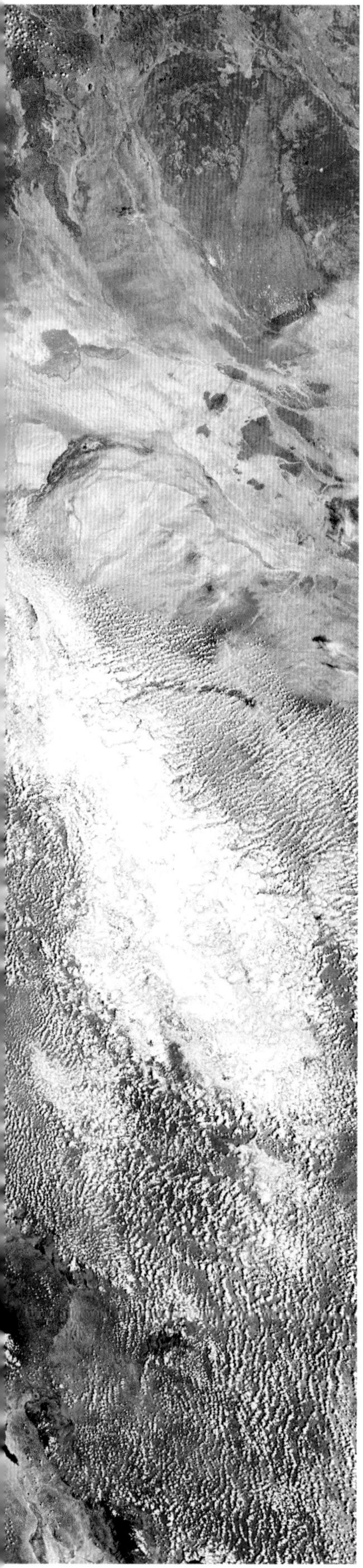

Lake Victoria
Located in East Africa's Rift Valley, Lake Victoria (captured here in a satellite image, bottom left) is a tectonic lake that formed when continental crust thinned close to a divergent boundary (see p.116). The largest lake in Africa, it lies mainly in Tanzania and Uganda but also extends into Kenya. The long, green lake at the top of the image is Kenya's Lake Turkana.

Large, wind-driven wave crashes against lakeshore

Lake Superior
Powerful storms blowing over the water's surface can create tall waves in large lakes, such as Lake Superior on the US–Canada border. The world's largest freshwater lake by surface area, Lake Superior has waves sometimes reaching several yards high.

lakes

Surrounded by land, and unconnected to the sea except by rivers or streams, lakes consist of a shallow depression called a basin that fills up with water from rain, melting snow and ice, rivers, or groundwater. Lake basins form in various ways. Most commonly, they are made by glaciers, which leave scoured depressions after they have retreated. Tectonic lakes form when movements of Earth's crust—such as at diverging plate boundaries (see pp.116–117) or along faults (see pp.126–127)—create a shallow depression. The water in lakes is relatively still compared to that of rivers, but it is still affected by currents driven by waves, wind, temperature changes, and river water entering the lake. Some lakes are entirely human-made.

TEMPERATURE LAYERS

In summer, many lakes have three distinct temperature zones: a relatively warm top layer (epilimnion); a colder bottom layer (hypolimnion); and a transition layer in between (thermocline), in which temperatures change rapidly. In the fall, seasonal mixing—called turnover—occurs, and the different layers combine until the entire lake is the same temperature. In winter, the layers become distinct again, although the temperatures are reversed: it is coldest at the icy surface and warmest on the lake bed. In spring, turnover reoccurs, during which the water is circulated from top to bottom. The warmer water is brought to the top, where the surface gains further heat from the sun.

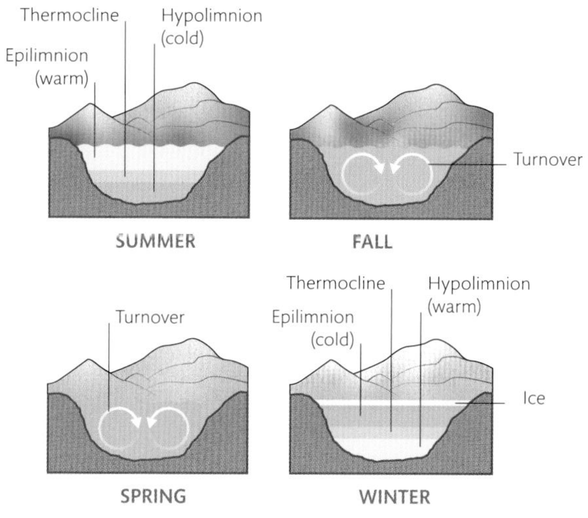

Towers shaped by water
This karst landscape in China is the result of millions of years of erosion of carbonate rocks by water. This process left behind an array of towering pinnacles that are now topped by lush vegetation.

karst landscapes

From giant pinnacles to densely cracked rock pavements, karst landscapes contain some of Earth's most spectacular features. These landscapes are sculpted by the action of precipitation and groundwater, which is slightly acidic and slowly dissolves carbonate rock such as limestone or dolomite. First, cracks in the rock start to appear. The cracks widen both on the surface and underground, where a complex system of caves and streams develops at the level of the water table (the line below which rock is saturated with water). Mineral-rich groundwater drips into the caves, forming icicle-like deposits called stalactites and mounds on the cave floor called stalagmites. When cave roofs collapse, towering spires are left behind.

HOW KARST LANDSCAPES DEVELOP

When water flows over carbonate rock, it slowly erodes it, opening up cracks called grikes and hollows called kamenitzas. The blocks surrounded by the grikes are called clints. Water flowing through the cracks dissolves the rock further, opening up large underground cavities called caverns. As water continues to flow through the cavern, it becomes larger. Eventually, the roof of the cavern may become too thin and falls, leaving behind the walls, which form towers in the landscape. Vegetation becomes established in soil inside cracks in the rock.

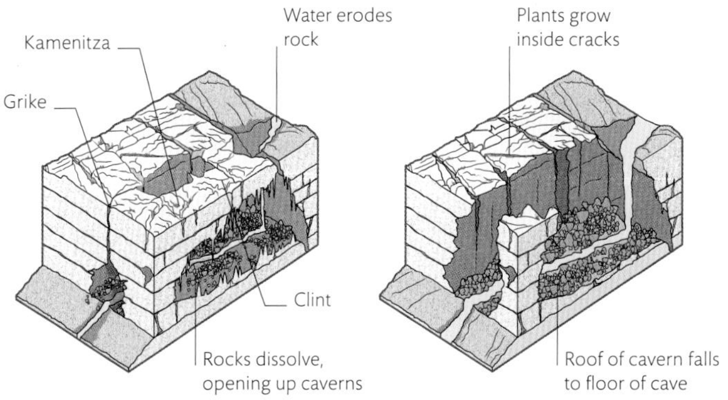

1. CAVERN FORMS

2. CAVERN ROOF COLLAPSES

caves

Caves are underground cavities that typically form in karst landscapes (see pp.166–167). Over time, acidic groundwater and precipitation seep through cracks in limestone (or sometimes dolomite), enlarging the cracks until they widen to form underground passages. Many cave systems descend steeply to the water table and then continue sideways as flooded, tubular passages, with the water sometimes reemerging at the surface as a spring. Eventually, if the water table drops, the caves dry out and sometimes the roofs collapse to form sinkholes. Dry caves can still widen as more precipitation and groundwater seep in. If the water table rises, dry caves can fill with water once again.

Stalactites hang from cave ceiling

Stalagmites form mounds on cave floor

Underwater cave
A cave diver explores a cenote cave system—an underground space filled with water that contains an opening to the outside. This cave system, in Yucatan, Mexico, was dry during the Last Glacial Maximum, about 22,000 years ago, when the ice sheets of the most recent ice ages were at their greatest extent and sea level was about 400 ft (120 m) lower than today.

Cave formations
This cave (left) is filled with limestone deposits that precipitate from water seeping into caves. Among the deposits are stalactites and stalagmites (see panel, below). Thin, hollow stalactites are called soda straws.

STALACTITES AND STALAGMITES

Rocks that form from mineral deposits inside caves are called speleothems. Two of the most common types are stalactites and stalagmites. As mineral-saturated water enters a cave's roof and drips down, residue builds up and stalactites form. As water droplets hit the ground, the minerals they deposit create mounds that grow into stalagmites. Over a long period of time, the two may join to form a single pillar that extends from floor to ceiling.

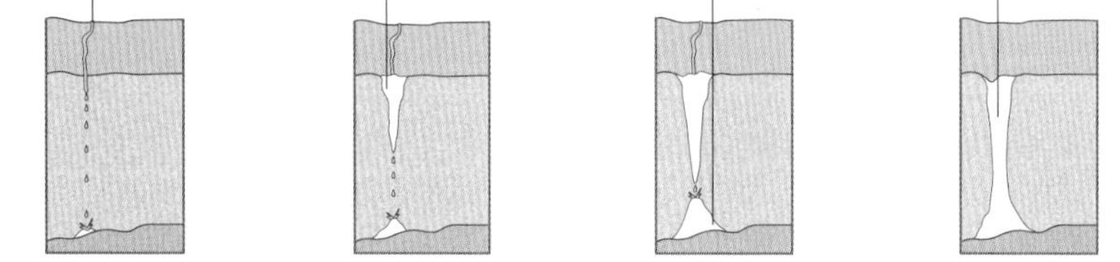

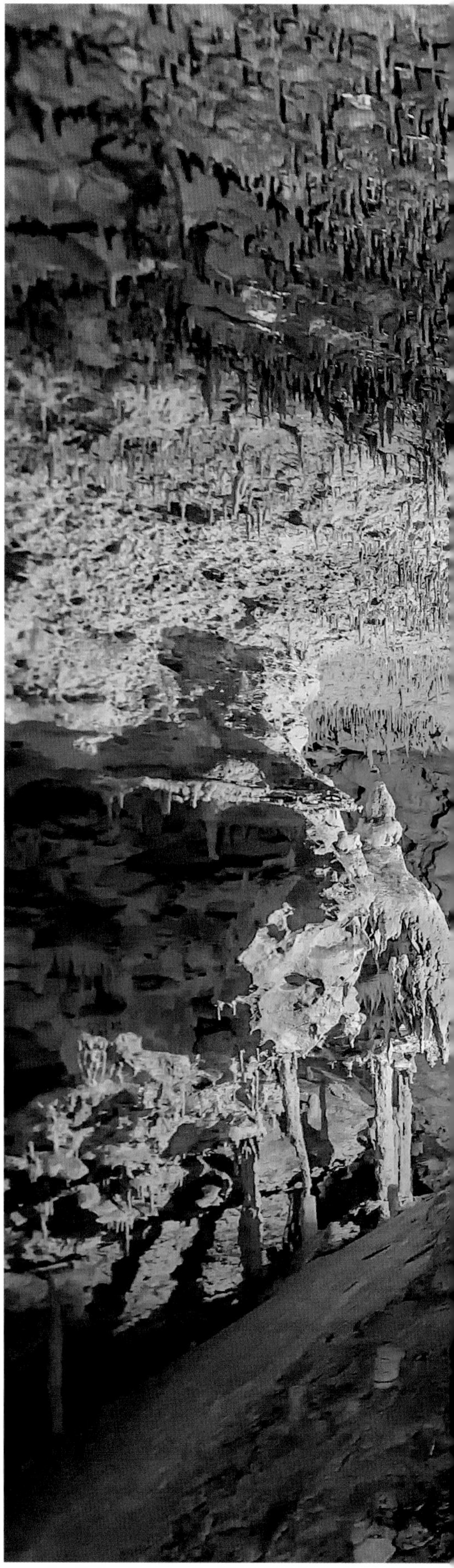

Confluence of glaciers
At 7.7 miles (12.4 km) in length, the Gorner Glacier (left) is joined by the smaller Grenz Glacier (right) as it descends the western slope of Monte Rosa in the Valais Alps, Switzerland. It forms the second largest glacial system in the Alps. The dark stripe of the medial moraine is clearly visible.

glaciation

Glaciers are moving bodies of ice that form from the accumulation and compaction of snow over thousands of years. Often dating from the last ice age, they are present in many mountain ranges, especially in cold regions. As they move downslope due to gravity and the pressure of their own weight, glaciers erode the landscape to form valleys, and in the process create deposits of rocks and sediments. Their abrasive action also leaves behind features such as striations, elongate scratches on rock surfaces, and grooves scooped into rocks in the landscape (see also pp.148–149).

GLACIAL EROSION AND DEPOSITION

Two glaciers, originating in semicircular depressions called cirques, may join to form a single glacier that flows into a preexisting valley. As they move, glaciers erode sediments by abrasion (when rock fragments on the underside scrape the surface of underlying rocks) and quarrying (when the glacier plucks out and incorporates bedrock from its base). The accumulated sediments left downslope are termed moraines. Terminal moraines are deposited at the lowest end of the glacier (snout), while lateral and medial moraines are deposited at its sides and in the center.

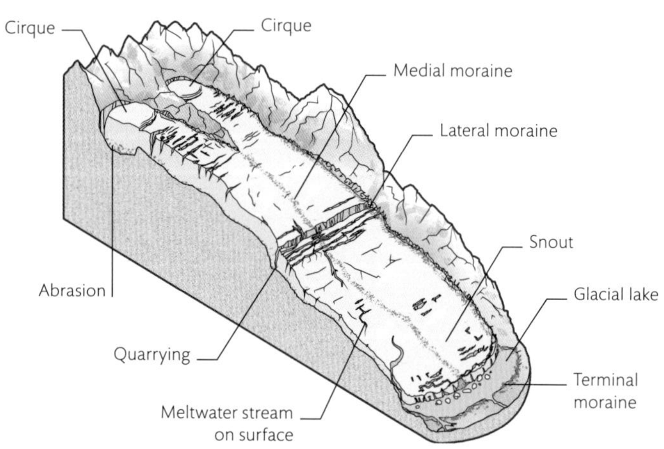

spotlight Glacier Bay

Glacier Bay is an arm of the Pacific Ocean that thrusts deep into a spectacular landscape in southern Alaska. The bay is 65 miles (105 km) long and surrounded by some of the highest coastal mountains in the world, which were thrown up by the collision of the Pacific and North American tectonic plates. To the west, the Fairweather Range reaches 15,000 ft (4,600 m) above sea level. Its high peaks trigger snowfall from the moist air blowing in from the Gulf of Alaska, feeding more than a thousand glaciers in the surrounding area.

Some of the glaciers flow directly into the sea as tidewater glaciers, while others terminate inland as valley glaciers. The Johns Hopkins and Margerie are two of the largest tidewater glaciers—each more than a mile wide—with ice fronts up to 200 ft (60 m) high, where icebergs calve off into the water. Most of the region's glaciers are retreating as the climate warms. The Grand Pacific Glacier no longer reaches the ocean.

Glacier Bay was not always a bay. In 1680, during the Little Ice Age, its southern half was a broad valley occupied by the Tlingit people. A huge glacier was advancing from the north, and by 1750 the whole valley lay under thousands of feet of ice. But the surge was short-lived. By 1880, the ice had retreated 45 miles (70 km) inland, leaving behind Glacier Bay. Today, the tributary glaciers have retreated farther up their valleys, which are now filled with water to form fjords.

The Lamplugh Glacier

The glacier is almost 1 mile (1.5 km) wide, 19 miles (30 km) long, and 165 ft (50 m) high at its terminus

Topeka Glacier
On the western side of Glacier Bay, the Topeka Glacier has carved a steep valley. Like 95 percent of the glaciers in Alaska, it is thinning and now no longer reaches the sea. It is estimated that the area has lost 11 percent of its glacial ice since the 1950s.

Glacier cave
Meltwater that runs through or under a glacier's ice can carve out a tunnel, or cave, between the ice and the bedrock. Unlike ice caves, which are formed in bedrock and contain ice all year round, glacier caves (here, in the Alps, left) are a seasonal feature. Increased melting due to climate change means glacier caves are becoming rarer.

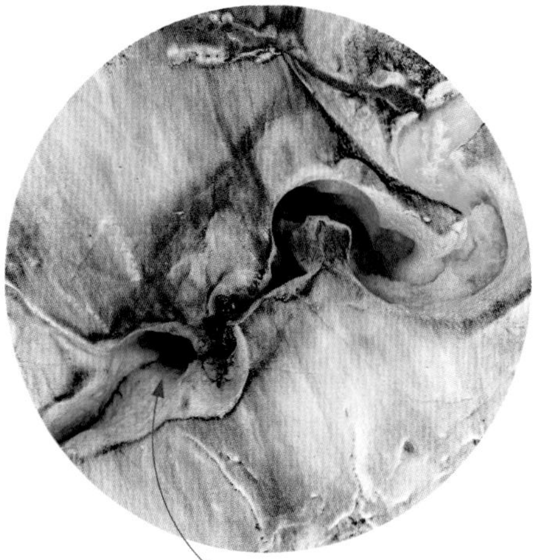

Glacial meltwater flowing on the ice is funneled into a moulin

Surface meltwater
This aerial view of the surface of Sawyer Glacier in Alaska shows a meltwater stream entering a moulin—a vertical shaft in a glacier. The meltwater travels through the moulin to the glacier's bed.

meltwater

When snow and ice held in ice sheets and glaciers melt, the liquid released is called meltwater. Streams of meltwater flow away from the glacier's terminus (the cliff-like edge where the glacier ends) and deposit debris—including rocks carried by the ice—downstream (see p.157). Meltwater streams can also create tunnels of water under a glacier, where they deposit sediment that can build up to form a long, sinuous ridge called an esker. Each year, glaciers melt naturally as part of the seasonal thaw. However, glaciers are melting at unprecedented rates worldwide as a result of climate change, contributing to a rise in sea level and increased flood risk.

MELTWATER LAKES

Lakes made of glacial meltwater sometimes form between the edge of a glacier and a moraine (a ridge of debris deposited by the glacier). When the melting glacier releases chunks of ice in a process called calving, the sudden input of material can cause waves, called seiche waves, on the lake's surface. The waves can move water over the moraine—a process known as overtopping—which may lead to flooding. Increased glacier melt due to global warming has led to an increase in size and number of meltwater lakes (also called proglacial lakes).

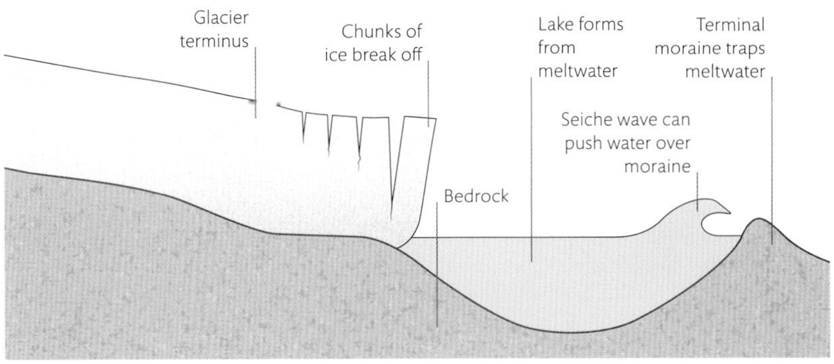

ocean and atmosphere

On Earth's formation, its most volatile ingredients jetted out of its crust as gas. Some escaped into space, some condensed into oceans, and some coalesced to form the planet's outermost, gaseous layers—its atmosphere. These fluid parts of Earth remain in constant motion and interaction—the sun's heat maintaining currents in the ocean and winds and weather systems in the air.

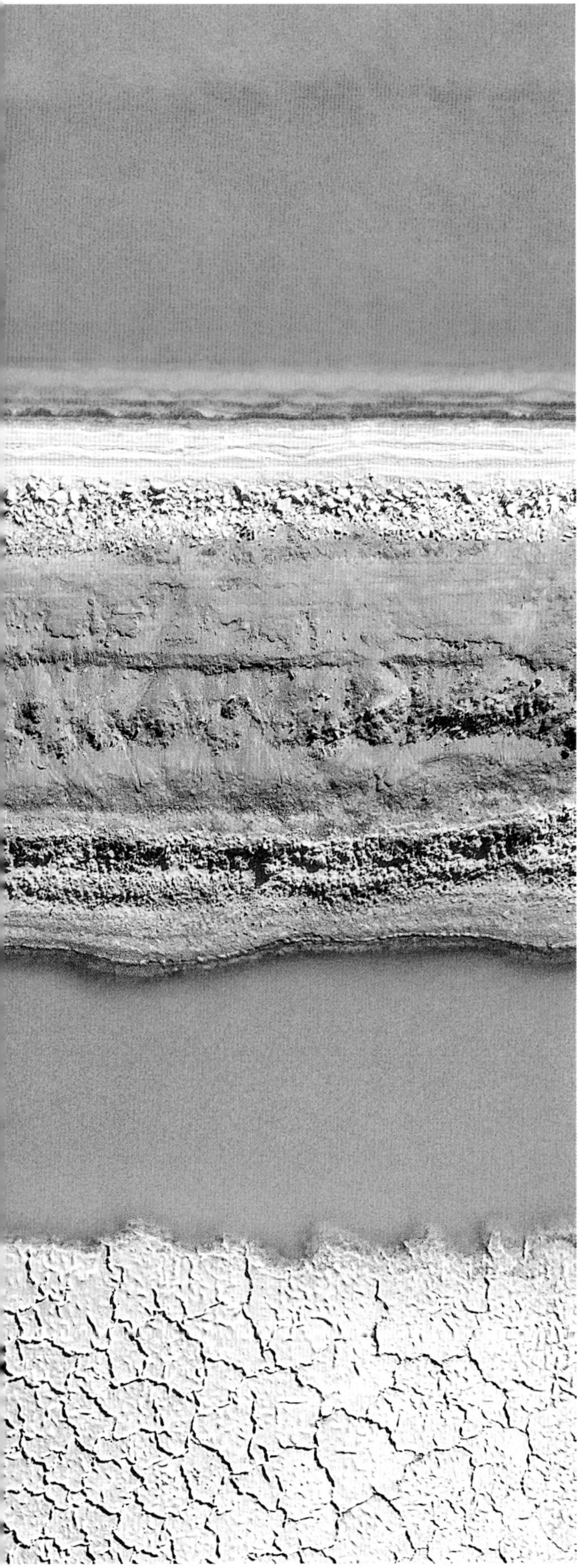

Aerial view of salt pans, France
Salt pans are shallow coastal ponds used to obtain salt from sea water. Sun and wind evaporate the sea water, concentrating the salts. Microscopic algae thrive in such extreme conditions. Different species bloom as salinity increases, changing the ponds' color from light green to vibrant red.

Salt pan harvest, Thailand
When enough water has evaporated and salinity is sufficiently high, the salt crystallizes out of solution, forming crusts of salt that can be harvested (right).

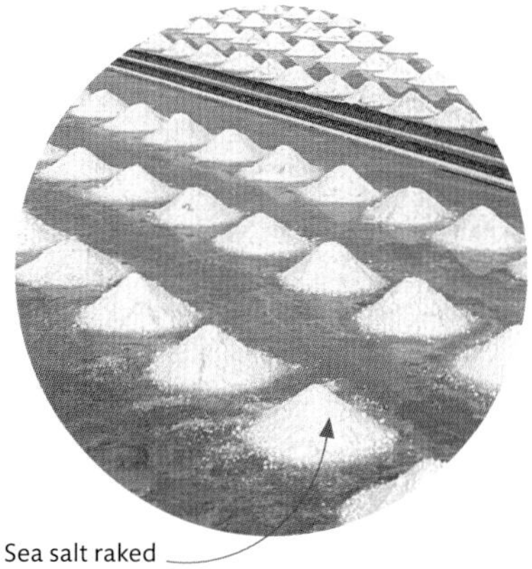

ocean chemistry

Sea water is a chemical cocktail of nearly 100 different elements, including life's building blocks—carbon, nitrogen, phosphorus, hydrogen, and oxygen—and even gold. Besides water (H_2O), the ocean contains 550 trillion tons of dissolved salts, the most common being chloride, sodium, sulfate, magnesium, calcium, potassium, and bicarbonate. The average concentration of salts (salinity) is 3.5 percent. Some marginal seas, such as the Mediterranean and the Red Sea, are saltier; others, like the Baltic, are less so. Sea water is slightly alkaline (pH 7.5–8.4), and it holds 60 times more dissolved carbon dioxide than the atmosphere.

THE SALT CYCLE

Ocean chemistry is in constant flux, but overall salinity remains unchanged due to a balancing process called the salt cycle. Rivers, volcanoes, and deserts add around 3 billion tons of dissolved chemicals to seas annually, and a similar quantity enters the ocean at mid-ocean ridges and from the dissolution of seabed minerals. An equal amount is removed by living organisms and the deposition of their remains on the seafloor, and by the subduction of marine sediments at oceanic trenches.

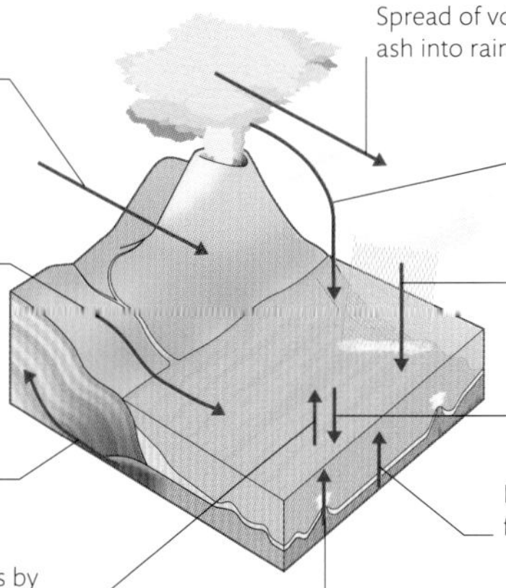

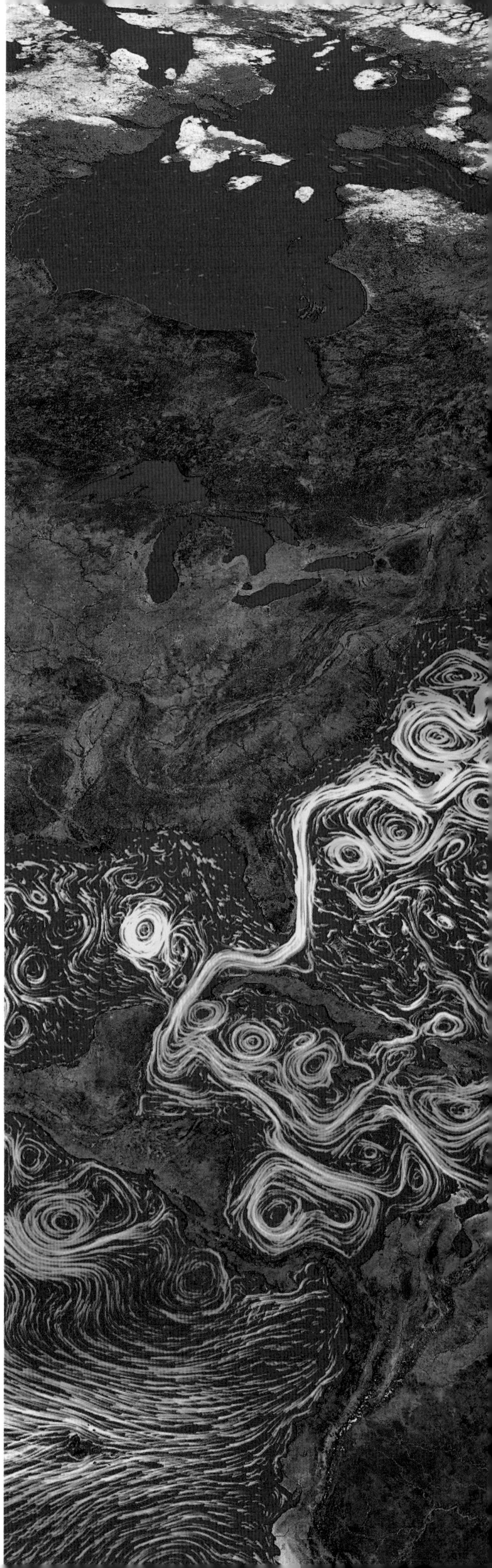

Atlantic eddies
Based on satellite readings of ocean temperatures, this graphic reveals the currents that flow across the Atlantic. The Gulf Stream, indicated in yellow (warm) and green (cool), flows from Florida to northwest Europe. It is fed through the Caribbean by the warm Atlantic Equatorial Current (shown in orange and red). Both major current systems flow in a series of giant eddies.

ocean circulation

The constant turmoil of atmospheric winds, powered by solar energy, drives the world's major ocean currents. Surface circulation consists of a series of wind-driven gyres (circular currents) around each major ocean north and south of the equator (the North Atlantic gyre is pictured here). The huge volumes of water at the core of these gyres exert great pressure. This combines with the effects of Earth's rotation (the Coriolis effect), the position of land masses, and the narrow gateways between ocean basins, to keep ocean waters in perpetual motion, moderating climate throughout the world.

OVERTURNING CIRCULATION

As surface currents flow toward the poles, so bottom water, driven by changes in temperature and salinity, diffuses slowly back toward the equator (see p.185). This overturning circulation transports vast amounts of heat, dissolved chemicals, and sediments around the world's oceans. It acts as a regulator of Earth's climate and locks carbon dioxide in the oceans.

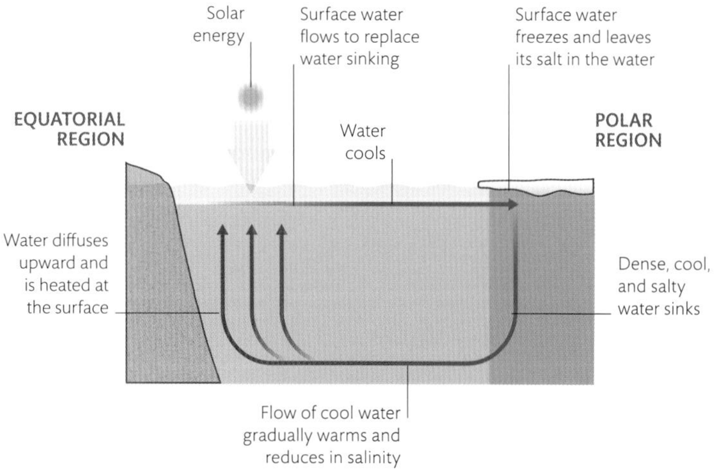

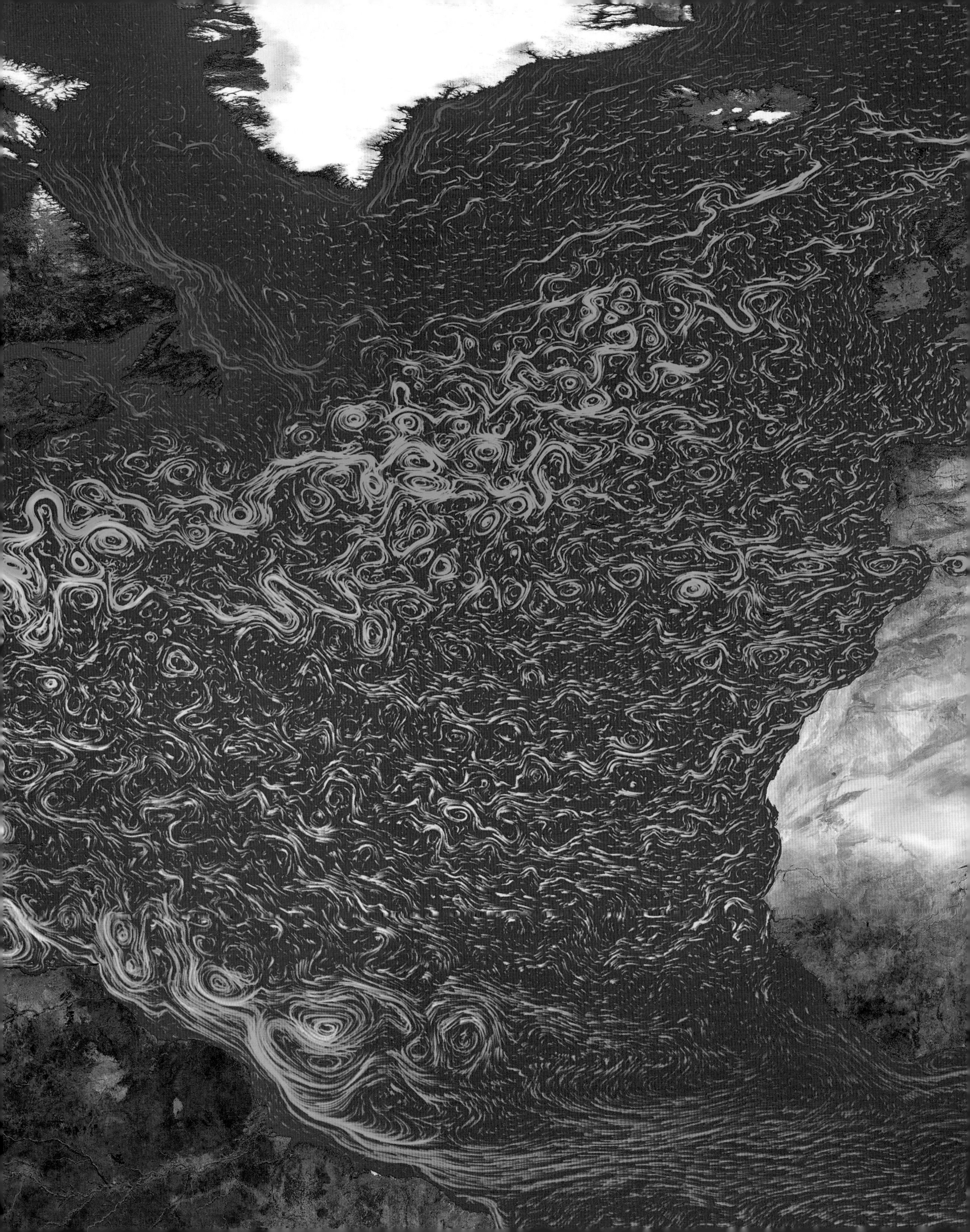

VERTICAL DIFFUSION

The slow upward diffusion of water from several hundred yards below the surface is called upwelling. The water rises to fill the space left by the offshore deflection of surface water, which is caused by a combination of wind direction, prevailing currents, and the Coriolis effect of Earth's rotation. Where winds and currents are reversed, the movement of water is toward the shore and the surface water is forced downward. This is known as downwelling.

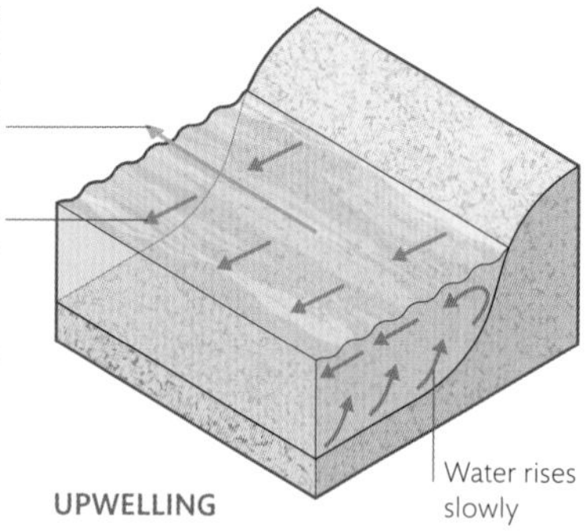

UPWELLING

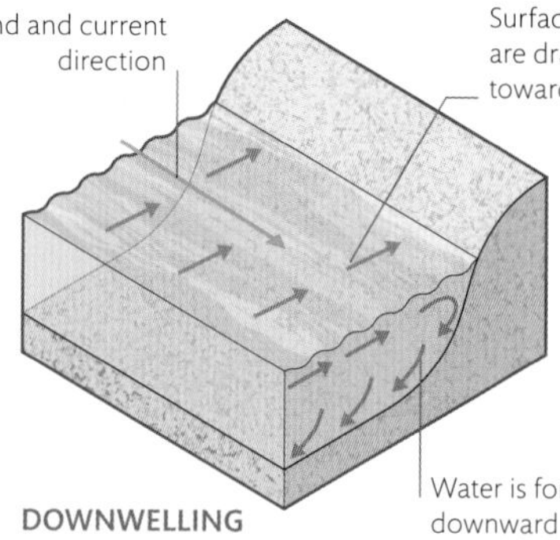

DOWNWELLING

upwelling and plankton blooms

Upwellings are rising currents that bring water rich in chemical nutrients up from the depths. At the surface, minute photosynthetic organisms called phytoplankton build their bodies from these nutrients, their growth fueled by solar energy. Phytoplankton, which include algae and cyanobacteria, are consumed by drifting animals known as zooplankton. Together, the two types of plankton support myriad marine grazers and predators.

Bloom of phytoplankton
In spring and summer, upwellings can trigger population explosions, or blooms, of phytoplankton. In this satellite view, a vibrant green phytoplankton bloom in the Baltic Sea is caught in swirling surface currents off Estonia. The central spiral eddy is about 20 miles (30 km) across. Such blooms can span hundreds or even thousands of miles.

Phytoplankton
Living out short lives in sunlit surface waters, phytoplankton are single-celled floaters and drifters. The majority, including most dinoflagellates, are microscopic; others, such as diatoms, are the size of sand grains. Many phytoplankton are master sculptors, building intricate skeletons out of calcium or silicon that they obtain from sea water.

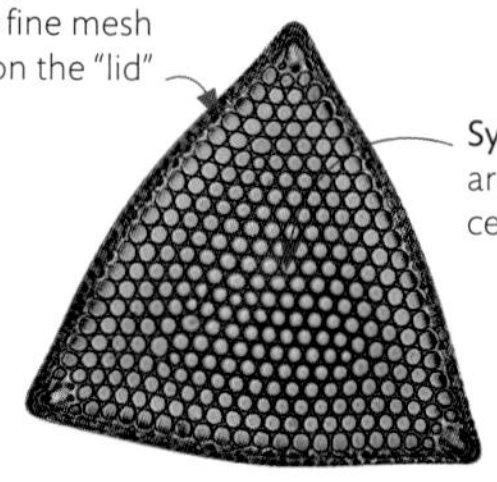

CENTRIC DIATOM
Triceratium faves

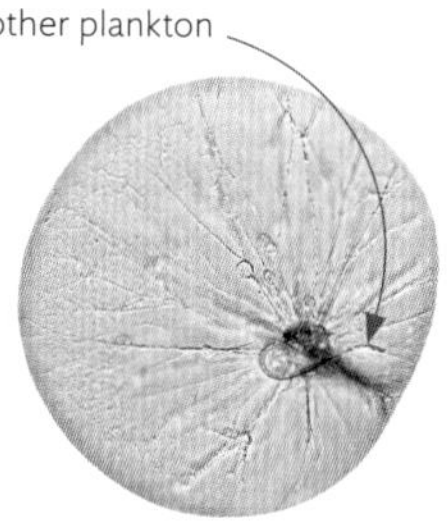

DINOFLAGELLATE
Noctiluca scintillans

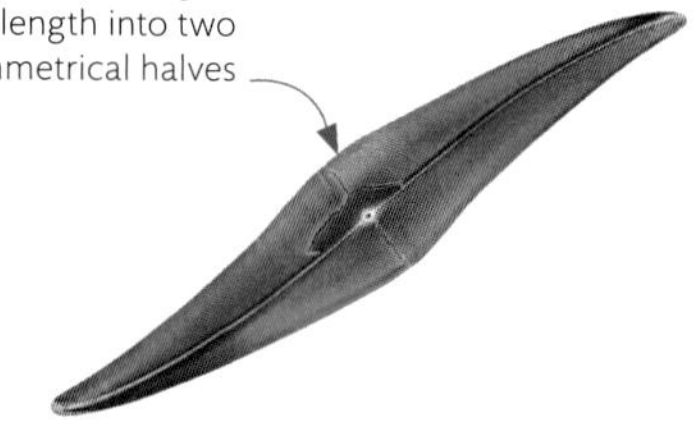

PENNATE DIATOM
Pleurosigma angulatum

Zooplankton
Wherever phytoplankton bloom, zooplankton abound, too, including some of the smallest of all marine animals. Zooplankton ride on tides and currents, although some are also able to swim very weakly. Holoplankton spend their entire lives as surface plankton; meroplankton are the larvae of organisms that complete their life cycles in deeper water.

CUSK-EEL LARVA
Brotulotaenia nielseni

OCTOPUS LARVA
Wunderpus photogenicus

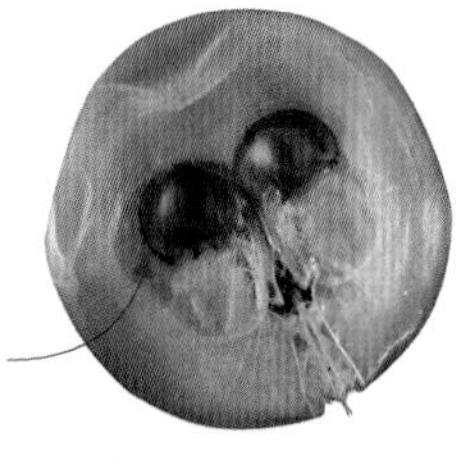

OSTRACOD
Gigantocypris muelleri

deep currents

Incised into the steep continental slopes that rim the ocean basins (see p.198) are narrow gullies, deep gorges, and tightly meandering channels. All but those in shallow water (main picture) are completely hidden from view. Some are deeper than Arizona's Grand Canyon, others snake across the seabed for more than 1,860 miles (3,000 km). These channels funnel sediment from land to the deep ocean in underwater flows, known as turbidity currents, up to a third of a mile thick and traveling at speeds over 60 mph (95 kph). Elsewhere, the ocean bottom is pockmarked with craters where methane gas has belched upward from buried organic-rich sediment, and scoured by huge submarine slides whose displacement may have triggered powerful tsunamis in the water above. Deep currents sculpt the seabed into ripples, dunes, and giant sediment waves, tracing the path of the thermohaline conveyor belt (see panel, opposite) across the marine underworld.

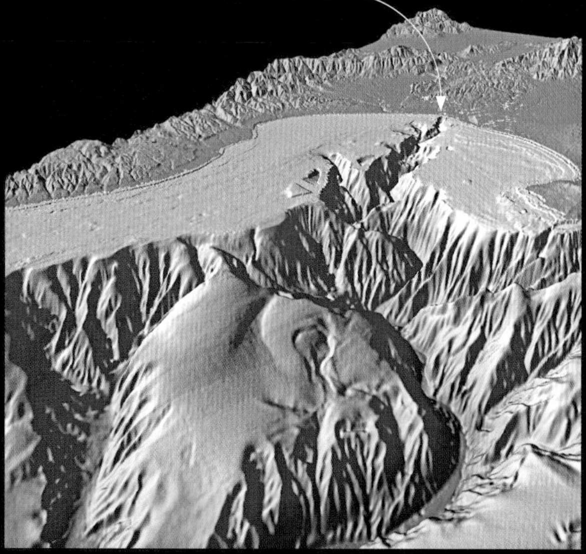

Seafloor relief, Monterey Bay, California
Monterey Canyon cuts a deep chasm across the Californian shelf before plunging down the steeply gullied slope. It extends for 290 miles (470 km) to the abyssal plain at a depth in excess of 13,000 ft (4,000 m).

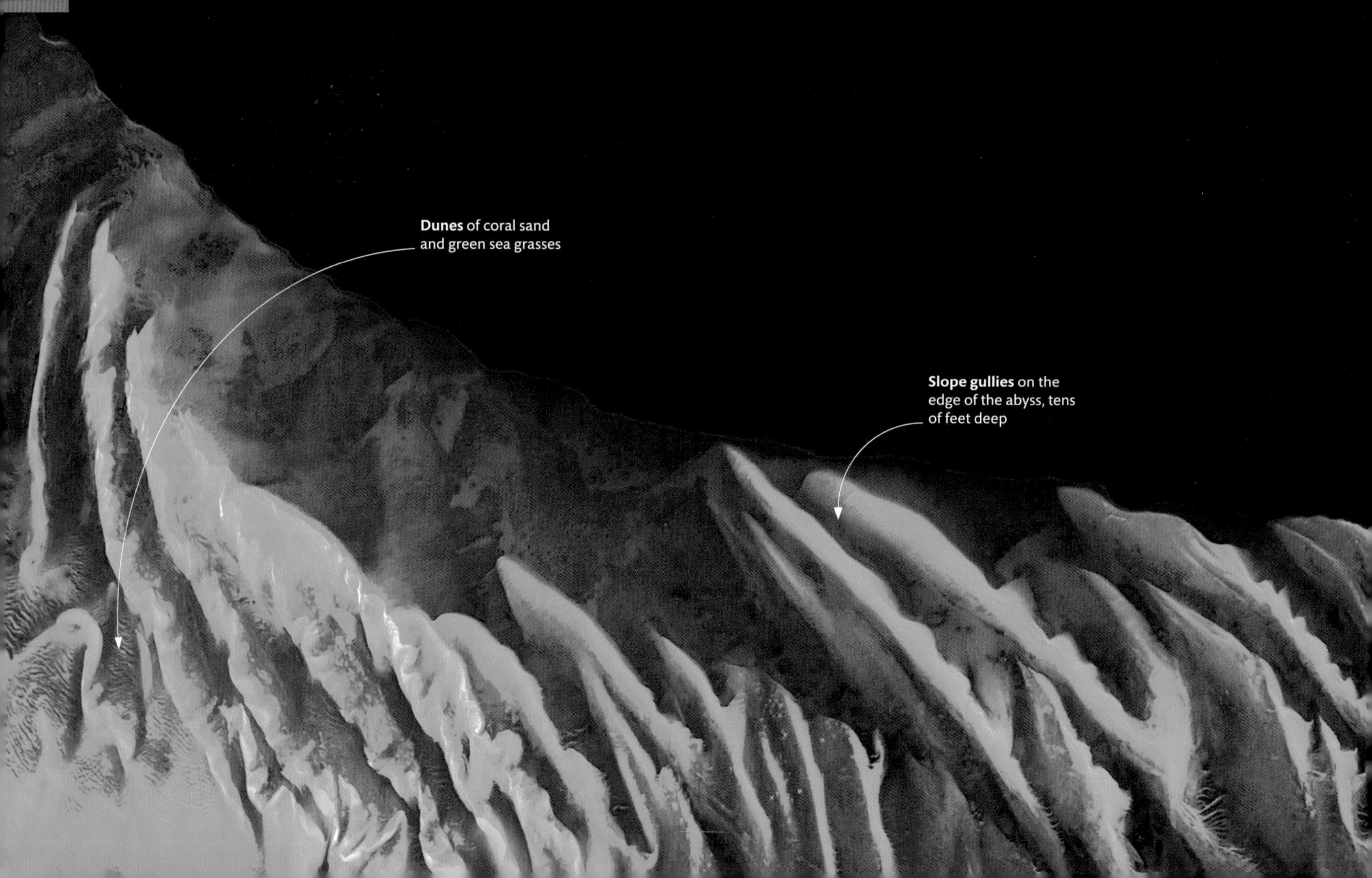

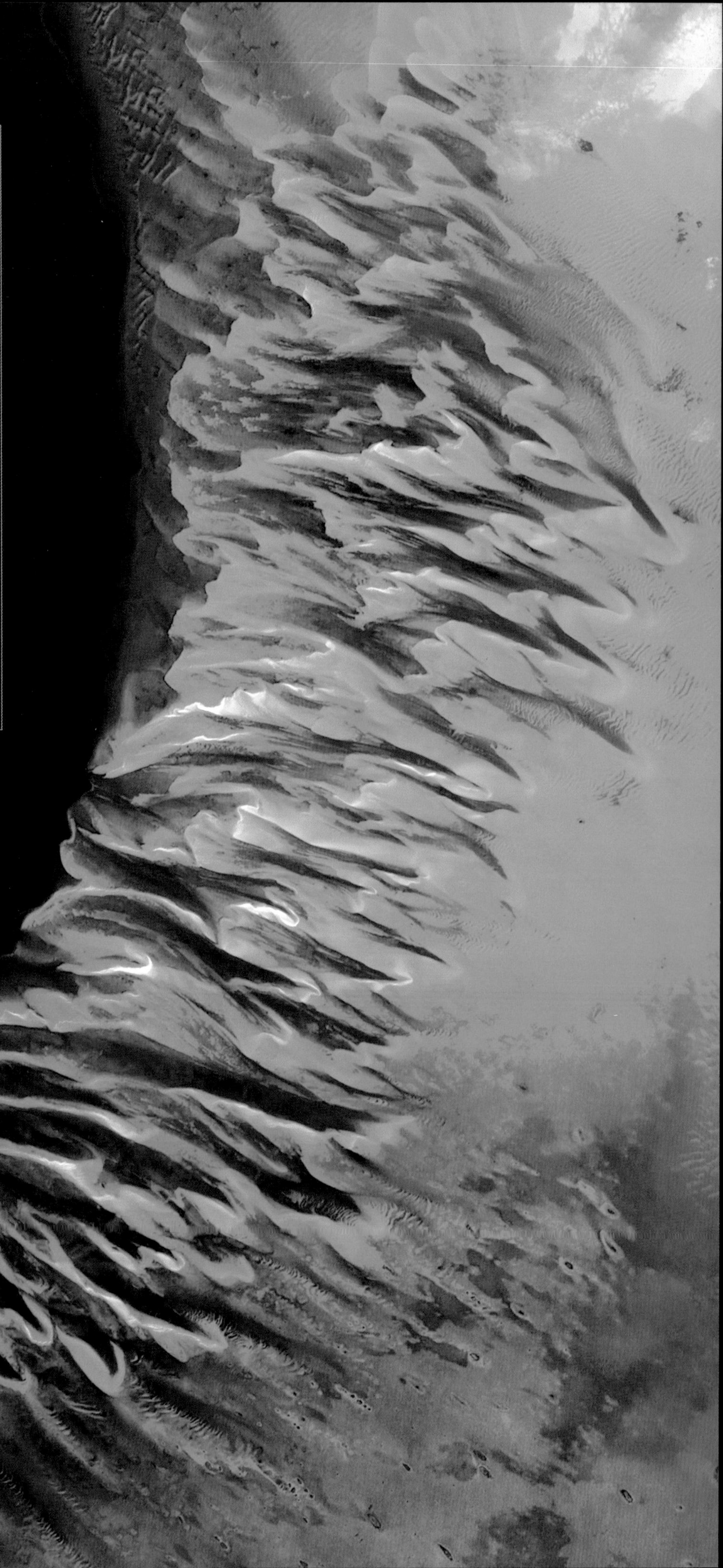

THERMOHALINE CONVEYOR BELT

Deep-ocean currents are part of the global pattern of thermohaline circulation. Driven by differences in water density—determined by temperature and salinity—thermohaline circulation is like a giant conveyor belt. As colder, denser, saltier water in polar regions sinks, it is replaced by warmer, less-dense, less salty surface water flowing poleward from the equator. In the deepest abyss, the conveyor-belt currents may cascade over silent waterfalls 2 miles (3 km) high or rage like storm-floods across the seabed, causing unseen turmoil.

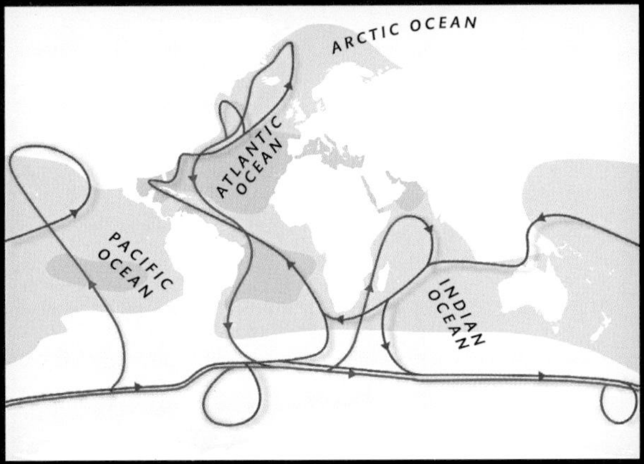

KEY

- Cold deep currents
- Warm surface currents

Salt content (%)

3.4 3.6 3.9

On the edge

This satellite image cuts through the clear waters to reveal the lip of Great Bahama Bank as it descends more than 6,560 ft (2,000 m) into a hidden blackness known as the Tongue of Ocean. The beautifully curved patterns on the bank's edge are gullies up to 1¼ miles (2 km) wide, carved into the seafloor by strong currents.

Surfing the tidal bore, Alaska
A tidal bore occurs when a strong tide surges up an inlet or river, traveling against the current as a wave or series of waves. At Turnagain Arm, Alaska, bores can be 10 ft (3 m) high and move at 15 mph (25 km/h).

Flowerpot Rocks, Bay of Fundy
The greatest tidal range—more than 50 ft (16 m)—occurs in Canada's Bay of Fundy. The sea covers and erodes the base of these sandstone stacks twice each day.

tides

The rise and fall of tides has occurred since the oceans formed around 4 billion years ago. Tides are waves with very long wavelengths that sweep around the planet. High tides are the crest of the wave and low tides the trough. Yet the reality of tidal motion is complex and variable due to frictional and topographic restrictions; the shape of ocean basins and marginal seas; and, in recent times, human changes such as the dredging of rivers. The difference between high and low tides is the tidal range, which varies from zero to as much as 40–50 ft (12–16 m) in confined inlets and estuaries.

MONTHLY TIDAL CYCLE

The monthly tidal cycle is affected by the gravitational forces of the Earth-Moon and Earth-Sun systems. When sun and moon are aligned (at full moon and new moon), the two force fields act in the same direction, combining to yield very high and low spring tides. When they are at right angles (at first and last quarter), they pull in different directions, giving less extreme neap tides.

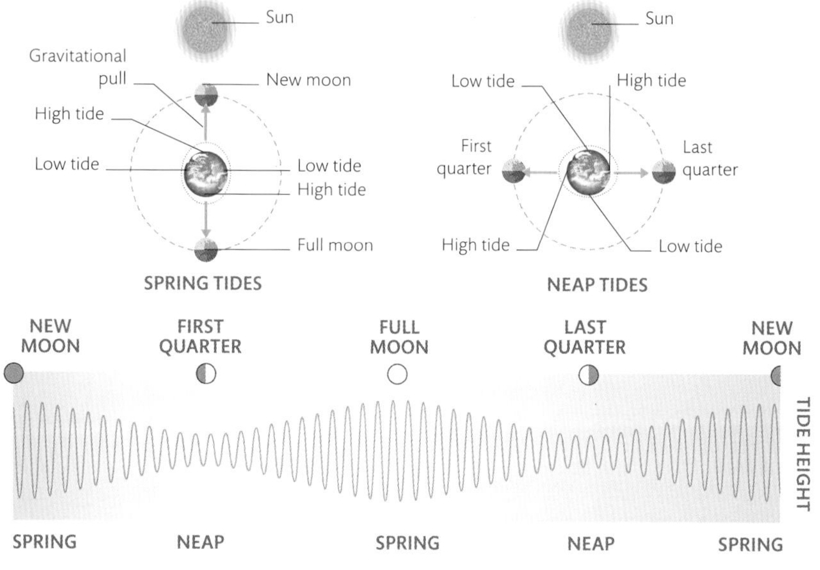

WAVE FORMATION

As winds blow across an area of sea, known as the fetch, waves develop. Small ripples, known as capillary waves, gradually build into chaotic choppiness. Beyond the fetch, wave interaction develops a distinctive regular pattern, called swell. As a wave enters shallow water, it slows against the seafloor. Its height increases until it spills forward as a breaker.

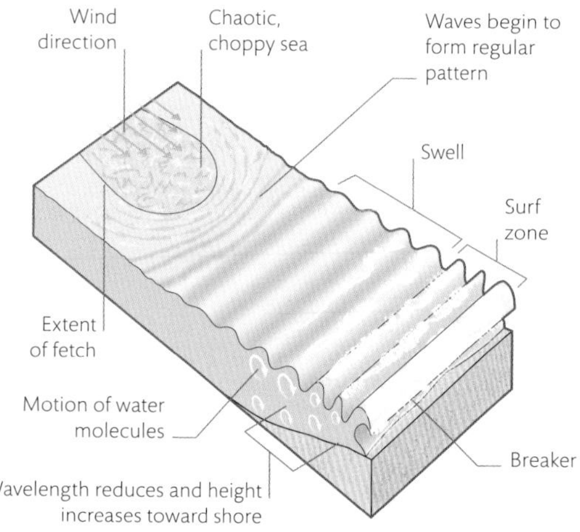

oceanic waves

As familiar and changeable as the color of the sea, waves pound the shore and erode the land with their relentless power. Almost all waves are caused by wind stress acting at the sea's surface. The greater the wind force, the bigger the wave, and the more energy is stored and then released when the wave breaks. Individual storm waves produce instantaneous pressures of up to 2.7 tons per sq ft (30 metric tons per sq m), enough to destroy cliffs, piers, and coastal buildings. Storms at sea generate a train—a distinctive series of waves of a similar wavelength that, if not disrupted, retains its unique signature as it travels thousands of miles across an entire ocean.

Ship steers to face wave head-on; a full cargo keeps it steady as it rides the rogue wave

Rogue wave can reach a height of more than 50 ft (15 m)

Rogue waves
Conditions at sea sometimes cause waves to merge and create a single, unusually large wave called a rogue wave. Rogue waves in the open ocean are isolated and unpredictable—and a danger even to large ships.

Wave power
A rebounding wave spectacularly clashes with an incoming breaker, whipping up the surf at Cape Disappointment in Washington state. Wave energy and erosive power become focused on coastal headlands such as this, when the wave train encounters land (see p.195).

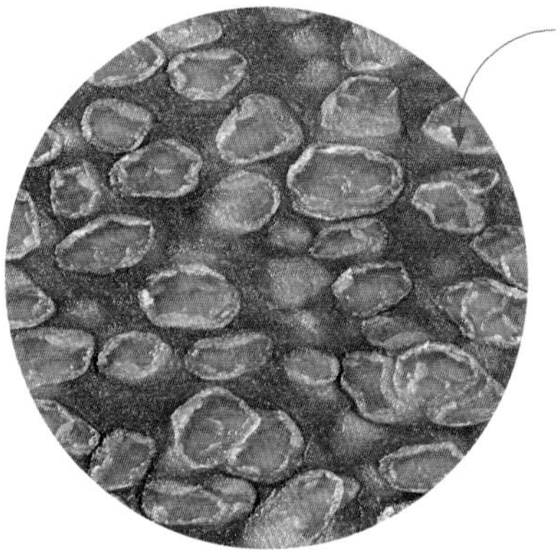

Pancake ice
Ice disks 1–10 ft (0.3–3 m) in diameter with upturned edges are called pancake ice. They form under conditions of moderate to high wave activity.

Ice floes
This jigsaw-like network of large ice floes, off Höfn in southeast Iceland, provides a temporary resting place for a seal, as well as a vantage point for hunting. Ice floes, which range in thickness from 1½–15 ft (0.5–5 m), repeatedly fuse into larger areas of sea-ice, and then break up again due to rough seas and partial melting.

frozen sea

Icebound, ice-cold, and inhospitable—such is the nature of polar seas. The North Pole sports an ice-covered ocean surrounded by land, the South Pole an ice-capped continent set amid frozen sea. In the dark polar winters, when temperatures drop below -22 °F (-30 °C), the sea freezes. Being less dense than sea water, the ice that forms floats. Around 7 percent of the ocean's surface—an area roughly equivalent to the size of North America—is covered by floating sea-ice at any given time. Fast ice is sea-ice that is anchored to the shore, while pack ice drifts on currents. With the continuous daylight of polar summers, much of the ice melts back into the ocean. Some ice remains year-round, but global warming is reducing the area of permanent sea-ice cover.

HOW SEA-ICE FORMS

Freezing conditions cause ice crystals to develop near the sea's surface. As the wind blows, the crystals fuse into a slushy layer—grease ice—that hardens and thickens as temperatures fall. Rough seas break this layer into rounded shapes known as pancake ice, which then coalesce into masses called ice floes. Year on year, thicker sea-ice builds up, especially near the shore.

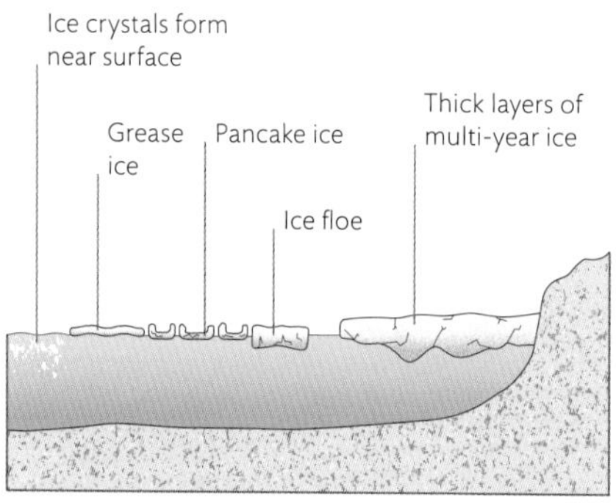

Deceptive appearance

The ice raft on which these walruses (*Odobenus rosmarus*) are resting is the above-surface portion of an iceberg in Hudson Bay, Canada. Most of an iceberg, around 90 percent, lies hidden below the waterline.

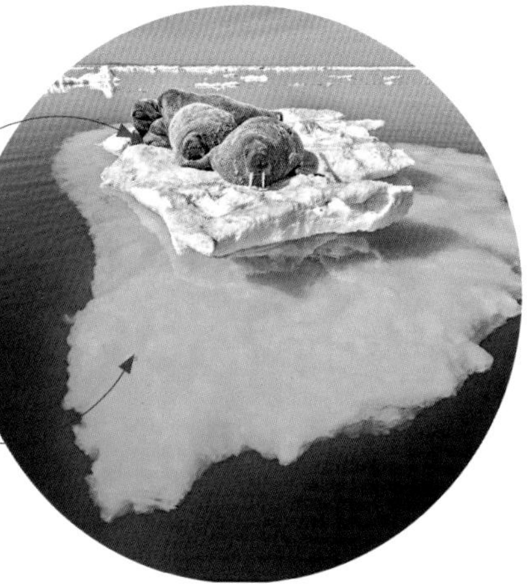

Walruses are insulated from the cold by blubber under the skin

Iceberg floats sideways, with its long axis parallel to the surface; it is only slightly less dense than the ocean, so it sits low in the water.

ice shelves and icebergs

Where flowing ice sheets and glaciers reach the coast they may extend over the sea as vast shelves of floating ice that tower above the waves. The ice in the shelves has accumulated from repeated snowfall over thousands of years and become progressively more compressed. As tiny air bubbles are squeezed out, the ice turns from white to a bluish hue. Marine algae frozen into the ice yield a vivid green, while rock flour—particles ground from rocks as the ice moved over land—produces reds, yellows, and browns. An iceberg is part of an ice shelf or glacier that breaks off and floats out to sea. Icebergs range from about 50 ft (15 m) in length to the size of a small country, such as Luxembourg.

Iceberg with chinstrap penguins

After a long winter, penguins such as chinstraps (*Pygoscelis antarcticus*) visit icebergs and await the food bounty triggered by the spring plankton bloom (see p.182). The melting icebergs themselves help fuel plankton growth by releasing nutrients picked up from land.

HOW ICEBERGS FORM

Floating ice shelves and glaciers are buffeted by fierce winds and powerful waves, but even greater strain is caused by tides rising and falling beneath the ice. Cracks form in the ice and open up, while existing crevices are further enlarged. Eventually, in a process known as calving, large chunks break off and crash into the sea, floating away as blocks of freshwater ice called icebergs.

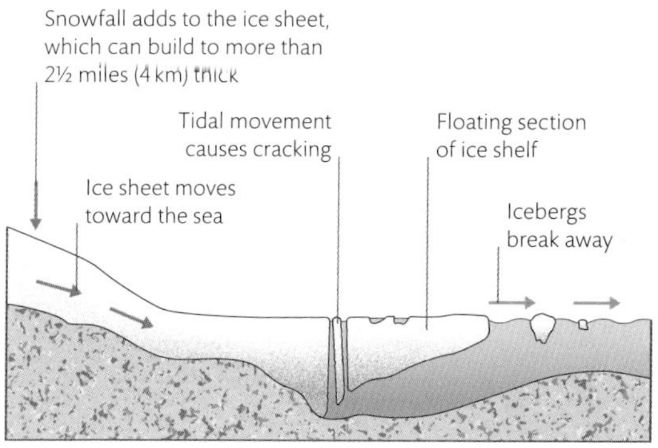

COASTAL EROSION

A typical high-energy coastline displays rapidly eroding cliffs, boulders, and headlands. When a wave front reaches the shore it refracts (bends) toward the shallower water adjacent to the headlands, where the energy becomes focused. Continuous erosion creates a negative sediment budget, with more material removed than added, and it carves out caves, arches, and stacks. Between headlands, beaches build up in sheltered bays, where more sediment is deposited than eroded.

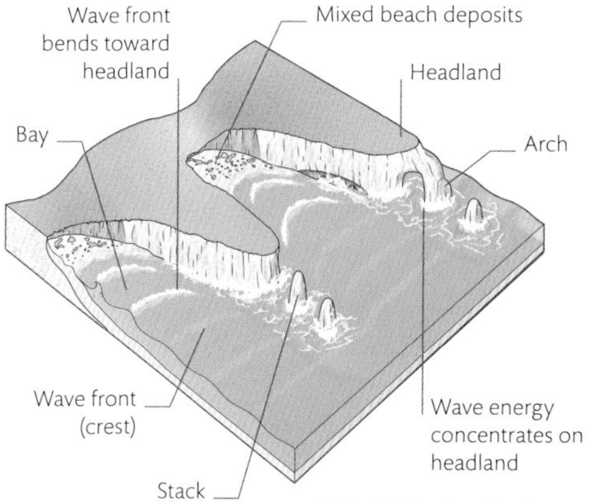

HIGH-ENERGY COASTLINE

where land meets the sea

A dynamic duel between land and sea is played out along the coast. Ocean water is the master sculptor and destroyer: it erodes cliffs, molds the rocky shore, rounds pebbles, and grinds sand to create long, golden beaches that are eventually washed out to sea. But the coastline is maintained by a sediment supply from rivers of around 22 billion tons per year, and more is added from windblown dust and glacial discharge. Coastlines are as varied as they are stunning, from intricate fiords and sea-pounded towering cliffs to vast muddy deltas, mangrove swamps, and white coral sands.

Skeleton Coast, Namibia
Four of the world's longest beaches, each stretching more than 60 miles (100 km), lie along the Skeleton Coast, where the Namib Desert meets the South Atlantic Ocean. The desert is around 60 million years old, and its dunes—some of which are 1,000 ft (300 m) tall—extend right up to the shore.

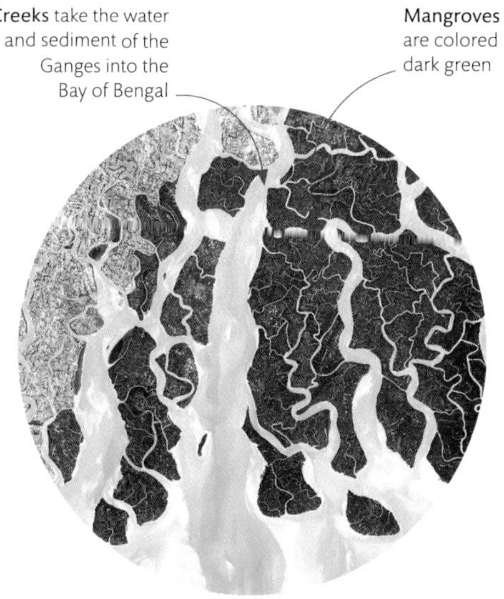

Ganges River Delta
This vast delta is part wild mangroves and part highly cultivated. Monsoon floods sweep its silty sediments out to sea, building a giant submarine delta 1,550 miles (2,500 km) long across the Bay of Bengal.

Uplifted beachfronts, New Zealand
In earthquake-prone regions, Earth's crust can be pushed either up or down by tectonic forces. This view (right) of Turakirae Head, New Zealand, shows a series of three ancient beach ridges snaking across the land above the present-day beach. The beaches were left behind between 160 years (the lowest) and 5,000 years ago (the highest). These represent the position of previous storm beachfronts that have been uplifted by major earthquake activity.

Rebounding land
Icecaps depress the underlying landmass, which slowly rebounds when the ice melts. Such a process, called isostatic rebound, created this raised shingle beach (left) at Loch Tarbert on the Isle of Jura, Scotland.

sea-level change

Sea level is the mean height of the ocean surface relative to land. This has varied throughout Earth's history in response to tectonic movements and climate change. When global temperatures were much colder, more water was locked up in icecaps and glaciers, causing sea level to drop by up to 400 ft (120 m) four times in the past million years. Broad continental shelves became dry land. Longer ago, the rate of ocean spreading and mountain uplift caused even greater change—sea level was 820–1,150 ft (250–350 m) higher and ocean covered 82 percent of the world. Global warming today is expanding sea water and melting icecaps. Sea level may rise 12–20 in (30–50 cm) by the century's end.

EVIDENCE FOR SEA-LEVEL CHANGE

This frontispiece of the ninth edition of Charles Lyell's 1853 *Principles of Geology* shows the Roman Temple of Serapis (now reinterpreted as a marketplace) in Pozzuoli, southern Italy. Lyell argued that a distinctive layer of boreholes made by marine *Lithophaga* bivalves in the temple columns indicated that they were once submerged under the sea and then later uplifted, thus supporting the idea of sea-level change and the uniformitarian theory of geology. Recent research broadly validates his conclusions.

TEMPLE OF SERAPIS, POZZUOLI, ITALY

SHALLOW SEA TO DEEP OCEAN

Continents are surrounded by gently sloping continental shelves. Beyond the shelves lie steeper slopes marked by deep canyons and drainage channels. Vast underwater rivers called turbidity currents and massive slides, or slumps, carry sediment down the slope. Deposited on the continental rise, sediment is reworked into sideways drifts by powerful contour currents, or spills across the deep-ocean basin as submarine fans.

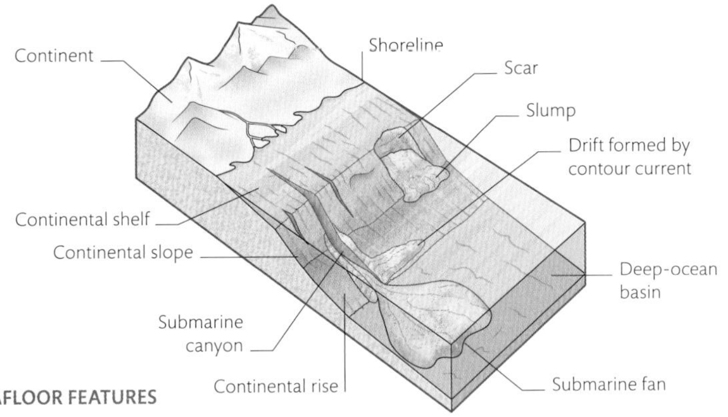

SEAFLOOR FEATURES

shallow seas

Shallow seas extend from the coast to the edge of the continental shelf, at about 300 ft (100 m) water depth. Bathed in sunlight and often rich in nutrients, they are ideal for photosynthesis and the growth of coral reefs and kelp forests. They also provide nursery habitats for juvenile organisms and feeding grounds for a host of marine species. Shallow seas are shaped both by their biogenic activity and by the dynamic effect of waves, tides, ocean currents, and coastal upwelling. Ripples and dunes covering the seafloor are testament to the transportation of sediment from land across the continental shelf.

Tentacles secrete an irritant that deters predators

Seabed engineer
Flame clams (*Limaria hians*) effect major transformations of the seabed. The nests they build from shells and stones combine to form a dense, reef-like structure that raises and stabilizes the seabed, creating a habitat that supports a rich community of organisms.

Kelp forest, offshore California
Forests of giant kelp (*Macrocystis pyrifera*) grow in nutrient-rich, temperate shelf waters, such as those off California. Anchored by holdfasts and kept afloat by gas bladders, the kelp can grow up to 2 ft (60 cm) per day. Many ocean animals find safe haven beneath the kelp forest's canopy.

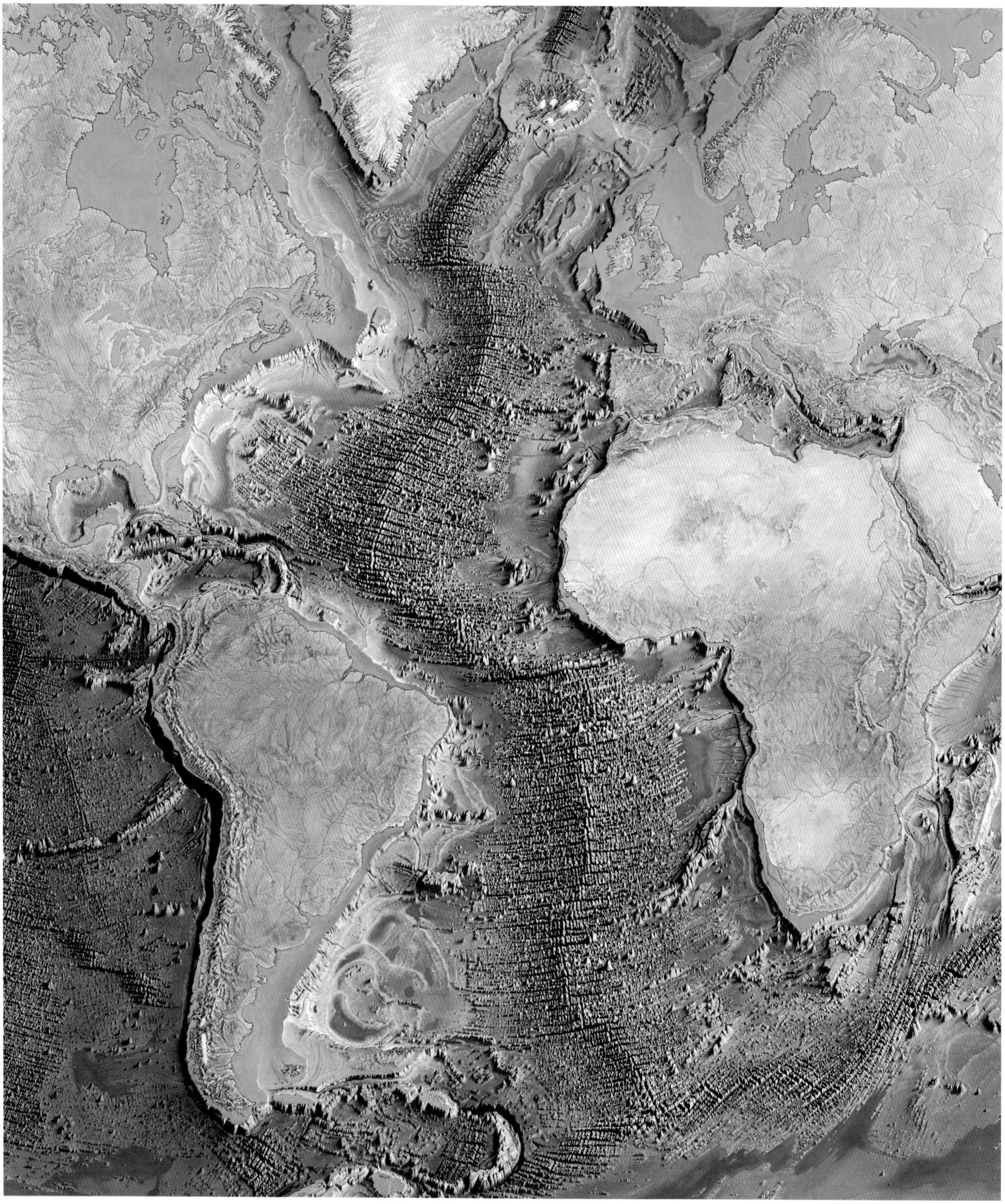

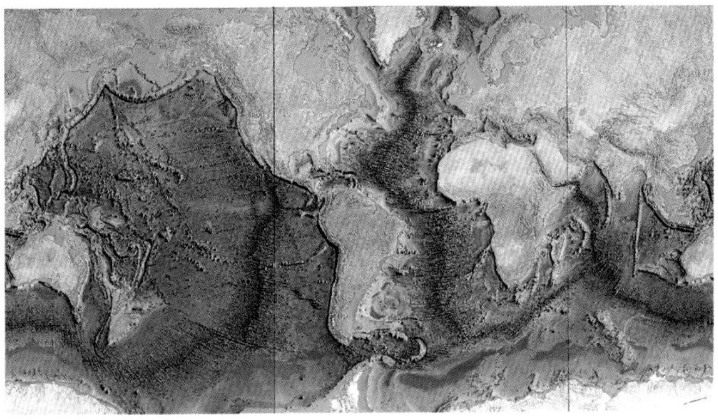

World ocean floor
Marie Tharp published this map of the world ocean floor in 1977, shortly after her colleague, Bruce Heezen, died at sea while on an oceanographic expedition in the Atlantic Ocean.

history of Earth science

seafloor mapping

Marie Tharp at work
Pictured here at the Lamont Geological Observatory in New York state, Marie Tharp painstakingly transcribed sonar readings to create her groundbreaking map of the world ocean floor.

Probably the most important map of the 20th century was a completely new view of the seafloor, painstakingly compiled from 1957 until 1977 by American geologists Marie Tharp and Bruce Heezen. For the first time, the ocean floor—which comprises 71 percent of Earth, and which had previously been largely unknown—was revealed to the public and scientists alike, leading to fundamental insights into the nature of our planet.

The Heezen-Tharp maps captured on paper the topography of the seafloor. The maps revealed the greatest mountain range on Earth—the mid-ocean ridge system—which rises 10,000 ft (3,000 m) above the seafloor and extends 37,000 miles (60,000 km) across. They showed dark chasms plunging to a depth of more than ⅔ mile (1 km). Deep canyons fringing the landmasses incise the flat shelves, feeding sediments eroded from land to gigantic submarine fans sprawled across flat abyssal plains. The maps revealed countless submerged seamounts as well as active volcanoes rising to the surface, constructing reef-fringed islands. Their work provided key evidence for plate tectonic theory, which was not widely accepted until the 1960s.

Tharp and Heezen used sonar readings of the seafloor to create their maps. Their map of the world ocean floor and its beautiful 1977 rendition painted by Heinrich Berann were true landmarks.

Today, our methods are ever more sophisticated and precise, using satellite imagery to refine the global ocean map, and high-precision deep-towed sonar to resolve myriad seafloor features. These include sediment waves, gas-escape pockmarks, deepwater coral reefs, and even the fault-displacement in the Sumatran trench, which was the cause of the devastating 2004 tsunami.

The map that shook the world
Tharp and Heezen's map, which revealed the Atlantic Mid-Ocean Ridge (opposite), revolutionized ocean science, but initially met with skepticism. Jacques Cousteau's subsequent exploration of the ridge in a submersible, intended to disprove their findings, confirmed the existence of the rift valley where seafloor spreading occurred.

> “It became clear that existing explanations for the formation of the Earth's surface no longer held.”
>
> HALI FELT, *SOUNDINGS: THE STORY OF THE REMARKABLE WOMAN WHO MAPPED THE OCEAN*, 2012

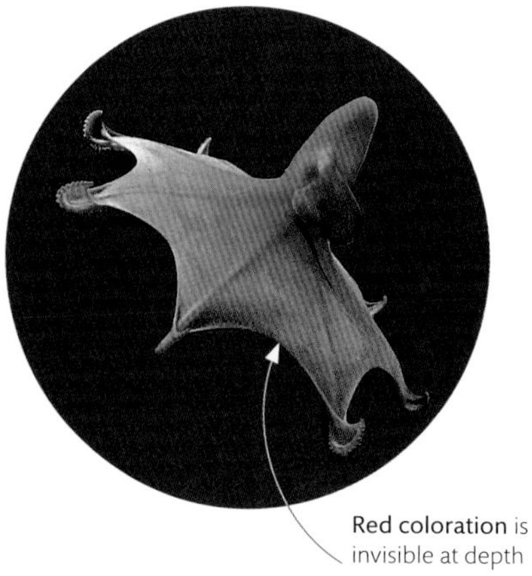

Hunters of the open ocean
Albatrosses are pelagic predators that spend many months, sometimes more than a year, on the open sea, before returning to remote islands to breed. These black-browed albatrosses (*Thalassalarche melanophrys*) scour the turbulent, nutrient-rich waters of the Southern Ocean for prey.

Denizen of the deep
The deep ocean is a dark, cold, high-pressure habitat. Dumbo octopuses (*Grimpoteuthis* spp.) live at depths of up to 13,000 ft (4,000 m) or more. They hover over the seafloor in search of bivalves, crustaceans, and worms.

open ocean waters

In the open ocean—a vast expanse covering two-thirds of the planet—primary productivity is low and the waters are sparsely inhabited. The ocean is stratified into layers of different temperature, salinity, and light penetration, and compartmentalized by currents. When such subtle barriers are disturbed by storm stirring, local upwelling, and the meeting of water masses, dissolved nutrients rise to the surface and promote phytoplankton blooms. Grazers gather to feed, but there is nowhere to hide. Many strategies have evolved to avoid predation: transparent bodies and countershading, mass migrations of zooplankton from the surface to deep water, and swimming in large schools.

OCEAN THERMOCLINE

The thermocline marks the base of a warm surface layer, 150–3,300 ft (50–1,000 m) thick, throughout temperate and tropical oceans. In the tropics, calm weather stabilizes the thermocline, inhibiting mixing so that surface waters are gradually depleted of nutrients. Storms in temperate mid-latitudes enhance mixing. This stirs the thermocline and brings nutrients to the surface: the green, phytoplankton-rich waters proliferate with life.

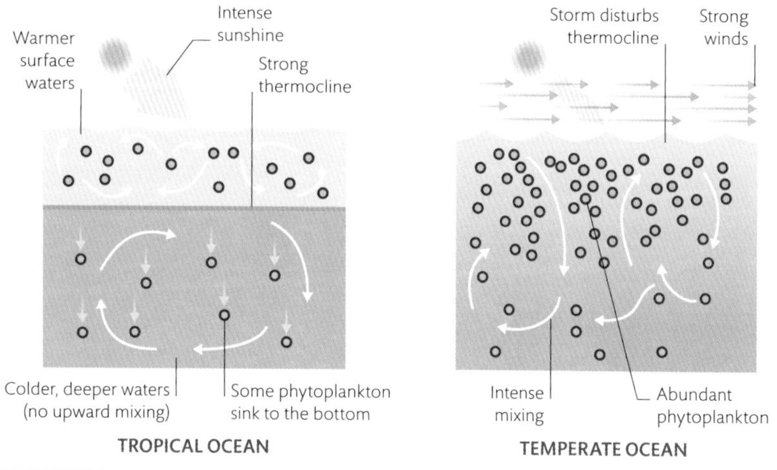

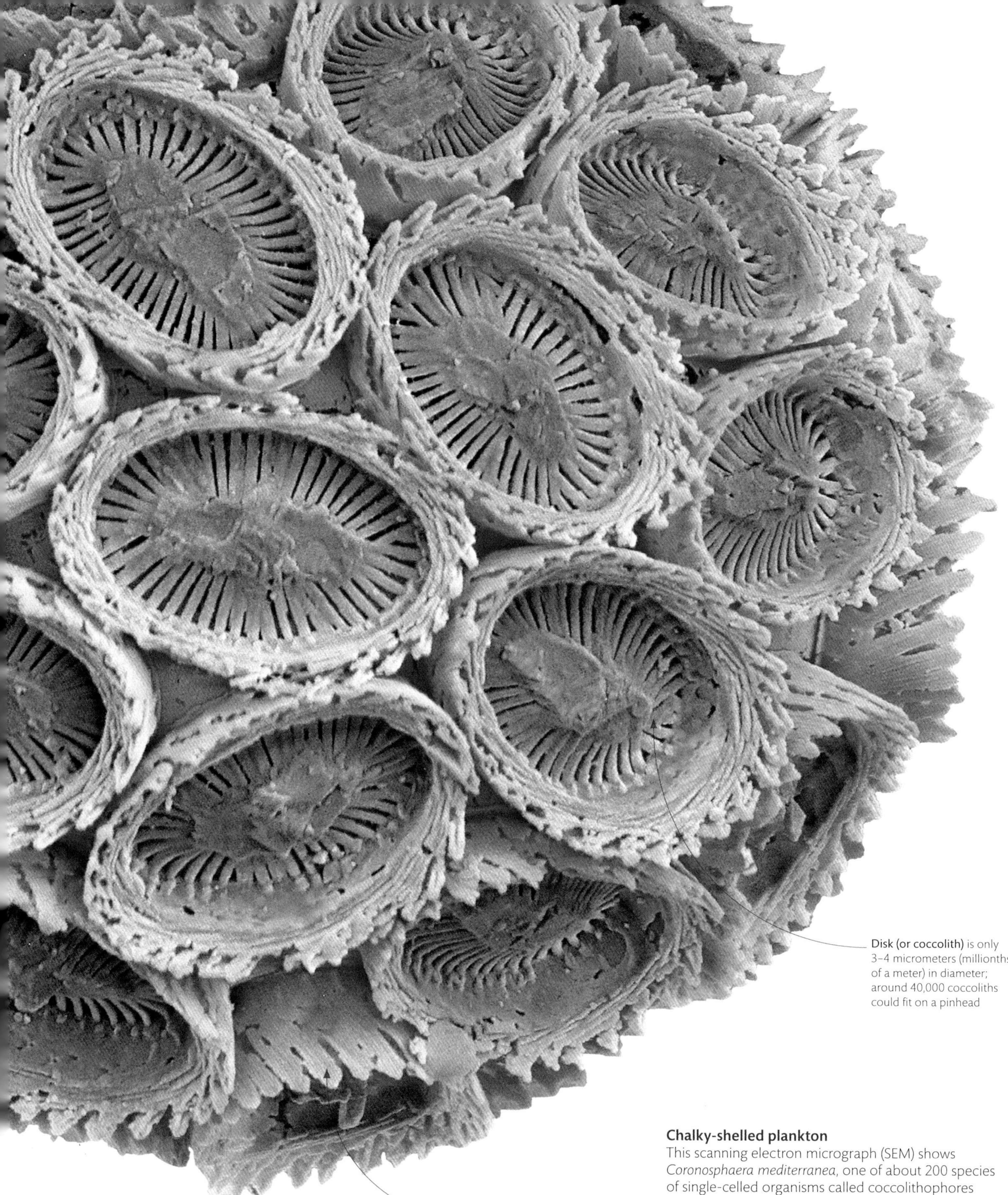

Chalky-shelled plankton
This scanning electron micrograph (SEM) shows *Coronosphaera mediterranea*, one of about 200 species of single-celled organisms called coccolithophores that form a vital part of the ocean's photosynthesizing plankton. The organism is surrounded by delicate patterned disks of chalky calcium carbonate, which sink to the seafloor when it dies. The disks contribute to the calcareous biogenic sediment.

patterns on the seafloor

The seafloor is almost entirely covered with sediment. On the flanks of mid-ocean ridges, the sediment is a thin veneer over oceanic crust, but along continental margins and beneath major deltas, its thickness can exceed 10 miles (16 km). Each sediment particle tells a different tale of process, climate, and environment. Some sediment grains are carried from land to seafloor by wind, rivers, glaciers, and underwater currents. Others, biogenic particles, are the remains of planktonic organisms such as foraminifera, coccolithophores, radiolarians, and diatoms (see p.182). Sediments can also contain mineral and metal deposits that precipitate out of concentrated brines or from volcanic emissions, and even space-dust that falls to the ocean floor.

Foraminiferan microfossils
The size of medium to coarse sand grains, foraminifera are single-celled protozoans, most with tests (shells) of calcium carbonate. Of more than 50,000 known species, the majority live on the seafloor; others are key members of the zooplankton.

BOTTOM DWELLING
(*Elphidium crispum*)

PLANKTONIC
(*Orbulina universa*)

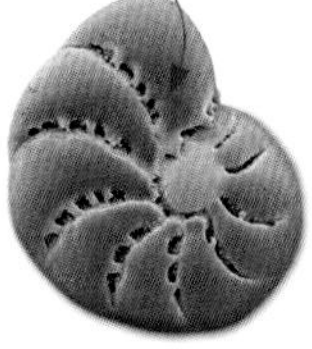

BOTTOM DWELLING
(*Cribroelphidium poeyanum*)

Radiolarian microfossils
Like foraminifera, radiolarians are zooplankton, or animal-like plankton. These tiny protozoan predators of the open ocean are about 1/12 in (0.03 to 2 mm) across. They have a glassy silica test with pores through which they project fingers of cellular material called pseudopodia.

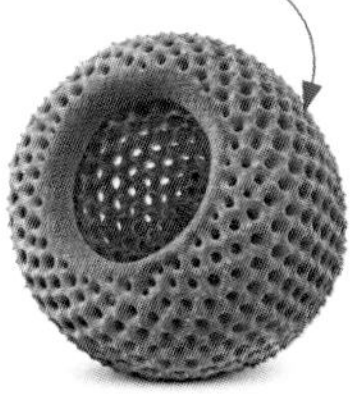

SPHERICAL OPEN FORM
(*Theocapsa* spp.)

SMOOTH NUT-SHAPED FORM
(*Tripospyris* spp.)

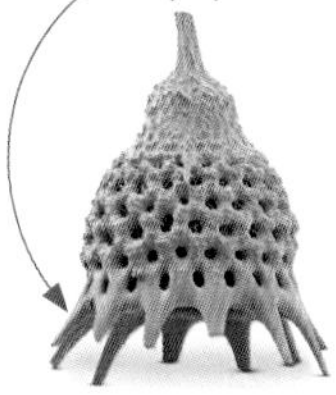

HELMET-SHAPED FORM
(*Anthocyrtidium ligularia*)

SEAFLOOR SEDIMENTS

Around land and near the poles, the seabed is blanketed by sediment from glaciers and other agents of land erosion. Siliceous sediments (made of silica) occur in belts where upwelling encourages blooms of silica-rich plankton. Calcareous sediments (made of calcium carbonate) also reflect the rain of microscopic tests from the plankton, but they are preserved only above a depth of 2½–3 miles (4–5 km). Below this level, more acidic waters dissolve the calcium carbonate, leaving just fine material called abyssal red clay.

ORIGIN OF SEDIMENTS ON THE SEABED

KEY
- Terrigenous (land-derived)
- Glacigenic (glacier-derived)
- Siliceous (biogenic—made by plankton)
- Calcareous (biogenic—made by plankton)
- Abyssal red clay

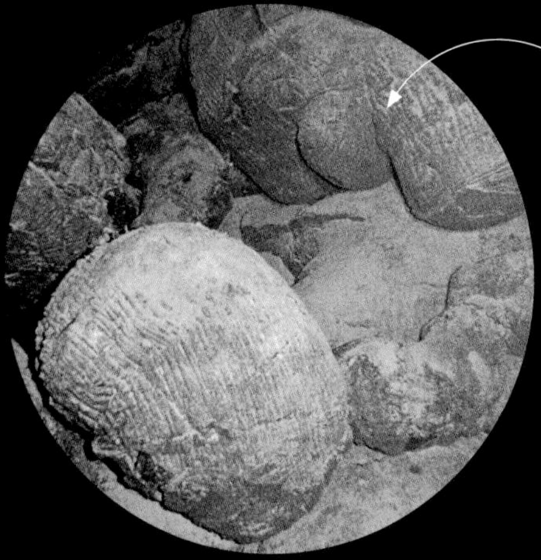

Rounded rock pillows, typically 1½–5 ft (0.5–1.5 m) in diameter

Pillow lava, Hawaii
When molten lava extrudes into cold sea water, a solid outer crust forms. Continued pressure inflates the crust with more lava before it cools and solidifies to create pillow-like masses on the seabed.

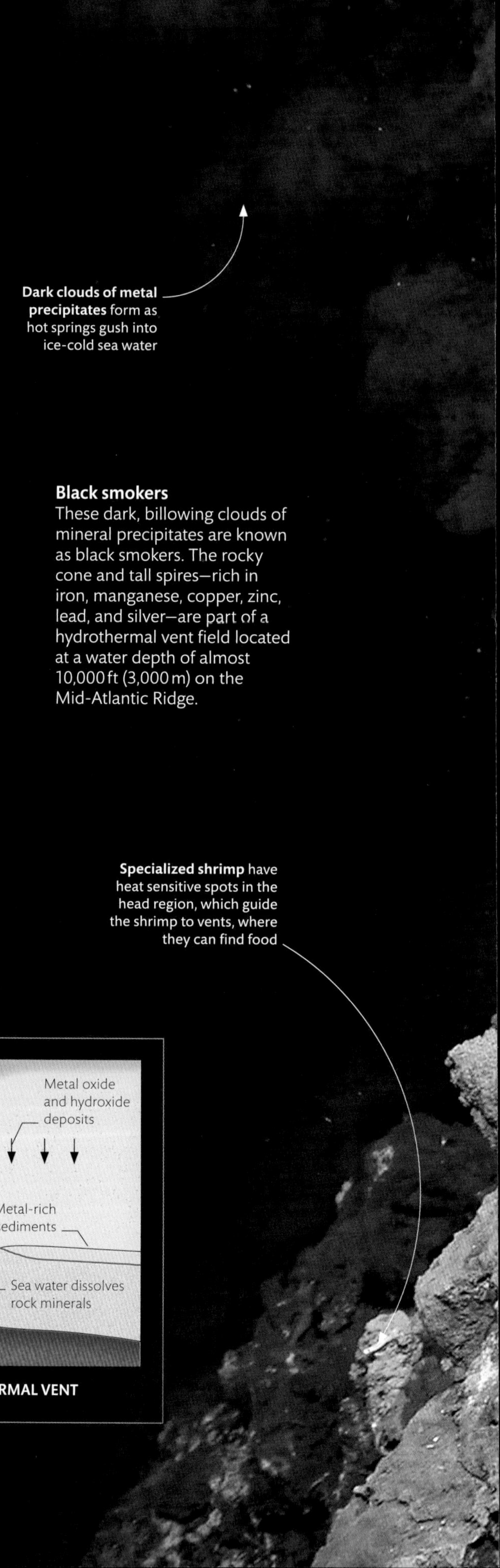

Dark clouds of metal precipitates form as hot springs gush into ice-cold sea water

ocean tectonics

Mid-ocean ridges (see p.117) are hotbeds of submarine volcanism. Here, rising basaltic magma drives apart the tectonic plates of the ocean floor and spews out onto the seabed as lava, forming new oceanic crust. Together, mid-ocean ridges host around 300 hydrothermal vent fields, where water superheated by magma under ground gushes into the ocean at 570–840 °F (300–450 °C). Dissolved minerals, leached from rocks below, precipitate on contact with cold sea water, creating black clouds of metal oxides and building tall chimneys of metal sulfides. Microbes thrive around the vents, deriving the energy they need for life by processing hydrogen sulfide from the jets of hydrothermal fluid and oxygen from the surrounding ocean. The microbes support a whole ecosystem of organisms, all adapted to life in these extreme conditions.

Black smokers
These dark, billowing clouds of mineral precipitates are known as black smokers. The rocky cone and tall spires—rich in iron, manganese, copper, zinc, lead, and silver—are part of a hydrothermal vent field located at a water depth of almost 10,000 ft (3,000 m) on the Mid-Atlantic Ridge.

Specialized shrimp have heat sensitive spots in the head region, which guide the shrimp to vents, where they can find food

HYDROTHERMAL VENTS

Cold sea water percolating into cracks and fissures on newly formed seafloor penetrates deep into the oceanic crust, leaching out a cocktail of chemicals as it seeps downward. When the water encounters hot magma, it is superheated and forced back to the surface, emerging from the vent as a hot spring. Dissolved minerals precipitate as black smokers, depositing metal-rich sediments and constructing a "cityscape" of chimneys, spires, and stacks. Communities of clams, mussels, barnacles, anemones, limpets, crabs, worms, shrimp, and fish surround the vents, entirely supported by microbes.

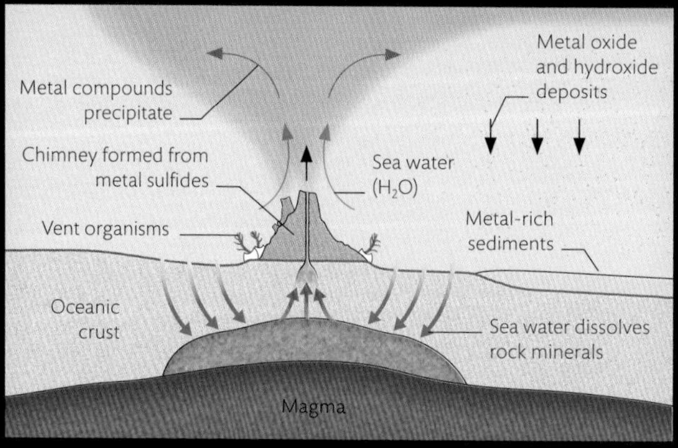

MINERAL DEPOSITION AT A HYDROTHERMAL VENT

islands in the ocean

Islands are like marine oases—havens for terrestrial and aquatic life. They can also tell us about oceans past and present. The hundreds of thousands of islands in the world's oceans are broadly classified into four main types. Volcanic islands and coral reefs are often smaller and ephemeral, subject to erosion, sinking, and biotic changes. Continental islands, which vary in size, are fragments splintered off the main landmass by a rise in sea level or by tectonic separation. Complex islands are larger and long-lived, formed by a combination of processes; some, such as Cyprus in the Mediterranean, have a core of deep-ocean crust, known as an ophiolite, which has been squeezed to the surface between colliding tectonic plates.

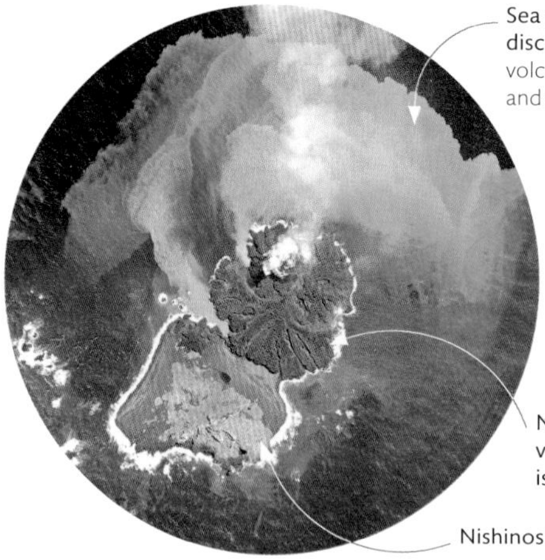

Merging islands
Japan's volcanic Nishinoshima Island grew by merging with a new island that arose after eruptions in 2013. Nishinoshima lies over a subduction zone (see pp.114–115) where the Pacific plate slides under the Philippine plate.

EVOLUTION OF VOLCANIC ISLANDS

Some volcanic islands, such as the Hawaiian and Galápagos Islands, emerge above hot spots from mantle plumes (see p.138); others, including those of Tonga and the Mariana Islands, form over subduction zones (see pp.114–115). Born from extensive outpourings of lava that build up on the seafloor, a volcanic island becomes explosive as it breaks the surface. If it survives early erosion, it may grow into a substantial cone, fringed in tropical waters by coral reefs. When the lava flow stops, the island subsides to form a seamount. As the island sinks, the corals continue growing upward toward the sunlit shallows, producing a ring-shaped reef at the surface called an atoll.

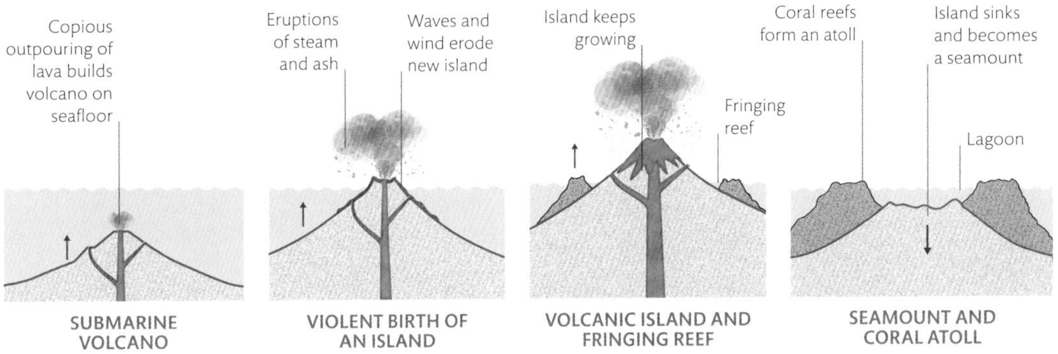

Coral atoll, Indonesia
Pulau Ndaa island, in the Banda Sea, is an atoll—a coral reef surrounding a lagoon. The atoll sits atop an inactive volcano that has subsided and become a seamount. If an atoll's coral reef grows fast enough to keep pace with the volcano's subsidence, the atoll remains at the surface; if not, it sinks with the volcano, disappearing beneath the waves.

LAYERS OF THE ATMOSPHERE

The atmosphere has distinct layers with their own characteristics and phenomena. The troposphere, the lowest layer, is the location of much of Earth's weather systems and commercial aircraft; the concentration of ozone gases peaks within the stratosphere; and the mesosphere and thermosphere play host to dazzling meteor showers and dramatic auroras. The Kármán Line marks the legal edge of space—the official upper edge of a country's airspace. The thermosphere and exosphere beyond are too thin to support aircraft.

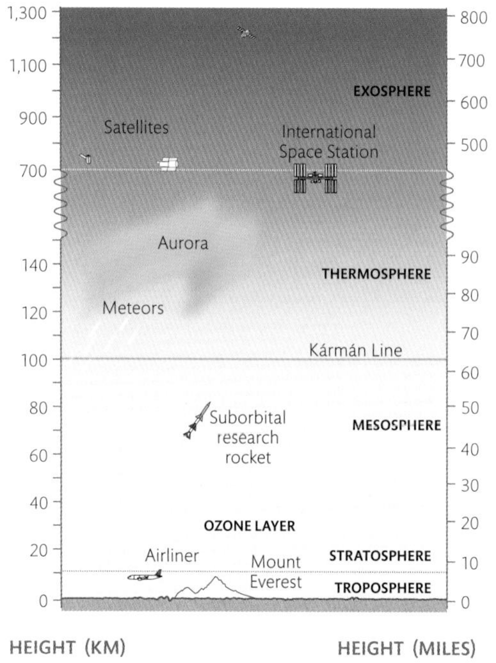

The Leonids comprise an annual meteor shower made by debris left by the Tempel-Tuttle comet

METEOR SHOWERS

Meteoroids are small rocky or metallic objects in outer space. If these objects enter Earth's atmosphere, they burn up, producing streaks of light called meteors. An estimated 25 million meteoroids, space dust, and other debris enter Earth's atmosphere each day.

THE EDGE OF SPACE

In addition to its key component gases, the lowest layer of the atmosphere (the troposphere) contains water vapor and weather systems driven by solar heating. The atmosphere gradually thins with increasing distance from Earth, until it is indistinguishable from the interplanetary medium—outer space.

Earth's atmosphere

First formed from gases spewed out by the volcanoes of primitive Earth (see pp.28–29), our planet's ocean-atmosphere couplet is unique in the solar system. Each of the atmosphere's layers comprises about 78 percent nitrogen, 20.9 percent oxygen, 0.9 percent argon, and traces of 10–15 other gases. In the lower layers, the gas ozone absorbs ultraviolet radiation from the sun, which would otherwise damage plant and animal DNA, and prevent plants from being able to carry out photosynthesis. Greenhouse gases, including carbon dioxide and methane, keep the planet at a habitable temperature by trapping infrared radiation. However, human activities have increased the amount of these gases in the atmosphere to dangerous levels, leading to global warming.

EARTH'S MAGNETIC FIELD

Earth is like a giant magnet. Its partially molten, iron-rich outer core is in constant motion, generating an extensive magnetic field around the planet that reaches thousands of miles into space. This magnetic shield, called the magnetosphere, protects Earth from harmful solar wind. Auroras occur where solar wind particles that penetrate the magnetosphere are pulled toward the magnetic poles.

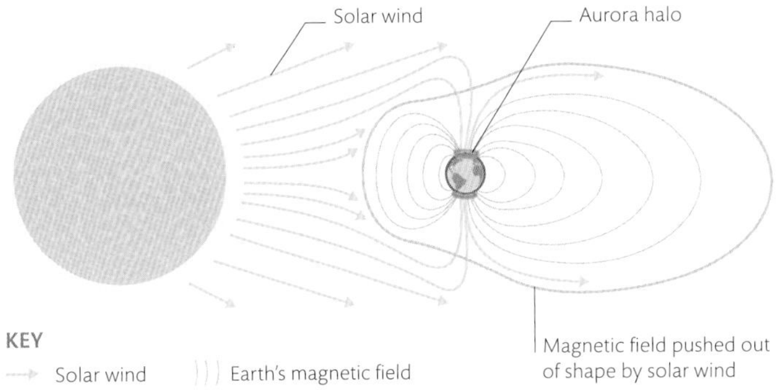

auroras

Earth's atmosphere is under constant bombardment from the solar wind—a flow of charged electrons and protons fired out from the sun. Most of these particles are deflected by the planet's magnetic field, but some penetrate it near the magnetic poles. When these particles collide with nitrogen and oxygen atoms in the atmosphere, energy is released in the form of light, creating a dazzling display of dancing lights across the sky. These light shows are known as aurora borealis, or northern lights, in the northern hemisphere and aurora australis, or southern lights, in the southern hemisphere.

Northern lights
The colors emitted by an aurora depend on the type of atom struck by solar wind particles and the altitude at which this occurs. Green and red shades, pictured here over Iceland, are most common, but yellow, purple, blue, and pink hues are also possible.

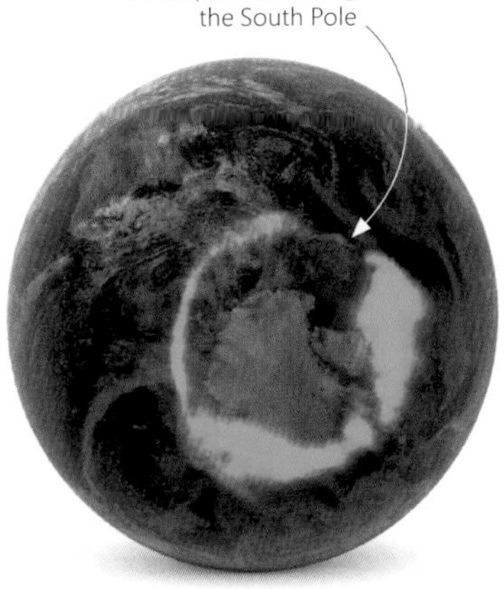

Aurora australis seen from space encircling the South Pole

Polar halo
Most auroras, like this one around Antarctica, occur in a band (halo) 3–6° wide located between 10–20° from the North or South Poles.

wind

From light breezes to hurricanes, air in the troposphere (see pp.210–11) is constantly moving around Earth due to differences in pressure and temperature. As the Sun heats different parts of the planet, the air around these warm spots heats up too. The resulting warm air rises into the sky, leaving an area of low pressure in its place. Cool air, which is more dense, sinks and fills the space. This constant circulation of air from regions of high pressure to regions of low pressure is what generates wind. These lower-atmosphere winds, called planetary winds or prevailing winds, form part of three great wind cells that dominate each hemisphere (see right). Local winds, which occur over a smaller area and time period, are the result of sea-breeze and land-breeze cycles in coastal areas.

Relentless winds
Prevailing westerly winds known as the Roaring Forties stunt and shape the growth of these trees at Slope Point on the southern tip of New Zealand. The powerful winds blow in the southern hemisphere between the latitudes of 40° and 50°, relatively uninterrupted by landmasses that could decrease their speed. The constant bombardment of air from one direction dries and kills the shoots and buds on one side of these trees, preventing new growth, while the other side is able to continue to grow as normal, creating the illusion of them bending in the breeze.

ATMOSPHERIC CIRCULATION

At the equator, intense solar heating warms the air, which then rises. More air is sucked in from either side of the equator to replace the rising air, causing winds, known as trade winds, to blow across the tropics. The rising air races poleward and cools rapidly, before sinking. This constant circulation generates a cell of air, called a Hadley cell, either side of the equator. Further pairs of cells occur at mid-latitudes and polar regions, causing prevailing westerlies and polar easterlies to blow.

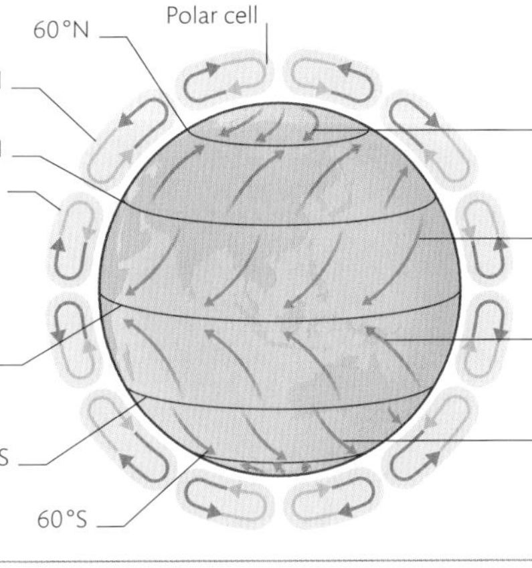

GLOBAL ATMOSPHERIC CIRCULATION

Haboob

A haboob, like this one approaching Khartoum, Sudan, is an intense wind storm common in arid regions around the world. The strong wind can approach with little or no warning, whipping up a wall of sand and dust that can reach up to 100 km (60 miles) wide.

Rossby waves are large-scale meanders in the jet stream

Wind speed color-coded from red (fastest), through orange, yellow, and cyan, to dark blue (slowest).

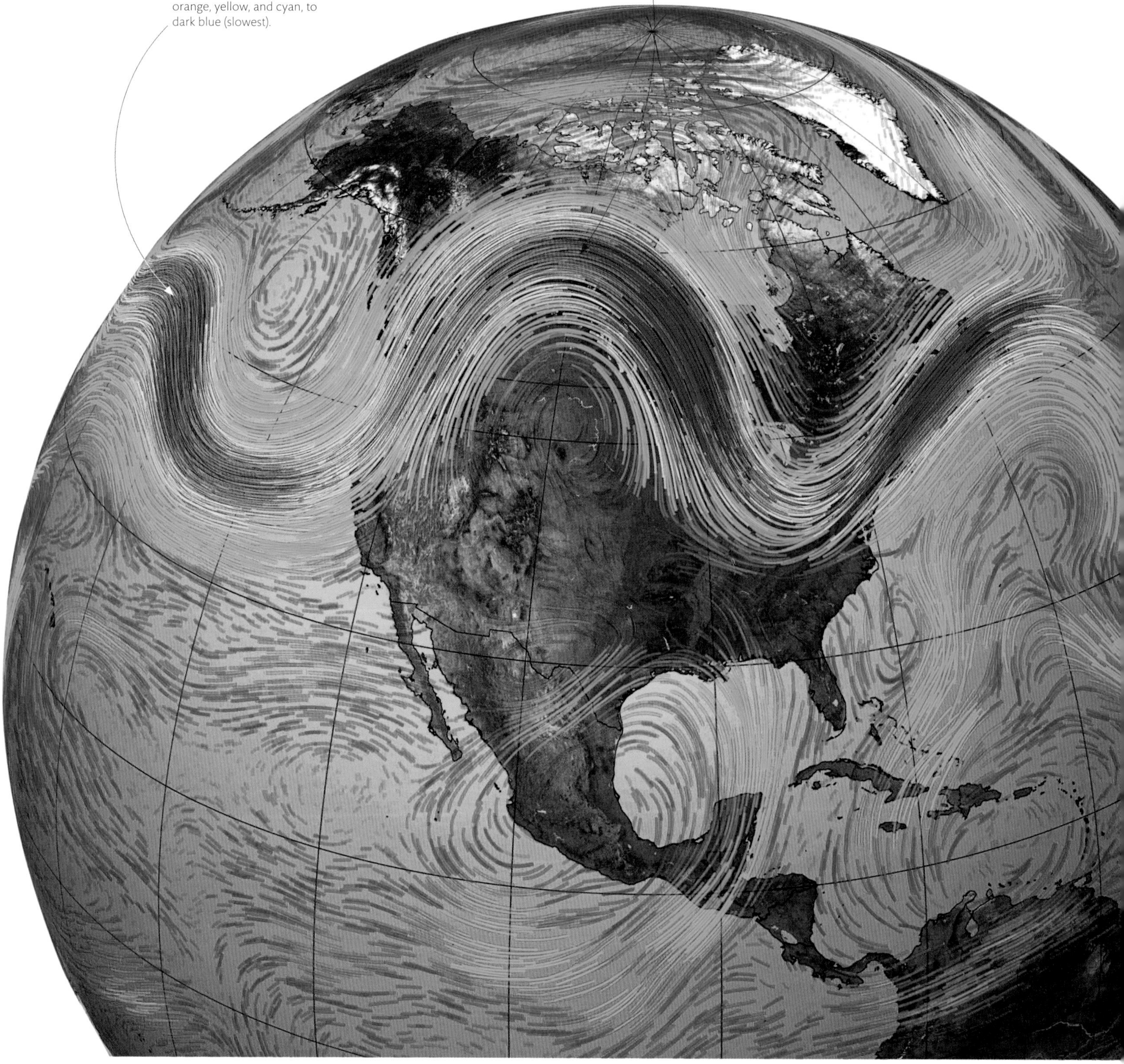

The ozone hole
Colored satellite maps (right) reveal depleted ozone levels in the atmosphere over Antarctica. Dark blue coloration shows areas of maximum depletion, known as the ozone hole. The hole continued to grow after 1987, reaching 11 million sq miles (28 million sq km) in 2000, but then stabilized and reduced.

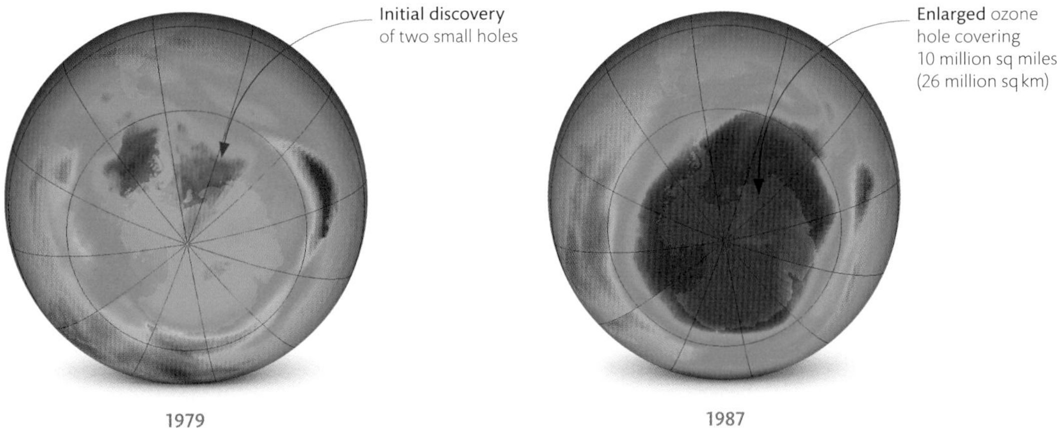

history of Earth science

atmospheric imaging

Eyes in space
Sensors aboard the historic satellite, Nimbus-7, detected abnormally low concentrations of ozone over the Antarctic in 1979–1980, helping to identify the ozone hole.

Visualizing the jet stream
This modeling of the jet stream, the high-speed belt of eastward-moving winds in the upper troposphere, is based on satellite observations of water vapor levels over North America. The position of the jet stream plays a major role in determining the weather below, causing frequent storms and other weather phenomena associated with low pressure.

Since the 1960s, the deployment of satellites carrying instruments to monitor Earth's atmosphere has allowed us to observe large parts of the world simultaneously, as well as to monitor specific parts over a longer time. Satellite imagery has made it possible to identify holes in the ozone layer, prompting action to address the problem, and has helped us understand the atmospheric changes that are likely to affect life on Earth.

Satellites are routinely used to inform all aspects of ocean and atmospheric sciences. They have become essential for weather observation and prediction, and provide warning of extreme events. Long term observations chart the effects of atmospheric pollution from volcanoes, wild fires, and urban traffic, and help us understand the nature and effects of natural climate oscillations, such as El Niño (see p.243).

To monitor weather, satellites are positioned in high geostationary orbits—nearly 22,400 miles (36,000 km). Most other scientific remote sensing is done from altitudes of 100–1,250 miles (160–2,000 km). These "low Earth orbits" have various inclinations to the equatorial plane, circling Earth every 90–120 minutes but covering a different track with each orbit. Complete coverage of the globe is possible in about 12 hours.

From the 1970s on, satellites have observed the atmosphere using instruments sensitive to infrared, the visible spectrum, and microwave bands. Some scan sideways to look at the upper atmosphere edge-on and can measure vertical variation in gases, temperature, and pressure. Others measure high-altitude winds by sensing Doppler shift.

To analyze atmospheric composition, mass spectrometry sensors identify gases by the way they absorb or emit particular radiation wavelengths. It was spectrometry data from satellites such as Nimbus-7 that was vital for identifying the thinning of the ozone layer. This led to the signing of the Montreal Protocol in 1987, which triggered action to stabilize and reduce the ozone hole. Another key contribution of satellite imaging is the measurement of levels of greenhouse gases to help us monitor the effect of efforts to combat climate change.

> "Satellite images… have revolutionized our understanding of the ocean-atmosphere…"
>
> PROFESSOR IAN ROBINSON, NATIONAL OCEANOGRAPHY CENTRE, SOUTHAMPTON, UK

weather systems

Sunshine or cloud, rain or snow, strong winds or still calm—these changing local weather conditions are the result of solar energy interacting with the oceans, atmosphere, and land. Sunlight heats Earth more in some places than others, leading to differences in air pressure and the generation of wind (see pp.214–215). It also drives the water cycle (pp.224–225). Different parts of the globe experience a variety of weather systems—tropical, monsoonal, temperate, and polar. Temperate regions are subject to mid-latitude depressions (or cyclones) that build and dissipate over a period of days, traveling from west to east one after another. These are areas of low-pressure generated by Rossby waves—giant meanders in the high-altitude jet-stream winds, created by Earth's rotation (pp.216–217).

East coast bomb cyclone
This satellite image of the east coast of the US shows the dramatic spiraling clouds of a bomb cyclone. This is caused by a dramatic drop in air pressure from outside to inside a weather system, created by the collision of cold continental air and warm oceanic air masses.

Mid-latitude cyclone
Driven by prevailing westerly winds, fast-moving cold polar air dives beneath the slower moving warm tropical air, which is pushed upward, producing an area of low pressure. In the northern hemisphere (pictured), air flowing into this area creates a counterclockwise spin. In the southern hemisphere, the spin is clockwise.

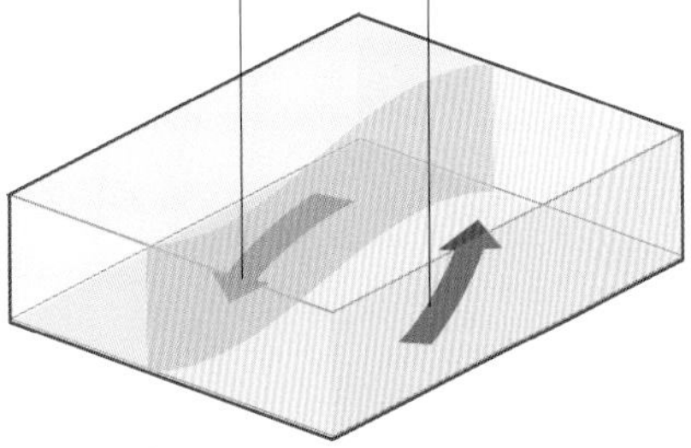

Stage 1
As a cold air mass advances toward a warm one, the air masses start to interact. This is the initial stage of the process of cyclone formation.

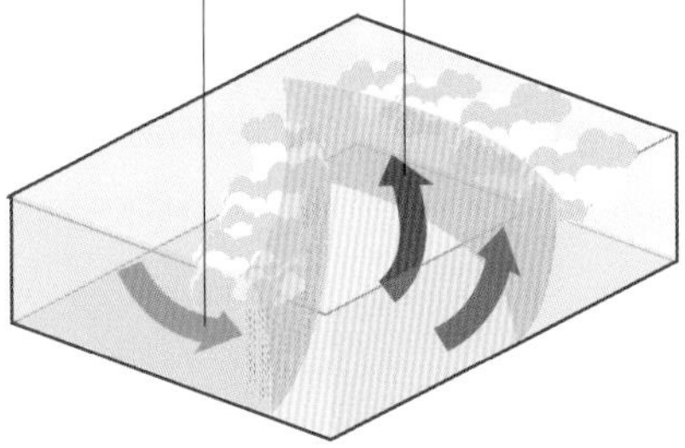

Stage 2
Cold air pushes warm air upward. The resulting rotation of both air masses starts to create a spiraling movement.

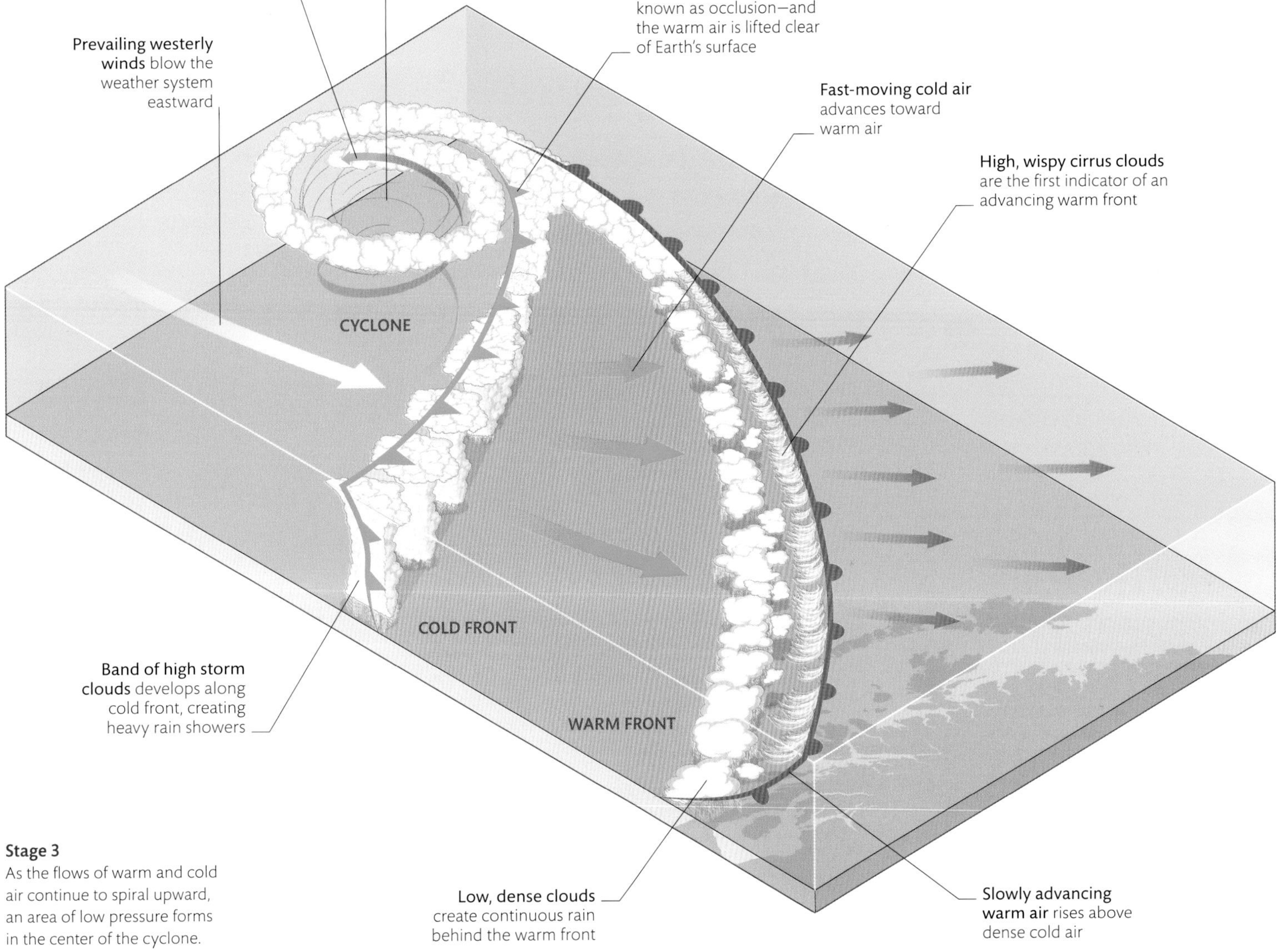

Stage 3
As the flows of warm and cold air continue to spiral upward, an area of low pressure forms in the center of the cyclone.

Eye of the storm
While there is a calm area of cool air that descends at the center of a storm—called the eye, seen here in Hurricane Florence in 2018—the strongest winds in a tropical cyclone spiral upward through the eye wall. The highest thunderclouds, powerful storms, and frequent lightning strikes surround the eye.

The calm eye region is typically around 20–40 miles (32–64 km) across.

tropical cyclones

Tropical cyclones are the largest, most violent storms on Earth, with sustained wind speeds of 75–155 mph (120–250 kph) and record gusts more than 250 mph (400 kph). Those that originate in the north Atlantic and northeast Pacific oceans are called hurricanes; those that begin in the northwest Pacific are called typhoons; and those that form in the south Pacific and Indian oceans are known simply as cyclones. The energy released by a single tropical cyclone is equivalent to about 10,000 nuclear bombs. Although they form over warm tropical seas, when these violent, rotating storms reach coasts and travel inland they can inflict widespread devastation—their combination of high winds and excessive rainfall tear down homes and trees, and cause catastrophic storm surges.

TROPICAL CYCLONE FORMATION

Tropical cyclones develop where sea surface temperatures exceed 80 °F (27 °C). Water evaporates from the overheated surface, rises, and condenses to form cumulonimbus thunderclouds that billow upward to the top of the troposphere (see pp.210–211), 6–10 miles (12–16 km) high.

Cool air rushes inward to fill the void created by the rising, humid air, and Earth's rotation causes the storm winds to spin counterclockwise in the northern hemisphere and clockwise in the southern hemisphere. Cool, dry air sinks through the cloud-free eye, a region of calm at the center of the storm.

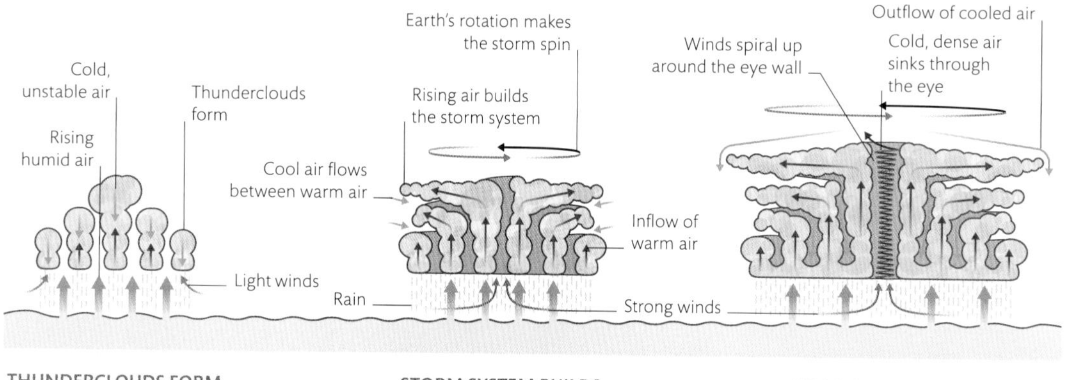

Atlantic hurricanes
Throughout the summer, sea surface temperatures build up in the tropical North Atlantic and Caribbean region, spawning a six-month hurricane season from June through November. Like many other hurricanes that hit the area, Hurricane Joaquin (seen here) caused widespread damage as it swept through the Caribbean in October 2015.

High-level to extreme-level clouds

While some types of cloud, including noctilucent and nacreous clouds, are found as high as the mesosphere and stratosphere (see p.210), most form in the troposphere. Within the troposphere, the highest-altitude clouds occur 23,000–40,000 ft (7,000–12,000 m) above ground. They are typically small, wispy or bobble-shaped clouds made up entirely of ice crystals. Multilevel cumulonimbus clouds (see pp.232–233) can tower through the entire height of the troposphere.

Veils and wisps visible at twilight, these extreme-level clouds form in the mesosphere

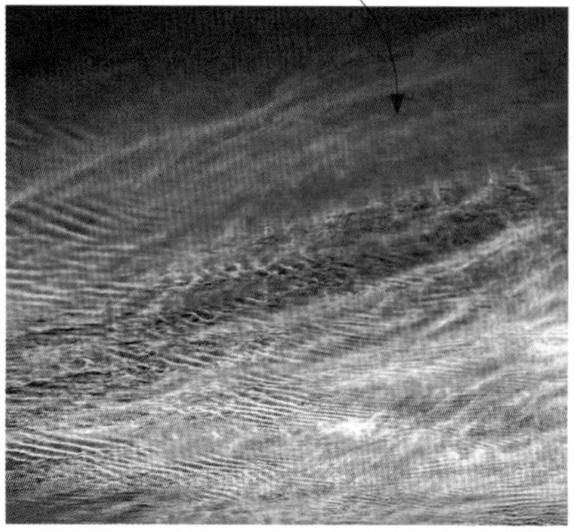

NOCTILUCENT

Very high-level stratospheric clouds visible at night and named for their resemblance to mother-of-pearl

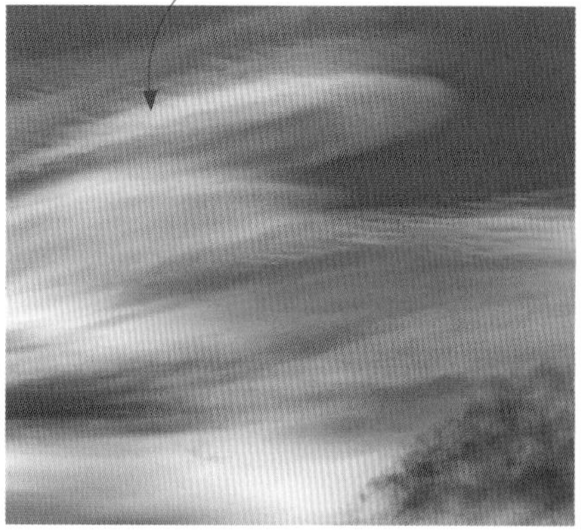

NACREOUS

Mid-level clouds

Lying 6,600–23,000 ft (2,000–7,000 m) above the ground, mid-level clouds are dominated by common, sheetlike altostratus and by the patches, rolls, and ripples of altocumulus. More unusual types include the striking lenticular cloud shapes (altocumulus lenticularis) and the distinctive rounded pouches that can protrude in a cellular pattern from the underside of clouds, called mammatus.

Sheetlike patches and streaks at a relatively high level in the sky

ALTOSTRATUS

Dispersed, bulbous patches, sometimes arranged in long parallel bands

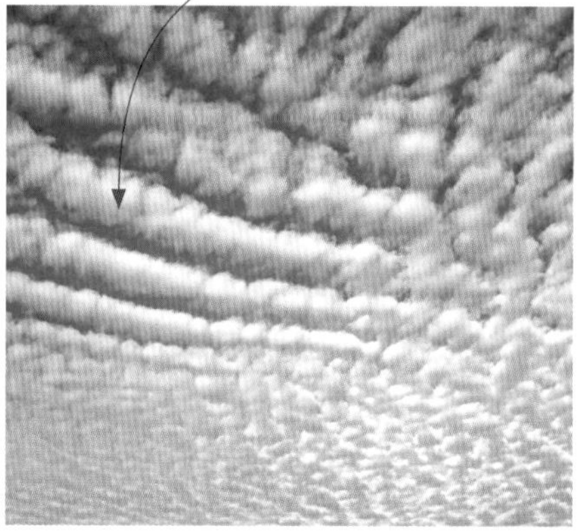

ALTOCUMULUS

Low-level clouds

On cloudy days, the whole lower troposphere, below 6,600 ft (2,000 m), can appear to be filled with low-lying, sheetlike stratus, fluffy cumulus, or mixed-form stratocumulus, as well as the bases of towering nimbostratus and cumulonimbus rainclouds. At this low level, the clouds are filled with water droplets. Other lenticular, tubular, and patchy clouds may also occur. Surface-level stratiform clouds are known as fog and mist.

Rare stratiform clouds with a rippling, wavelike underside, only added to the International Cloud Atlas in 2017

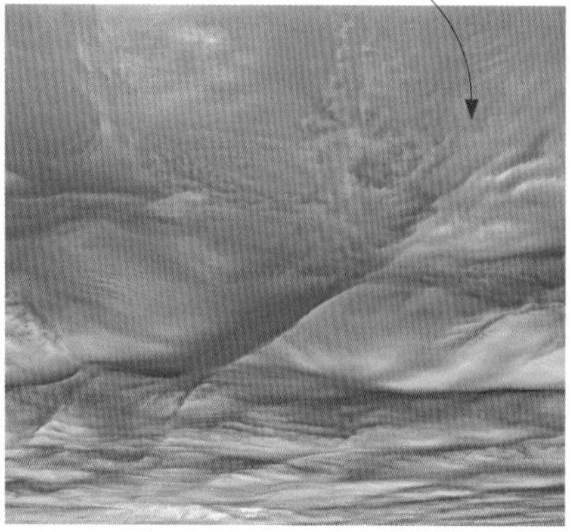

ASPERITAS

Rounded, bulbous clouds, sometimes arranged in parallel bands

STRATOCUMULUS

types of clouds

From iridescent wisps to bulbous pockets, the shapes, colors, and sizes of clouds vary wildly. Clouds form when moist air that has been warmed by the ground or driven over high ground rises upward. As air rises it cools, reducing its ability to hold moisture. At a temperature known as the dew point, this air becomes saturated and the water vapor within it condenses around tiny particles called condensation nuclei, such as dust, pollen, and spores. This condensation forms minute droplets or freezes into tiny ice crystals and becomes the base of a cloud. When water condenses it releases latent heat, which allows the air to continue to rise and build up thick banks of cloud. The type of cloud created depends on temperature, humidity, and air stability.

Fine, wispy clouds associated with fine weather, but may portend a coming storm

CIRRUS

Lens-shaped clouds, often very smooth and with a bulbous top

ALTOCUMULUS LENTICULARIS

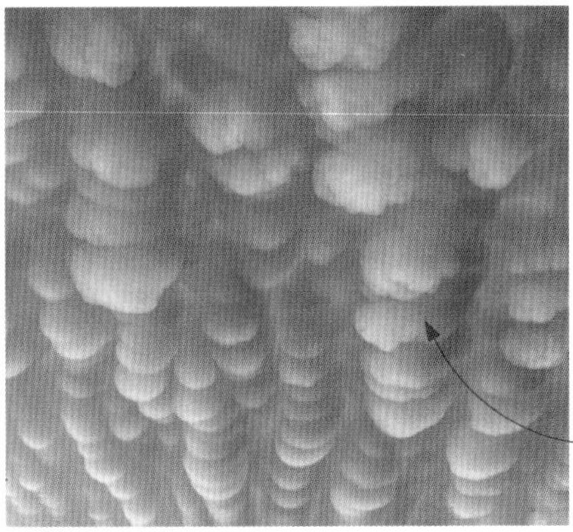

An unusual formation of globular pouches on the underside of a parent cloud

MAMMATUS

Fluffy, cotton-like clouds that often appear in small patches

CUMULUS

Low-level, sheetlike clouds that may extend across the whole sky

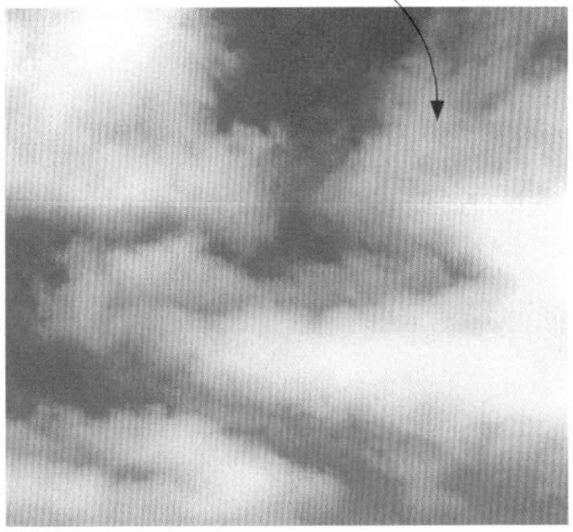

STRATUS

Spectacular roll clouds formed by downdrafts from cumulonimbus thunderclouds

ARCUS

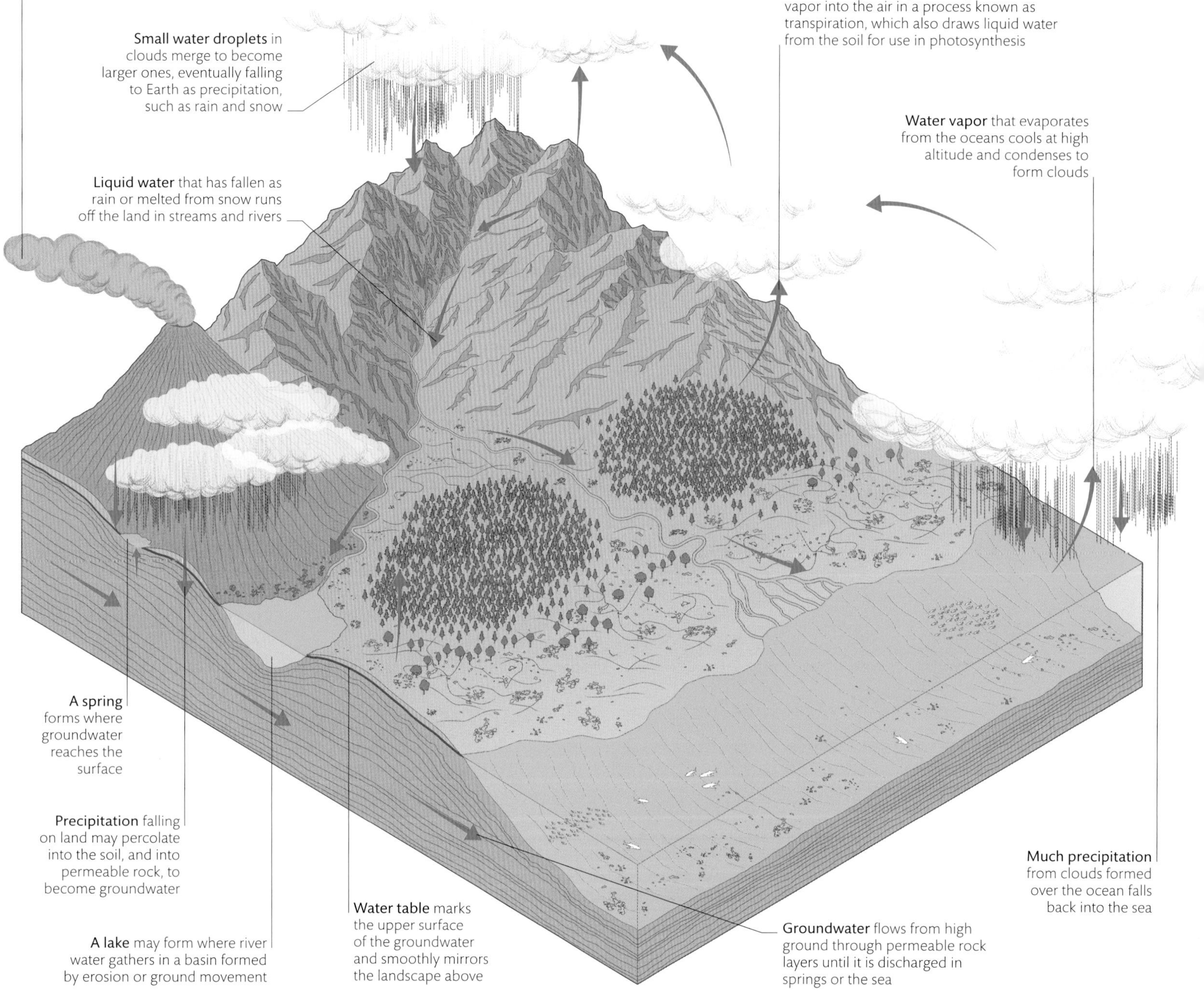

Perpetual transfer

Solar heat causes evaporation of water from oceans and land, and, by the process of transpiration, from plants. Water vapor rises, cools, and forms clouds. Eventually water in the atmosphere returns to Earth as precipitation. Melting of snow and ice, surface run-off in rivers, and infiltration into the ground all transfer liquid water from the land back to the oceans.

Trapped groundwater

Water from precipitation may seep though permeable rock into deeper rock layers, where it can become trapped by layers of impermeable rock. The pressure of water in the permeable rock may force the water out above ground, forming a natural spring. In some cases, humans can tap it to create an artesian well.

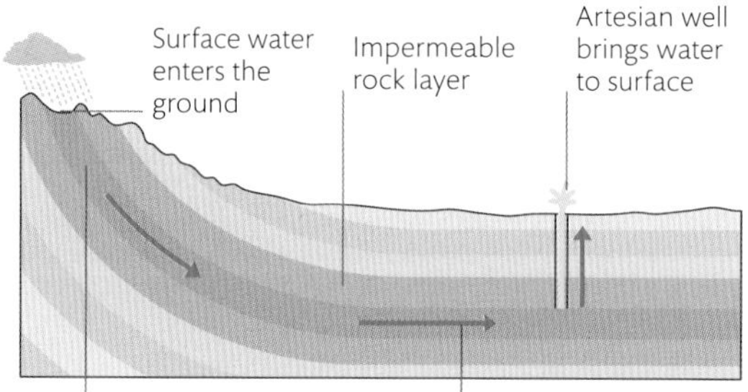

the water cycle

One of the most important systems that mold Earth's surface and regulate its atmosphere, the global water cycle (or hydrologic cycle) is driven ultimately by solar energy. It is vital for plant and animal life, drives the weather, and transfers energy through the oceans, atmosphere, and land. Water also lubricates plate movement and can trigger earthquakes. The total water volume of Earth is around 333 million cubic miles (1.4 billion cubic km), distributed on and below Earth's surface. The oceans hold nearly 96 percent of this total, glaciers and ice sheets around 3 percent, and subterranean groundwater about 1 percent. All the world's rivers, lakes, atmosphere, and biosphere together contain a mere fraction (less than 1 percent). A single water molecule may spend only a few days in the atmosphere or a few weeks in a river, but can be locked up in an ice sheet for a million years.

Seasonal water cycle
Summer snowmelt in eastern Greenland feeds a steep torrent of water and sediment from the Syltoppene glacier across a rocky fan delta into King Oscar Fjord more than 8,900 ft (2,700 m) below. Heavy winter snowfall restarts this cycle.

RAINDROPS

The shapes of raindrops depend on their size, but they are not the pear-shaped teardrops of popular perception. Very small drops up to 0.04 in (1 mm) in diameter are more or less spherical. Flattened by air resistance as they fall, larger droplets are more ovoid in shape. As they grow to 0.18 in (4.5 mm), raindrops begin to split apart and eventually become two separate spherical droplets once again.

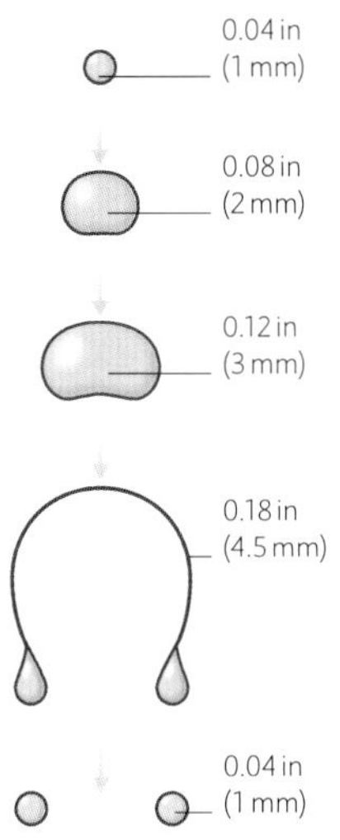

EVOLUTION OF A RAINDROP

Downpour
A ship flees the deluge of a large cumulus congestus raincloud over the Black Sea in Europe. The seemingly impenetrable curtain beneath the cloud is in fact the downpour of falling rain spattering the sea surface. Where violent updrafts of air are too strong, or high temperatures cause evaporation, the rain may never reach Earth's surface.

falling **rain**

When water vapor in the atmosphere condenses around minute particles called condensation nuclei (see pp.222–223), the tiny droplets that form gather into clouds. The constant motion in clouds causes these droplets to coalesce and grow until they are heavy enough to fall to the ground. As the principal source of fresh water, rainfall is vital to life on Earth, and an unimaginable volume—more than 18,000 trillion cubic feet (500 trillion cubic meters)—falls on the planet every year. Not just a valuable water source, rainwater also helps reduce greenhouse gas buildup, as carbon dioxide from the atmosphere dissolves in the water.

WINTER PRECIPITATION

At low temperatures, usually in winter, cold air causes snow instead of rain to form in the clouds. The type of precipitation that hits the ground, however, depends on the warm and cold air masses that it passes through on its way, as shown below. The red front depicts a warm air mass moving through cold air.

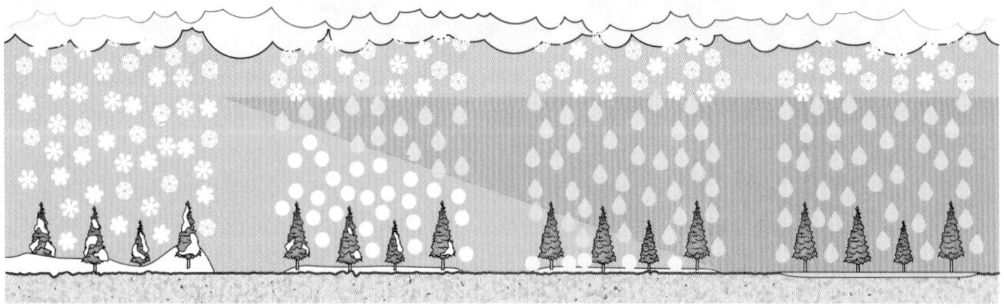

Snow
Snow falls through cold air all the way to the ground

Sleet
Snow partially melts in warm air and refreezes before reaching the ground

Freezing rain
Snow melts in warm air and refreezes upon contact with the cold ground

Rain
Snow melts in the warm air and cannot refreeze before reaching the ground

Sunseekers and surfers flock to the tropical Hawaiian Islands in their millions each year, but, perhaps surprisingly, the islands boast one of the wettest places on the planet: the summit of Mount Wai'ale'ale on the island of Kauai. This extinct shield volcano rises to a height of 5,148 ft (1,569 m) and boasts lush green rainforest and boggy swampland nestled inside its deep crater, due to the high rainfall the mountain experiences.

spotlight Mount Wai'ale'ale

With a name that means "Rippling Water" or "Overflowing Water," Mount Wai'ale'ale has one of the highest rates of rainfall on Earth, typically ranging from 31.2 ft to 37.4 ft (9.5 m to 11.4 m) annually. It rains for between 335 and 360 days a year, and in 1982 alone, a staggering 56.8 ft (17.3 m) fell on the peak.

As the northernmost of the Hawaiian Islands, Kauai is surrounded by ocean and exposed to moisture-laden trade winds that bring rain from the northeast. Helped by the mountain's conical shape, incoming humid sea air is quickly driven up Wai'ale'ale's steep eastern walls, rising as high as 3,000 ft (900 m) over a distance of just 2,600 ft (800 m). As it rises and cools, this air condenses into clouds. Perhaps the most important factor that makes Wai'ale'ale one of the wettest spots not only in Hawaii, but also the world, is its height. The mountain's summit lies just below the trade wind inversion layer—the level at which trade winds cannot rise. This means that, unable to rise any farther, the water-laden clouds are funneled and squeezed into a small space and forced to drop a large portion of their rain on one concentrated spot: Mount Wai'ale'ale's peak.

Dubautia waialealae

Endemic to Mount Wai'ale'ale, this rare plant's compact dome shape and dense whorls of leaves help it to thrive in the mountain's wet climate

Weeping Wall
Mount Wai'ale'ale's abundant precipitation feeds many waterfalls across the mountain, including these tumbling cascades that spill down a steep flank of the volcanic crater. One of the mountain's most famous sights, this lush, verdant wall is known as the Weeping Wall or Wall of Tears.

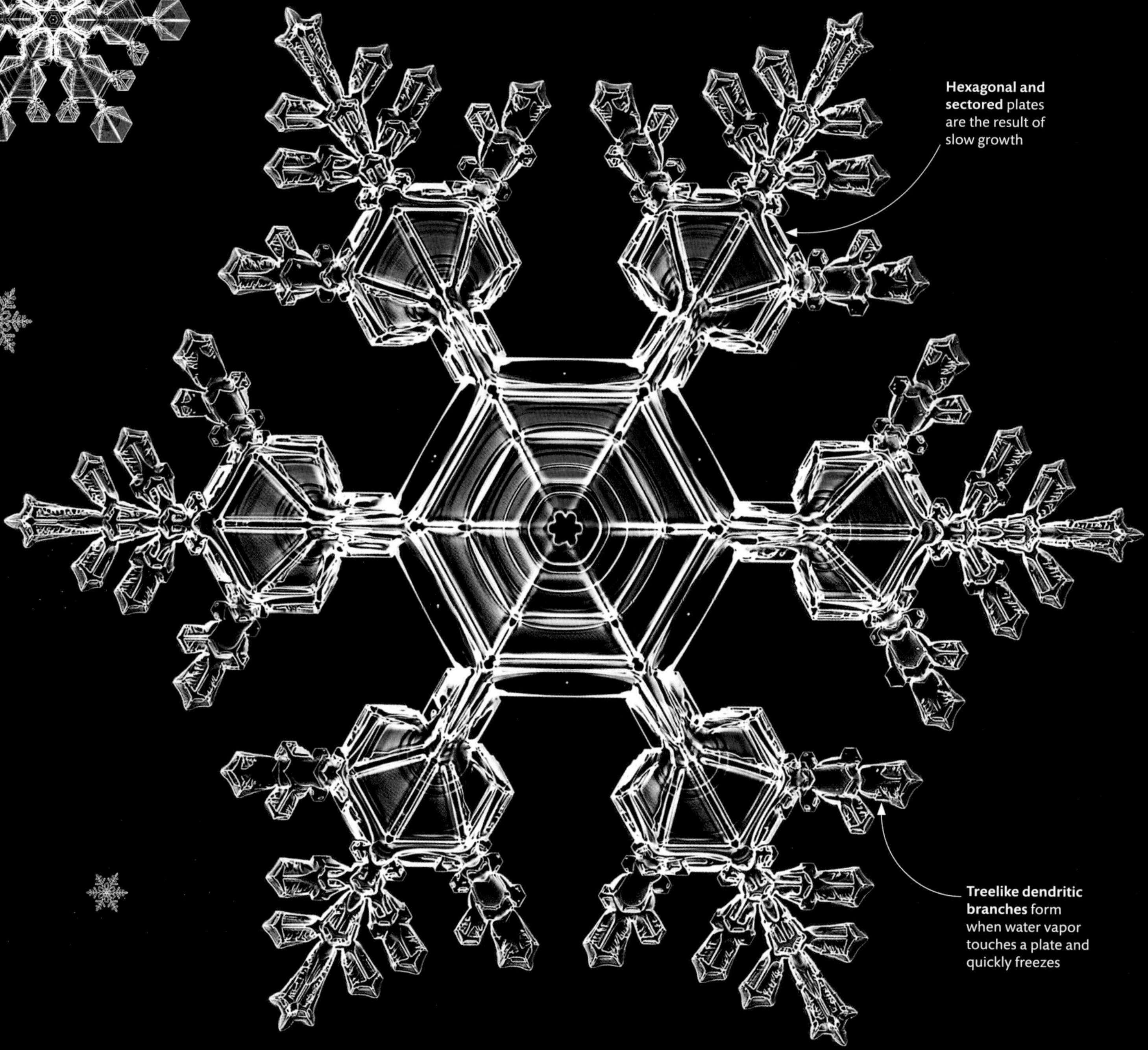

snow

In cold conditions, water vapor that condenses onto dust and other tiny particles can create minute ice crystals. Snowflakes grow as supercooled cloud droplets freeze onto these primary ice crystals, building them up into unique and often elaborate forms that are typically 0.08–0.4 in (2–10 mm) in size. Snowflakes are always hexagonal, due to the internal structure of the primary ice crystals they grow from. As they move through the turbulent air, they collide and stick together, gradually building up into the fluffy aggregate snowflakes that eventually land on the ground.

This snowflake has an almost entirely dendritic form, telling us it formed quickly throughout its growth

SNOWFLAKE MORPHOLOGY

The forms snowflakes take are partly dependent on the temperature and moisture content of the air. Solid plates and prisms form at low humidity, whereas more elaborate dendritic and delicate needle shapes form when humidity is higher. The largest flakes and heaviest snowfall occur between 32°F (0°C) and 23°F (-5°C), when the air holds more moisture. At much colder temperatures the air is drier, resulting in less snow and smaller flakes.

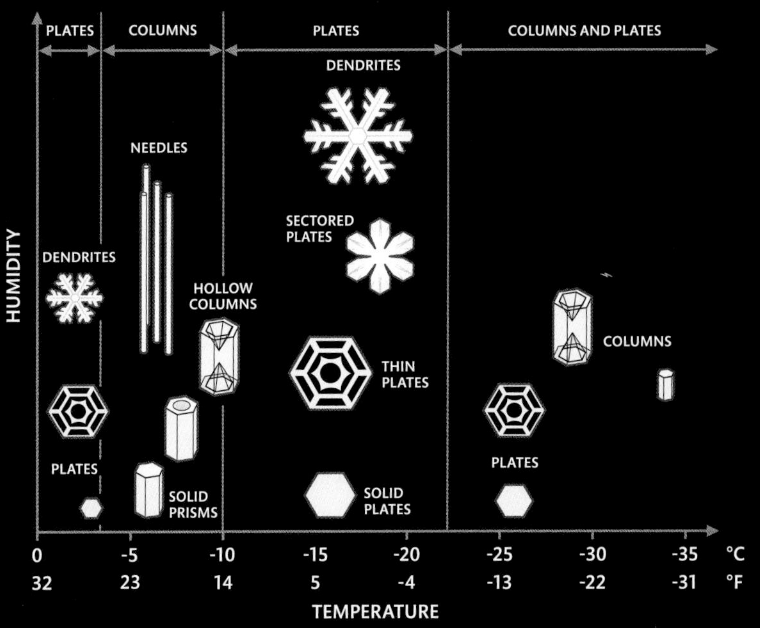

Snowflakes

There is a magnificent array of delicate tracery in even the most common of snowflakes, as illustrated by these light micrographs. Slow growth of a snowflake can lead to hexagonal and sectored plates, while finer dendritic branches occur when growth is faster and less stable. Exposure to varied conditions can lead to snowflakes made up of several different forms.

A simple hexagonal plate forms the center of this snowflake

A very thin hexagonal prism

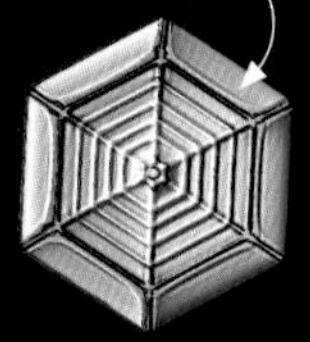

HEXAGONAL PLATE

A plate with broad arms and ridges

SECTORED PLATE

Thin, hairlike crystal

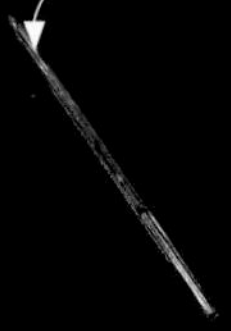

NEEDLE

Hollow region at each end

HOLLOW COLUMN

The result of plate growth on a column

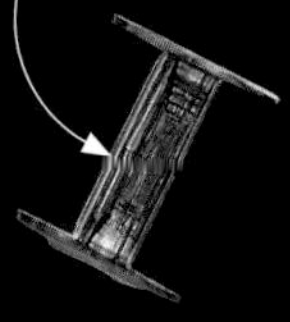

CAPPED COLUMN

Multiple columns growing together

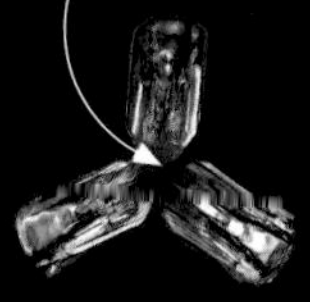

BULLET ROSETTE

Snowflake forms

Crystals of ice that are symmetrical and hexagonal, snowflakes' final forms depend on the conditions in which they grow. While dendritic shapes are most familiar, snowflakes may also take the form of plates, needles, columns, or rosettes.

Supercell
Sometimes the updrafts and downdrafts within a cumulonimbus separate and rotate around each other to form a more extensive storm front called a supercell. This supercell in Nebraska sports forked lightning, a spiraling tornado, and extensive precipitation.

Hailstones
Made within the strong updrafts of cumulonimbus clouds, hailstones are rounded, solid ice balls formed by the freezing and accretion (accumulation of layers) of supercooled water.

Wet-growth hail forms when smaller ice-crystals collide and stick together

thunderstorms

Generated inside towering cumulonimbus clouds, thunderstorms have a dark, menacing base that lies just a few hundred yards off the ground and multilevel, billowing plumes that can reach the top of the troposphere (the lowest layer of Earth's atmosphere) 7½ miles (12 km) up. These storm clouds lead to sudden, brief deluges of rain, hail, or snow, accompanied by lightning flashes, thunder, and even tornadoes. They can store as much energy as 10 Hiroshima atom bombs, causing considerable damage when it is unleashed. Thunderstorms are most common in the warm, moist equatorial zone, but occur everywhere on Earth except for the polar regions.

FORMATION OF A THUNDERSTORM

When warm, moist air rises, it condenses into water droplets. If an air mass is highly unstable, this rise, called convection, can be rapid, leading to the formation of cumulonimbus clouds. Rapid convection releases heat, which causes further updrafts of warm air that can reach 100 mph (160 km/h), and associated downdrafts of cool air. If the cloud meets the tropopause, a flat, anvil-shaped top forms. Sometimes high-speed updrafts will break through the surface, creating an overshooting top.

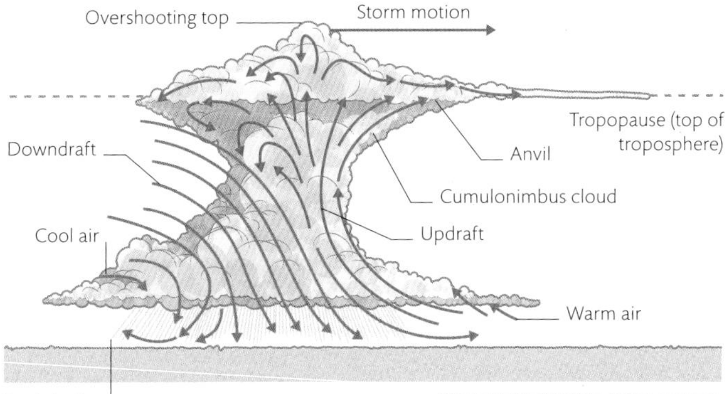

THUNDERSTORM STRUCTURE

The Great Plains of North America cut a broad swathe from Texas to Canada. A land of weather extremes, these flatlands are home to Tornado Alley—a region famous for its high number of tornadoes. Although this region has more tornadoes than anywhere else in the world, scientists warn that the most intense, dangerous, and long-lived tornadoes often devastate regions to the south and east of the Great Plains.

spotlight Tornado Alley

A tornado begins near the top of a cumulonimbus cloud, where rising warm air starts to rotate slowly, creating a funnel shape in the center. Downdrafts of cool air can force this spiraling funnel downward, so that it protrudes from the base of the cloud. As the radius of the funnel decreases, it rotates more and more quickly, and when it touches the ground, it becomes a tornado. Tornadoes can last anywhere between a few seconds to over an hour. Meteorologists use the Enhanced Fujita scale to rate their intensity from EF0 to EF5. The fastest EF5 tornado winds ever recorded reached a staggering 300 mph (480 kph). Tornadoes can span over $4\frac{1}{3}$ miles (7 km) in width and tear across land for more than 185 miles (300 km).

The loosely defined area of Tornado Alley, which sweeps across the US from Texas in the south to South Dakota in the north, records around 1,000 tornadoes each year. The region makes the ideal cauldron for the development of severe, tornado-spawning storms, as warm moist air traveling north from the Gulf of Mexico meets cold dry air from Canada and the Rocky Mountains, creating unstable atmospheric conditions.

Complete devastation or near-miss is the lottery faced by residents of Tornado Alley

Violent air
In the northern hemisphere, tornadoes typically rotate counterclockwise due to Earth's spin. Clockwise-rotating tornadoes, called anticyclonic tornadoes, are rare. This anticyclonic tornado near Simla, Colorado, has swept up many tons of dust and debris from the plains and is about to devastate a hapless ranch.

Natural glass from electricity
Fulgurites are a type of natural glass formed by the fusion of sand and silt under intense heat, such as when lightning strikes the ground. This blue fulgurite was created when a fallen tree channeled electricity from power lines into the ground over several hours, heating it and forming glass.

electricity in the air

On average there are three million lightning strikes worldwide per day, with about 70 percent occurring over the tropics. Lightning is the dramatic result of a sudden discharge of static electricity from storm clouds. When static charge builds up an electric potential of several hundred million volts, the surrounding atmosphere can no longer insulate against its escape, leading to cloud-to-cloud lightning sheets or jagged lightning bolts that stretch to the ground. Typically two fingers wide and 1.2–1.8 miles (2–3 km) long, bolts emit dazzling light caused by superheating air to 54,000 °F (30,000 °C) in less than a second. This heat makes the air rapidly expand, resulting in a sonic shock wave, or thunderclap.

Volcanic lightning
During major eruptions—like this one from Iceland's Eyjafjallajökull volcano—ash, water, and gases are spewed high into the atmosphere. Turbulent movement of these particles within the ash cloud can lead to static electricity buildup and lightning, just as in rain clouds.

WHY LIGHTNING STRIKES

As they are pulled up and down by the air currents within a storm cloud, particles of rain, snow, ice, and soft hail (graupels) collide. Positive ions rub off particles and transfer to upward-moving smaller particles, such as ice crystals, while negative ions transfer to the heavier graupels that accumulate near the cloud base. A buildup of opposite charges eventually leads to lightning within and between clouds, and directly to the ground below.

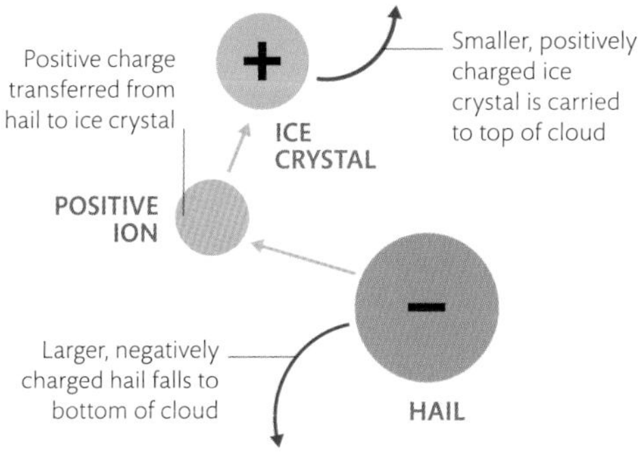

Red sprites
Fainter, faster, and significantly larger than lightning flashes, red sprites are elusive, short-lived optical events triggered far above active thunderstorms. Their red color is due to light emitted from electrically ionized nitrogen. Their tendrils can become green-blue as they descend. These sprites were captured from the McDonald Observatory in Texas.

Gigantic jet
Seen here from the International Gemini Observatory in Hawaii, gigantic jets flare upward from highly charged thunderclouds. The white and blue colors of a gigantic jet's initial burst end in a red fountain near the thermosphere (see p.210).

sprites, jets, and elves

Eyewitness accounts of pilots who glimpsed these unexplained lights were initially dismissed by meteorologists, until they were first caught on video in 1989. Part of a family of large-scale electrical discharge phenomena that occur high above major thunderstorms, called Transient Luminous Events (TLEs), red sprites, jets, and elves light up the atmosphere like fireworks. There are an estimated several million TLEs per year, but they are rarely witnessed from land, because they occur at altitudes of 25–60 miles (40–100 km) and mostly last only a millisecond. Sprites, which produce delicate hanging tendrils below a halo, are the most common.

TRANSIENT LUMINOUS EVENTS

TLEs are triggered by normal lightning discharge to the ground through the troposphere (see p.210), but mainly occur in the stratosphere and mesosphere as electrically induced forms of luminous plasma (charged particles). Sprites move downward from the upper mesosphere, while the expanding red glows of elves occur higher still. Jets (cloud-to-air electrical discharges similar to lightning) extend tens of miles upward. Other TLEs, including trolls, pixies, ghosts, and gnomes, are still poorly understood.

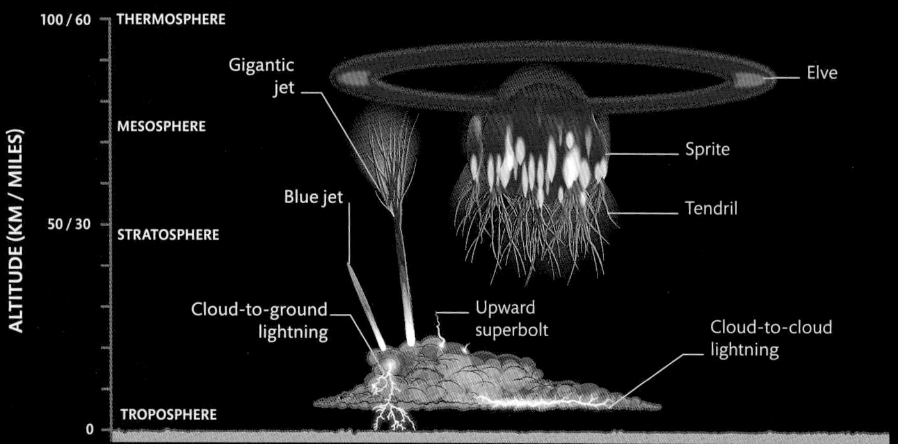

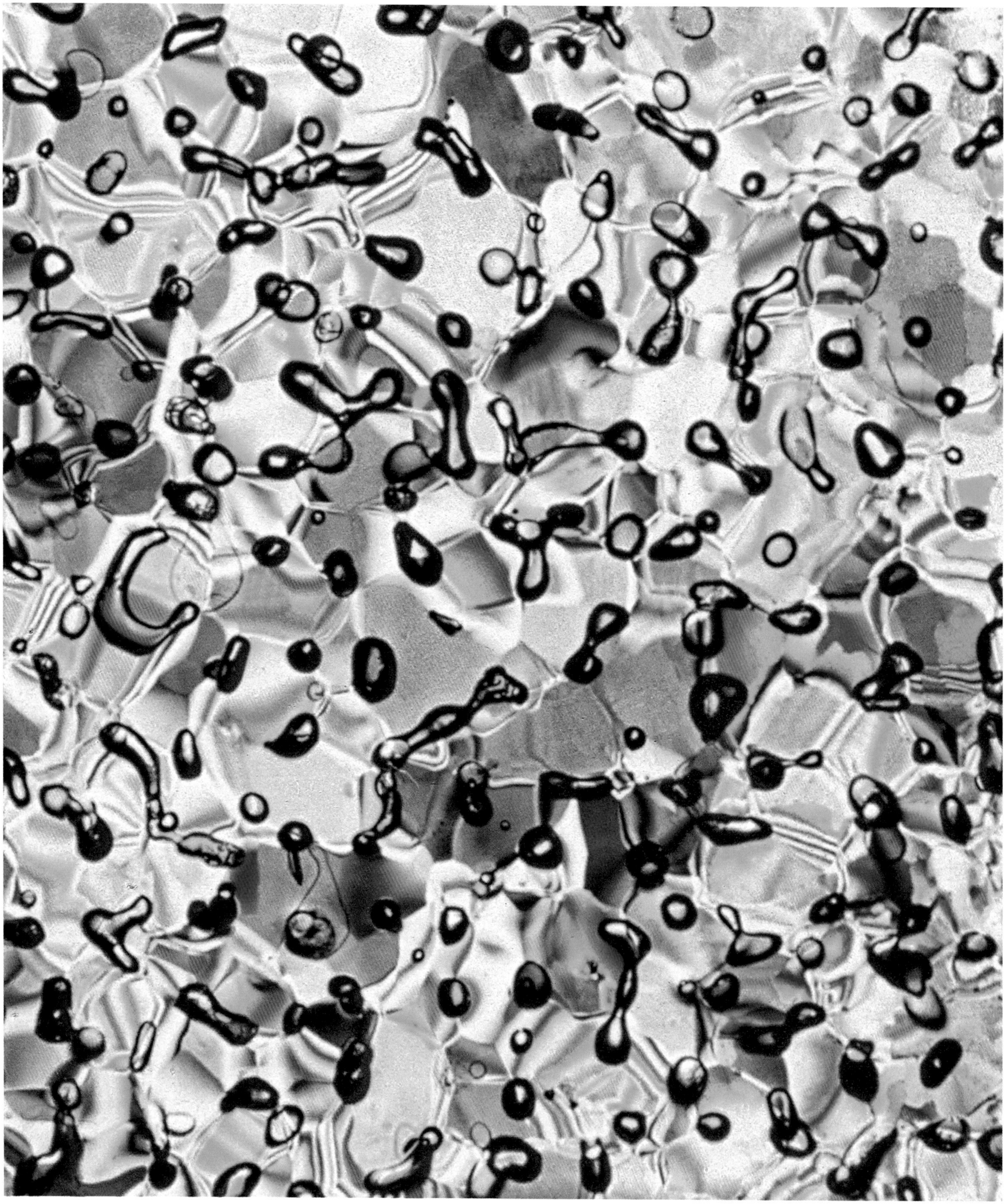

Ice cores from Greenland
Narrow cylinders of ice are extracted by rotating drills, or thermal drills, which melt the ice. The cores are stored in giant freezers and then examined and sampled. Here, three sections extracted from the Greenland ice sheet contain information revealed by ice from different depths.

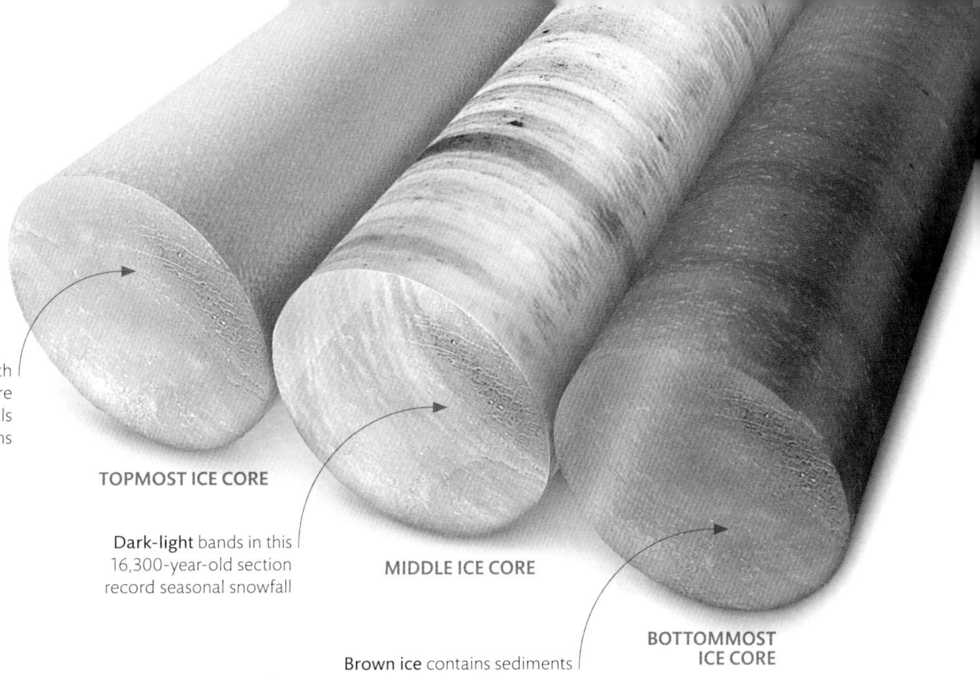

history of Earth science

ice-core analysis

Ice cores contain tiny bubbles of atmospheric gases from the distant past that have been trapped in ice and buried beneath glaciers and ice sheets, and yield a climate record over a period of up to 800,000 years. A few cores contain ice of more than 2 million years old. The deepest cores from Antarctica penetrate over 3,000 m (10,000 ft) and contain clear evidence of the correlation between levels of greenhouse gases and climate change.

Researching the Greenland ice sheet
Ernst Sorge (1899–1946), a German glaciologist, is pictured here during an expedition to Greenland in 1930–1931. On this mission, he dug a shaft 49 ft (15 m) deep to make the first records of the density and temperature of buried ice.

Scientists have long been curious to discover the secrets that lie under the thick layers of ice that occur in glaciers and the polar regions. Louis Agassiz (1807–1873), a Swiss-born American pioneer of ice age theories, used iron rods to drill into Alpine glaciers. Open-hole drilling, vertical shafts, and trenching were used to investigate the Antarctic ice in the early 20th century. Similar methods were used by Ernst Sorge (see inset, above left) for his research into the Greenland ice sheet in the 1930s.

The first ice cores (cylinders of ice) drilled from deep in the ice layer were recovered from Greenland and Antarctica in the 1950s. Since then, large numbers of ice cores have been drilled and stored for research. Scientists can detect the levels of key elements in these ice cores to date the samples accurately. Different forms of oxygen occur in ice formed during and between glacial periods. Such evidence helps to give a precise record of past climates, including snow accumulation, temperature changes, glacier retreat, and variation in ocean and atmospheric composition. Particulate matter in ice cores can also provide evidence of major volcanic eruptions and wild fires.

Analysis of ice cores from glaciers around the world is also used to further advance our understanding of natural climate variation.

Air bubbles from the past
This ultra-thin sliver of Antarctic ice shows hundreds of tiny air bubbles trapped between ice crystals. The bubbles contain nitrogen, oxygen, argon, carbon dioxide, and methane, and provide vital data on the composition of the atmosphere at the time they became trapped. Polarized lighting of the ice causes the colorful effect.

> “The era of intense cold, which preceded the present… has been only a temporary oscillation of the Earth’s temperature.”
>
> LOUIS AGASSIZ, 1837

The current ice age

During icehouse periods, or ice ages, Earth cycles between colder periods (glaciations), with extensive polar icecaps, and warmer interglacials, with less ice and higher seas. Today we live in an interglacial of the current icehouse period. The ice has receded since the last glaciation, which peaked 22,000 years ago. Even so, the Antarctic icecap is 6,900 ft (2,100 m) thick. It buries mountains, leaving just their peaks visible as so-called "nunataks".

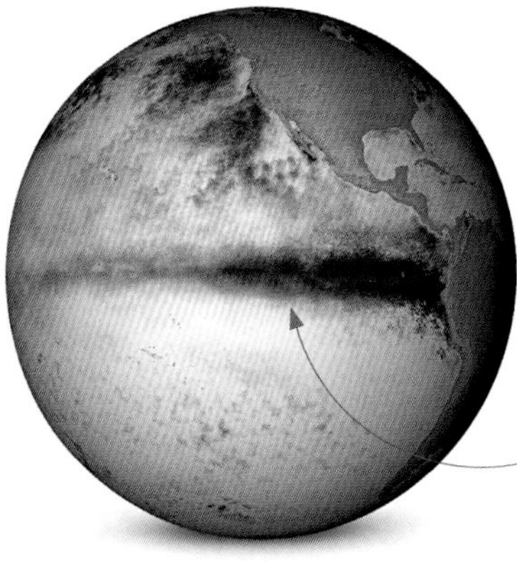

The surface water of the eastern Pacific experiences excess warmth (red), leading to drier, hotter weather across North America

El Niño events

Every few years, a natural periodic oscillation of the ocean-atmosphere system significantly warms Pacific surface waters. Known as El Niño, the phase profoundly affects Earth's climate and can warm the planet by 1.3 °F (0.7 °C).

natural climate change

About 100 million years ago, palm trees and crocodiles lived near the North Pole; now the region is covered with thick ice. Earth's climate cycles between two states: greenhouse and icehouse Earth. Icehouse periods, also known as ice ages, feature lower temperatures, continental ice sheets, and icecaps in the polar regions, whereas greenhouse periods experience an increase in global temperatures and the retreat of ice. These natural changes to the planet's climate are driven by variations in solar radiation, the position of continents and oceans, and greenhouse gases.

MILANKOVITCH CYCLES

Fluctuations in Earth's orbit around the sun and rotation on its axis lead to additional cyclical changes in the climate. Named after Milutin Milankovitch, the Serbian mathematician who first described their effect, these cycles are responsible for the alternating glacial and interglacial conditions during the current icehouse period. Also known as the Quaternary Ice Age, this began just over 2.5 million years ago.

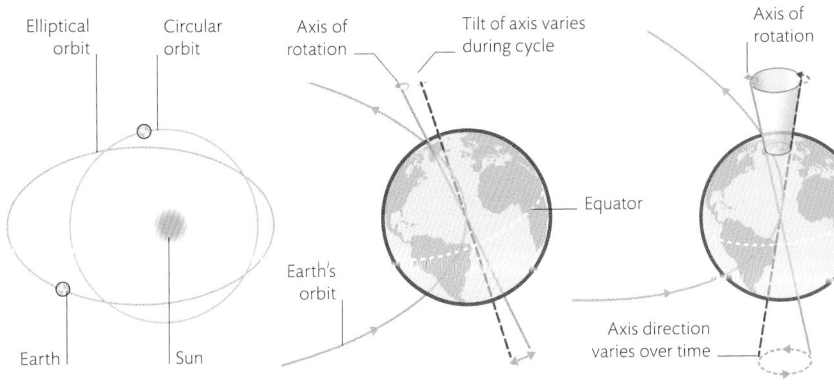

ECCENTRICITY
Earth's orbit varies from nearly circular to distinctly elliptical (eccentric) over about 100,000 years.

OBLIQUITY
The tilt (obliquity) of Earth's axis varies between 24.5 ° and 22.1 ° over a period of about 42,000 years.

PRECESSION
Earth's axis wobbles slightly as it spins over a period of nearly 26,000 years. This is termed precession.

living planet

The living part of Earth, its biosphere, forms a thin film on its surface measuring less than two-thousandths of the planet's diameter. This veneer has nevertheless had a profound influence on the planet's development over the last 4 billion years. Living processes have transformed Earth's atmosphere, and living organisms have left a fossil record that allows us to trace the path of evolution and to reconstruct geological history in fine detail.

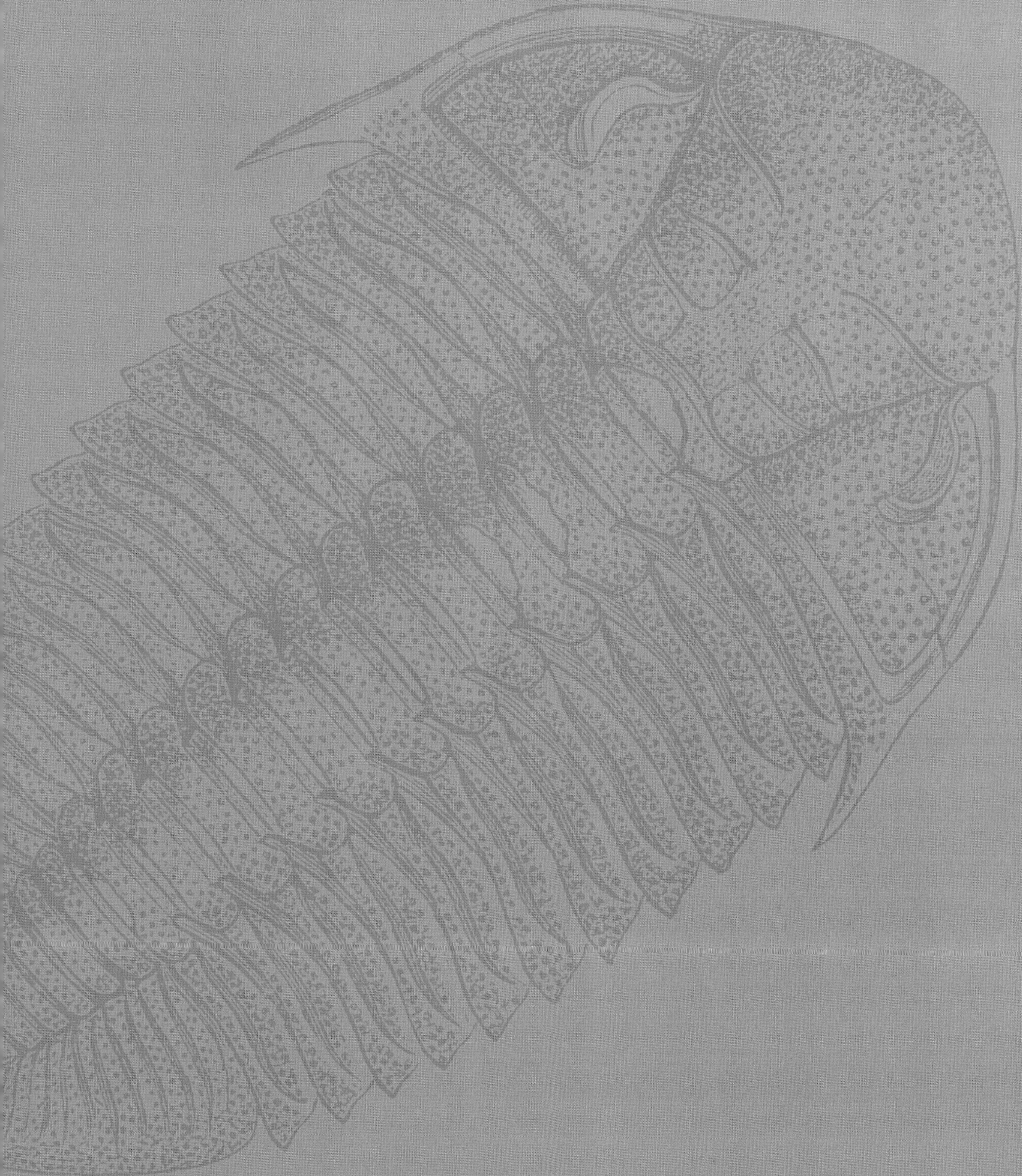

the biosphere

The parts of Earth that are habitable by life are confined to its surface—in a layer of rock, water, and air called the biosphere. The thickness of this layer accounts for less than 0.2 percent of the diameter of the planet. This extraordinary living coating is the only example yet discovered in the universe. At its deepest, it follows ocean trenches that plunge as far down into the crust as the highest mountains that rise above it. At these extremes, life is sparse and specialized to cope with hostile conditions. But closer to sea level—on land and in the oceans—life teems in forests, deserts, rivers, lakes, and the vast marine realm. Almost all of this life is fueled by sunlight, as food-making plants and algae nourish complex food chains.

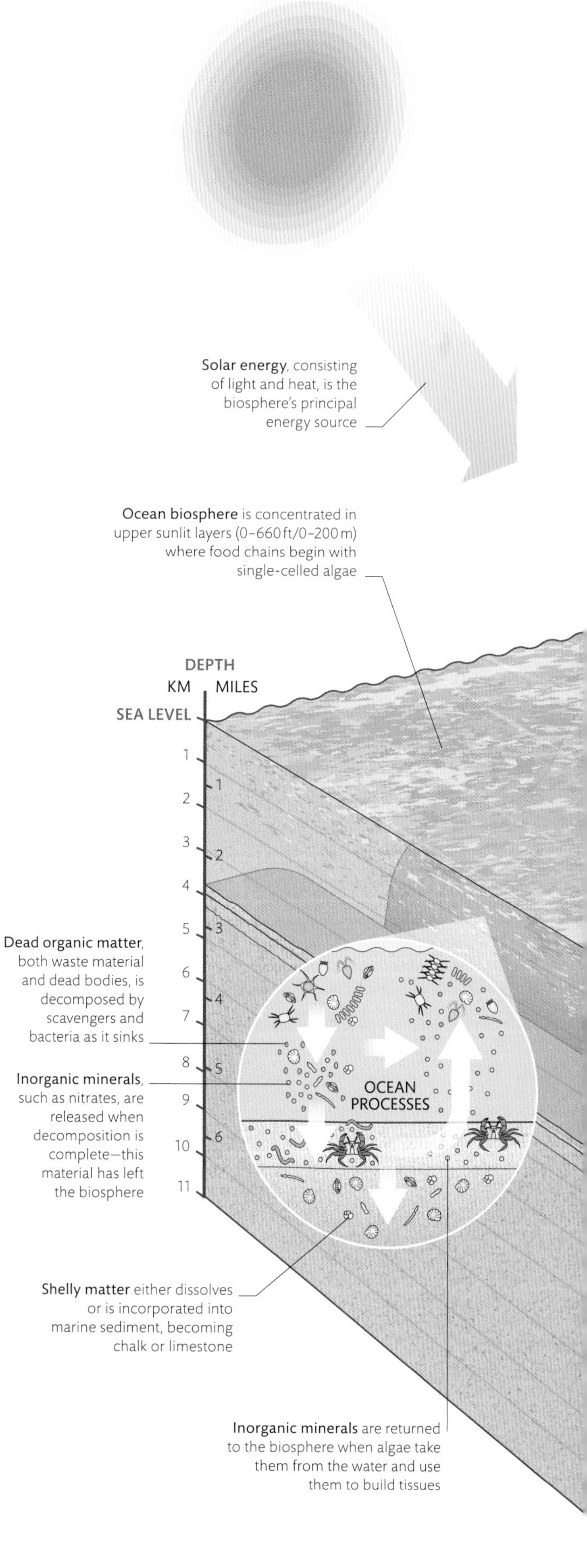

Life fueled by minerals, not sunlight
In parts of the deep ocean floor, life flourishes without the sun's energy. Here, volcanic hydrothermal vents emit hot streams of mineral-rich sea water. Bacteria release energy from the minerals to make food, while invertebrates graze on these bacteria and are, in turn, food for predators, such as fish.

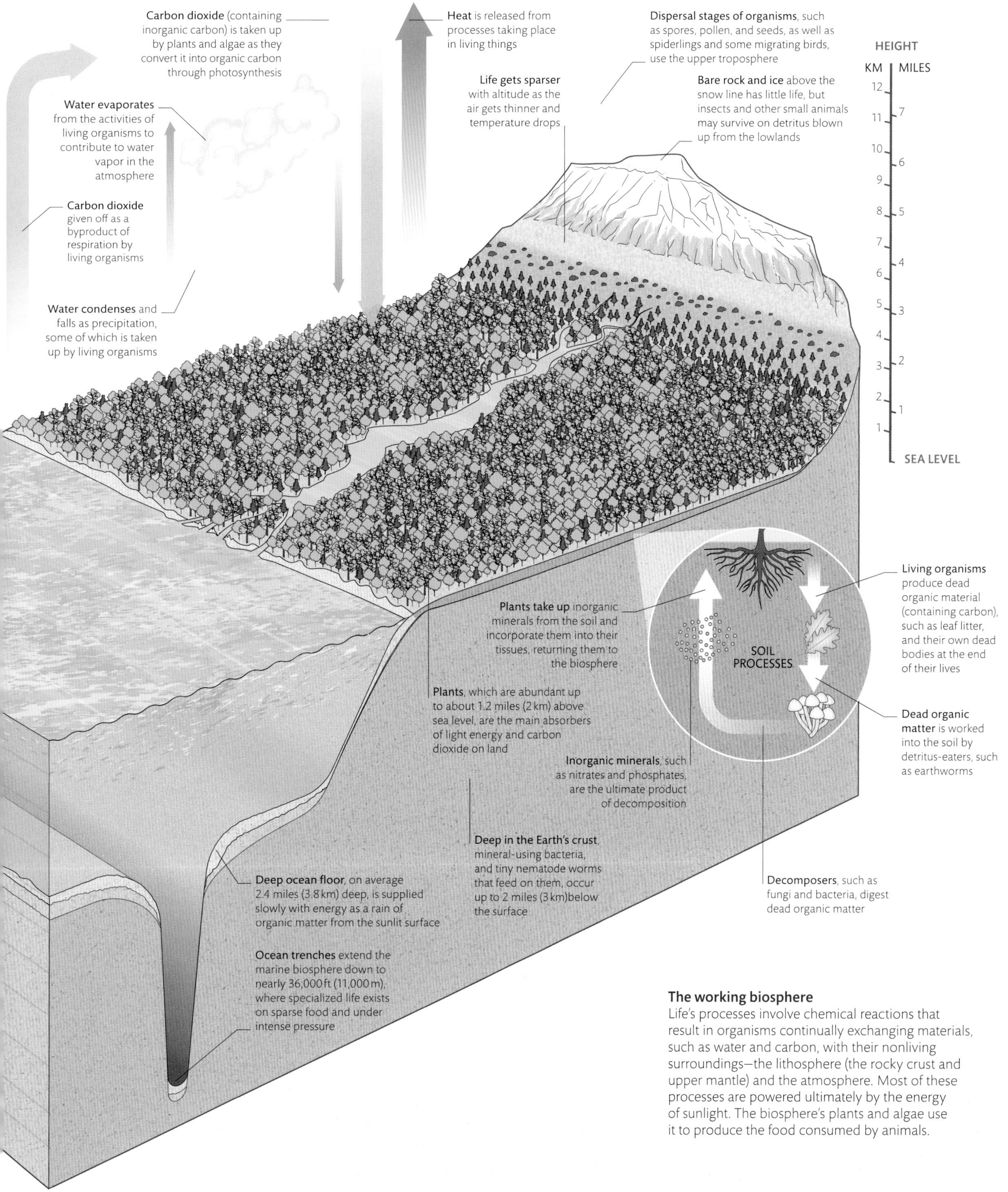

The working biosphere

Life's processes involve chemical reactions that result in organisms continually exchanging materials, such as water and carbon, with their nonliving surroundings—the lithosphere (the rocky crust and upper mantle) and the atmosphere. Most of these processes are powered ultimately by the energy of sunlight. The biosphere's plants and algae use it to produce the food consumed by animals.

millions of species

Earth is the only planet known to have life, and there is a breathtaking variety of color, shape, and movement in its surface biosphere (see pp.246–247). Every species has developed its own way of surviving and reproducing, and each one is descended by evolution from a single ancestor that was just a cell some 4 billion years ago. Today, the diversity of microbes, plants, and animals is at its richest in ancient forests and sun-drenched coral reefs, but life forms can exist anywhere they can feed and take in oxygen—and have found a way to do so—from the highest mountains to the deepest ocean trenches.

Pollination partners
Insects and flowering plants are the most species-rich elements of Earth's biodiversity. As these marbled white butterflies, *Melanargia galathea*, drink nectar from greater knapweed, *Centaurea scabiosa*, and transfer pollen, they continue a partnership of insects and flowering plants that evolved millions of years ago and led to the planet's astonishing diversity of both.

Prehistoric diversity
This 420-million-year-old slab of rock is a snapshot of early biodiversity. It carries fossils of a reef community, ancestors of modern marine animals.

HOW SPECIES EVOLVE THROUGH GEOGRAPHICAL SEPARATION

Over many millions of years, life on Earth changes through a process of evolution. As this happens, populations split and their characteristics can diverge so much that they become different species. The diversity of species at any one time comes from a balance between new species generated and those that become extinct because their populations do not survive.

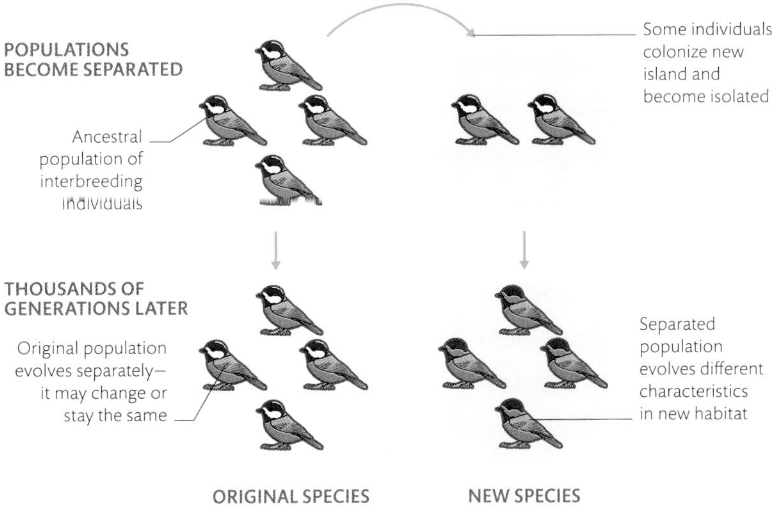

early life

For the first billion years of its existence, Earth was lifeless. But given the vast timescale available, the rare chemical reactions that were possible on its mineral-rich surface became likely—perhaps inevitable. In turn, these reactions produced the first complex organic matter, and eventually, self-replicating molecules enveloped in oily membranes became the first living cells. No one knows exactly where the first life appeared, although a strong theory points to volcanic vents on the ocean floor. But by at least 3.4 billion years ago, life had colonized shallow seas, leaving strange fossilized rocky mounds called stromatolites.

Living stromatolite
Stromatolites, like these in Shark Bay, Western Australia, still grow today. They are confined to places that are inhospitable to animals that graze the microbes that produce them, such as areas where the water is too salty.

Active microbes remain on the surface layer

Lower section is formed by remnants of former microbial surface communities

Dome-shaped layers built up over remains of algal tufts

Lower layers formed from algae that died as new algae grew above them, blocking out the light

FROM BIOFILM TO STROMATOLITE

Even today, many kinds of microbes including bacteria and algae—produce sticky slime that keeps their cells amassed into thin colonies, or biofilms. Photosynthetic microbes called cyanobacteria need light, so continually migrate upward, leaving dead layers of the microfilm beneath. The slime binds sediment, which accumulates in layers that build up over thousands of years and cement together to form rocky stromatolites.

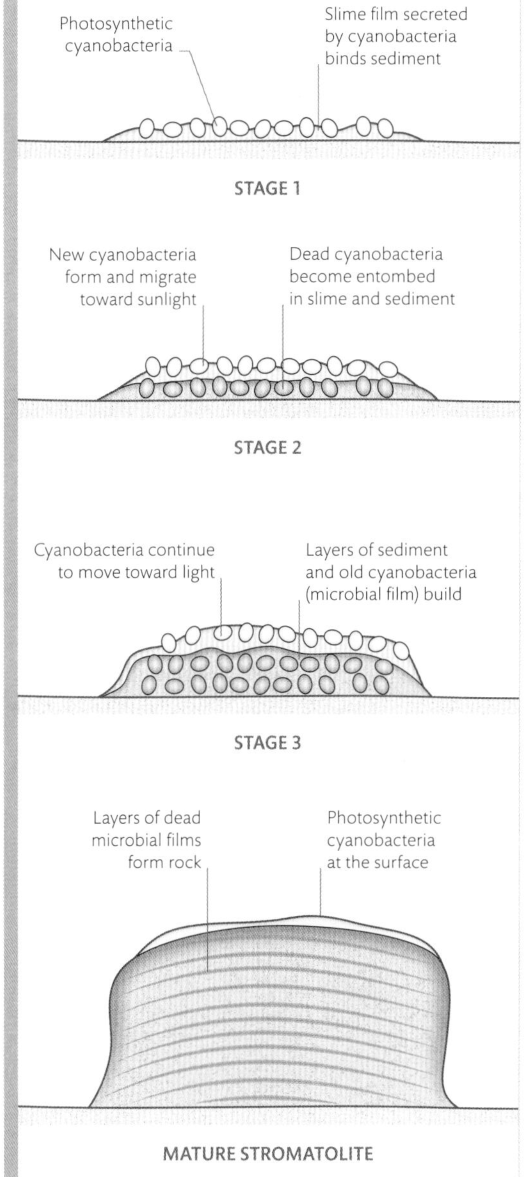

From microbes to marble
This slab of Cothan marble from the dawn of the dinosaurs—billions of years after the first stromatolites formed—shows that simple microbes have been sculpting rocks throughout life on Earth. The dark, blue-black bushy streaks were once sticky tufts of living algae that bound mounds of mud; over thousands of years, they solidified into limestone, then crystallized to marble.

Banded iron formation
Iron ore develops from iron oxides—the products of the reaction between soluble iron salts and oxygen. As oxygen-releasing microbes waxed and waned over time, bands were formed in sedimentary rocks on the seabed; the bands correspond to the seasonal changes in the activity of microbes. Earth's shifting crust has brought the rocks, such as this one found in the iron mining districts of Western Australia, to the surface.

HOW PREHISTORIC LIFE OXYGENATED THE WORLD

The first photosynthesizing organisms were microbes that built rocky mounds called stromatolites (see pp.250–251) found in sunlit shallow seas. Iron salts welling up from the deep ocean reacted with the oxygen to deposit bands of solid iron oxide, or rust. As iron salts in the water declined, the oxygen from photosynthesis effectively overspilled into the atmosphere, eventually rising to the 21 percent level of today.

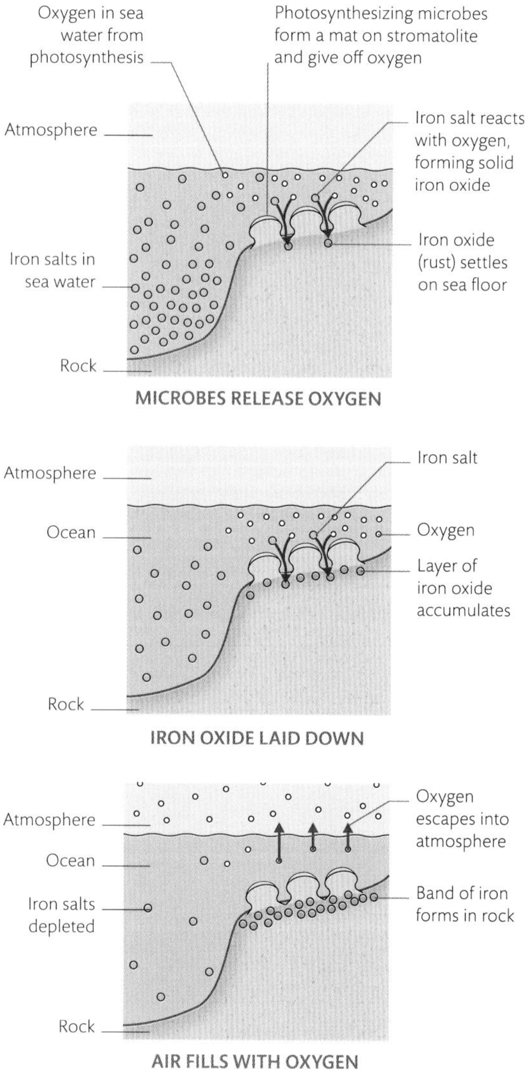

Lighter colored bands are composed mainly of scattered grains of chalcedony, a silica-based mineral containing iron oxide

Darkest bands have the richest concentration of iron and are made of iron oxide minerals called magnetite and hematite

Golden bands, known as "tiger eye", carry an intermediate amount of iron

Tiny cells of *Nostoc* form chains linked by branchlike structures

Bubbling oxygen
Slimy underwater colonies of modern bacteria, called *Nostoc*, produce bubbles of oxygen. Today these photosynthesizers—along with more complex algae and plants—keep air and water oxygen-rich.

life transforms Earth

More than 2 billion years ago, the first green photosynthetic microbes bubbled up oxygen, producing some of Earth's most striking rock patterns, and affecting the course of evolution forever. Some of Earth's richest concentrations of iron ore are the result of this event at the dawn of life. The oxygen reacted with dissolved iron particles floating in sea water, producing minerals deposited under prehistoric oceans as solid red ore. When the soluble iron in the oceans was depleted, the oxygen from the microbes escaped into the air—helping to create the breathable atmosphere that life relies on today.

life branches out

As the first single-celled organisms evolved into bigger, multi-celled bodies, life began to have a bigger physical impact on the world around it. Animals have nerves and muscles, so their reactions and movements are faster and strong enough to churn sediment and stir water as they search for nutrients in the water column, or from dead matter on the ocean beds. The first simple animals flourished on the seafloor around 600 million years ago, during a period called the Ediacaran, named after the Ediacara Hills in South Australia, where their fossils are especially abundant.

Wide head had sense organs concentrated in the animal's direction of movement

Forward movement
Just 1½–2 in (3–5 cm) long, *Spriggina* was a complex animal with distinctively different ends to its body. It must have moved in one direction—head first.

HOW ORGANISMS CHANGED THE SEABED

The movements of the first muscle-powered animals changed conditions beyond the flat seabed, making other parts of the ocean more habitable for life. Some animals bulldozed the surface, while others burrowed down. Together they stirred up ocean sediment, which helped oxygenate the mud below and nourish the waters above.

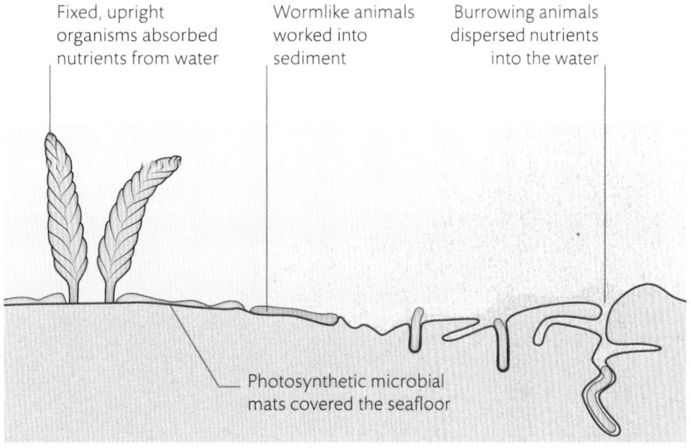

CROSS SECTION OF SEAFLOOR

Life on the seafloor
The ribbed, leaflike impressions on these *Dickinsonia* fossils indicate they were organisms that lived on the seafloor. The prominent ribs suggest that their bodies were supported by a tough hornlike material. Trace fossils on the rock could be track marks, evidence that they moved from place to place.

CRINOID
Order Comatulida

BRACHIOPOD
Lingula hians

Cambrian origins
The Cambrian explosion was a time of intensive evolution, when many animal body plans originated. Today's crinoids—relatives of starfish—are very similar to their Cambrian counterparts. Some animals, such as brachiopods, have changed so little in half a billion years that they are described as "living fossils."

Cambrian explosion

As the first wriggling animals churned mud and used ocean minerals to build the first shells, life began to affect the cycles of sediment and rocks beyond the living biosphere. Shell-building and reef-forming animals amassed calcium and silica in their bodies, and many of them used muscle power to swim upward into the water column. Early in a period of time called the Cambrian, about 539–520 million years ago, an explosion in the development of these hard-shelled animals found the prehistoric seas teeming with the distant ancestors of sea creatures alive today—including shrimplike arthropods, snails, and sea urchins.

Shelly fauna
Exoskeletons and shells preserve better than soft bodies, and it is the predominance of shelled fauna from the Cambrian that helps explain the rich diversity of its fossil record. Seen here are the cone-like tubes of *Archotuba* (center)—which may have enclosed a soft, tentacled anemone—and a jointed-legged arthropod (top), called a trilobite (see pp.278–279).

CAMBRIAN OCEAN DIVERSITY

Fossils of the ocean-dwelling animals of the Cambrian reveal body components, such as mouthparts and limbs, that are important clues not only about their feeding habits but also about how they moved. Their diversity points to a community of swimmers and crawlers, scavengers and predators, which were all linked in the first complex food chains. By dispersing upward from the ocean floor, where the first animals originated, they colonized the open water—some drifting in the currents as the earliest plankton, while others swam against the currents. In this way, animal life expanded the biosphere into three dimensions of the ocean.

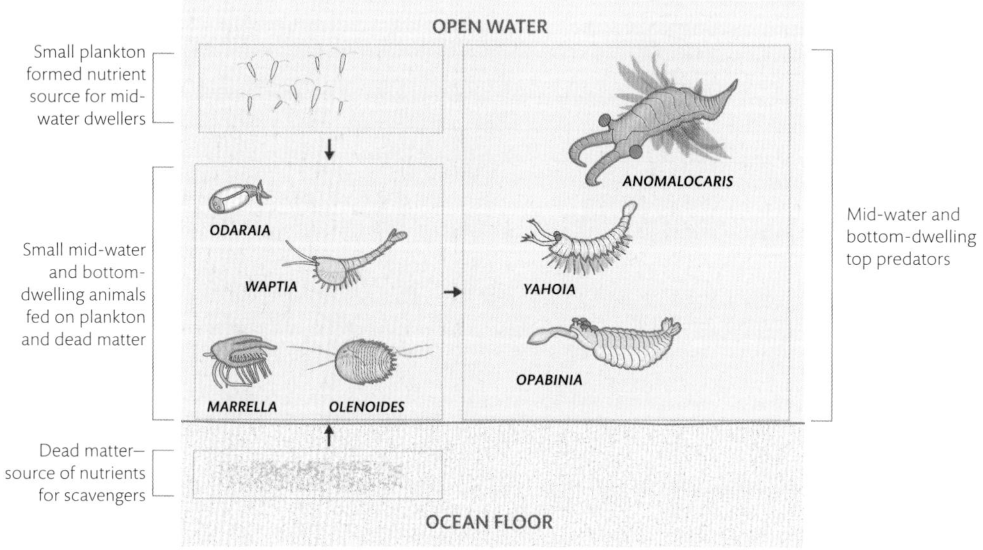

THE CAMBRIAN OCEAN FOOD CHAIN

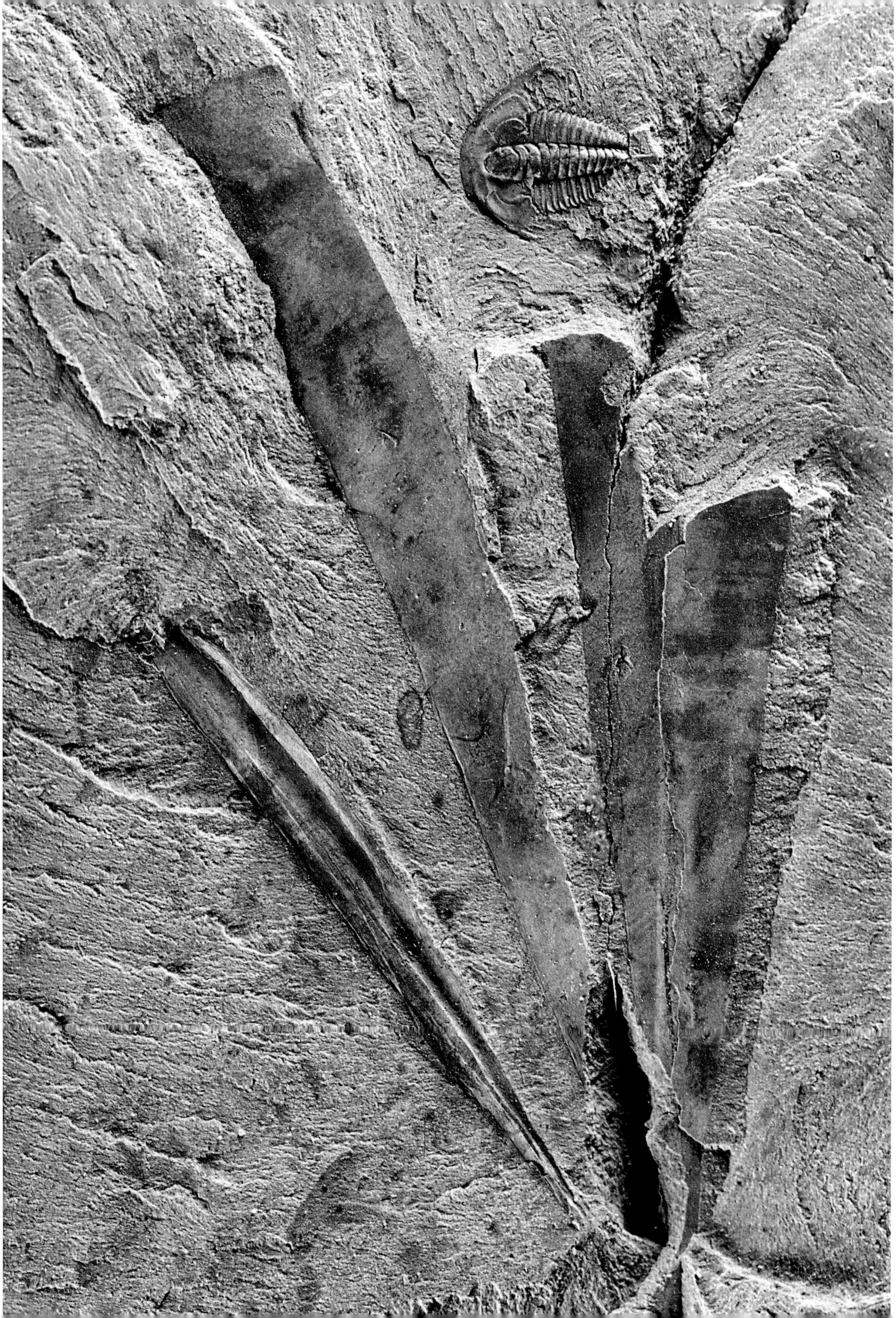

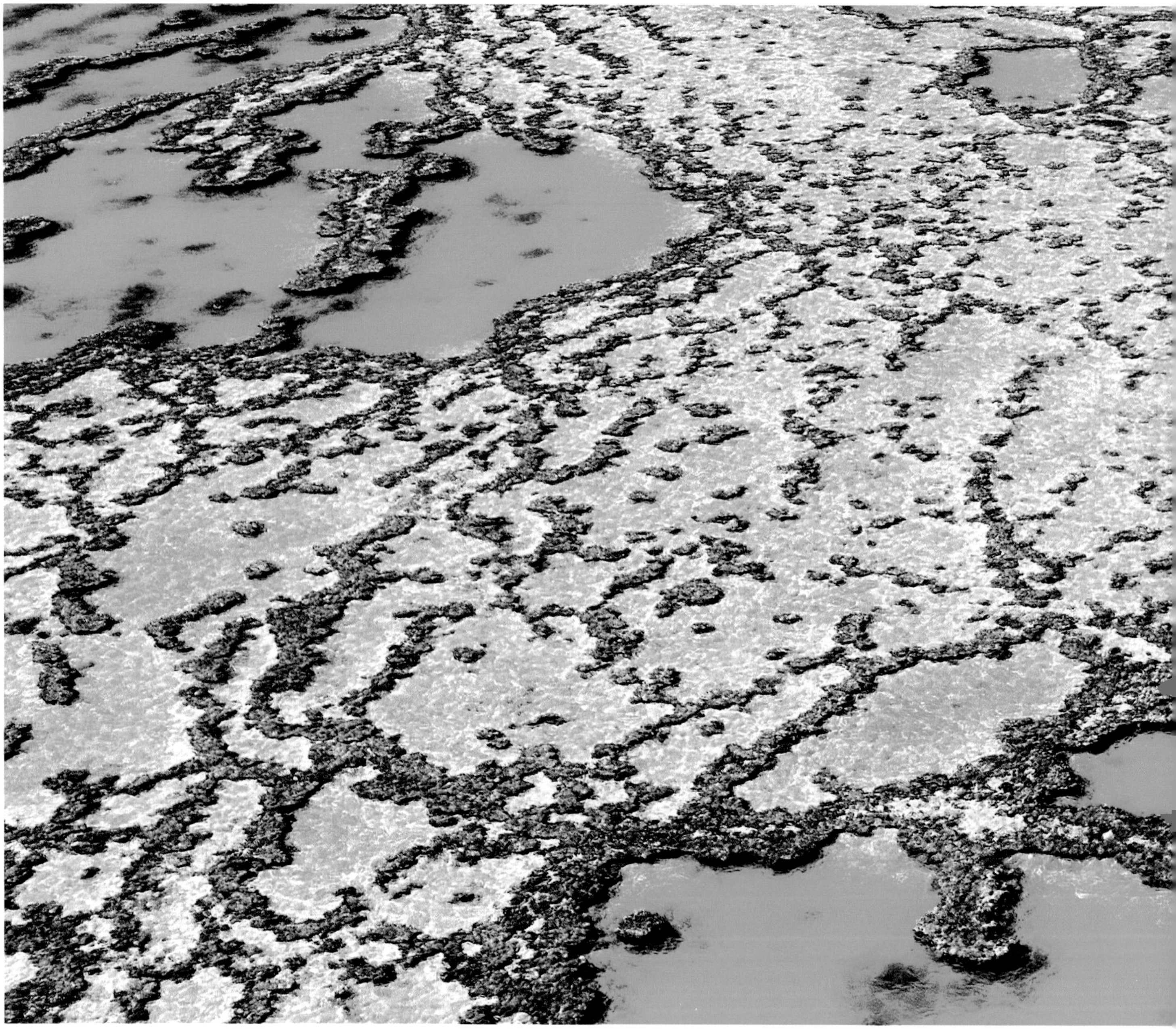

making reefs

Ocean reefs are the biggest structures made by living things and can be seen from space. They are constructed by organisms depositing sedimentary rock over millennia. Today's reefs are built by coral polyps using minerals extracted from sea water. These colonial animals are cousins of anemones and jellyfish—a group whose ancestry stretches back half a billion years. But the history of prehistoric reefs is far richer—once stromatolites (see pp.250–251) were grazed almost to oblivion, organisms as diverse as microbes and sponges and, later, rugose corals, succeeded them as the main reef-builders.

Great Barrier Reef
Stretching halfway down Australia's east coast, the Great Barrier Reef is Earth's largest modern reef system. Its complex geography of underwater ridges and islands creates a mosaic of underwater habitats. Like reefs of the prehistoric past, it is a "hot spot" of marine biodiversity: a place where huge numbers of species colonize and evolve.

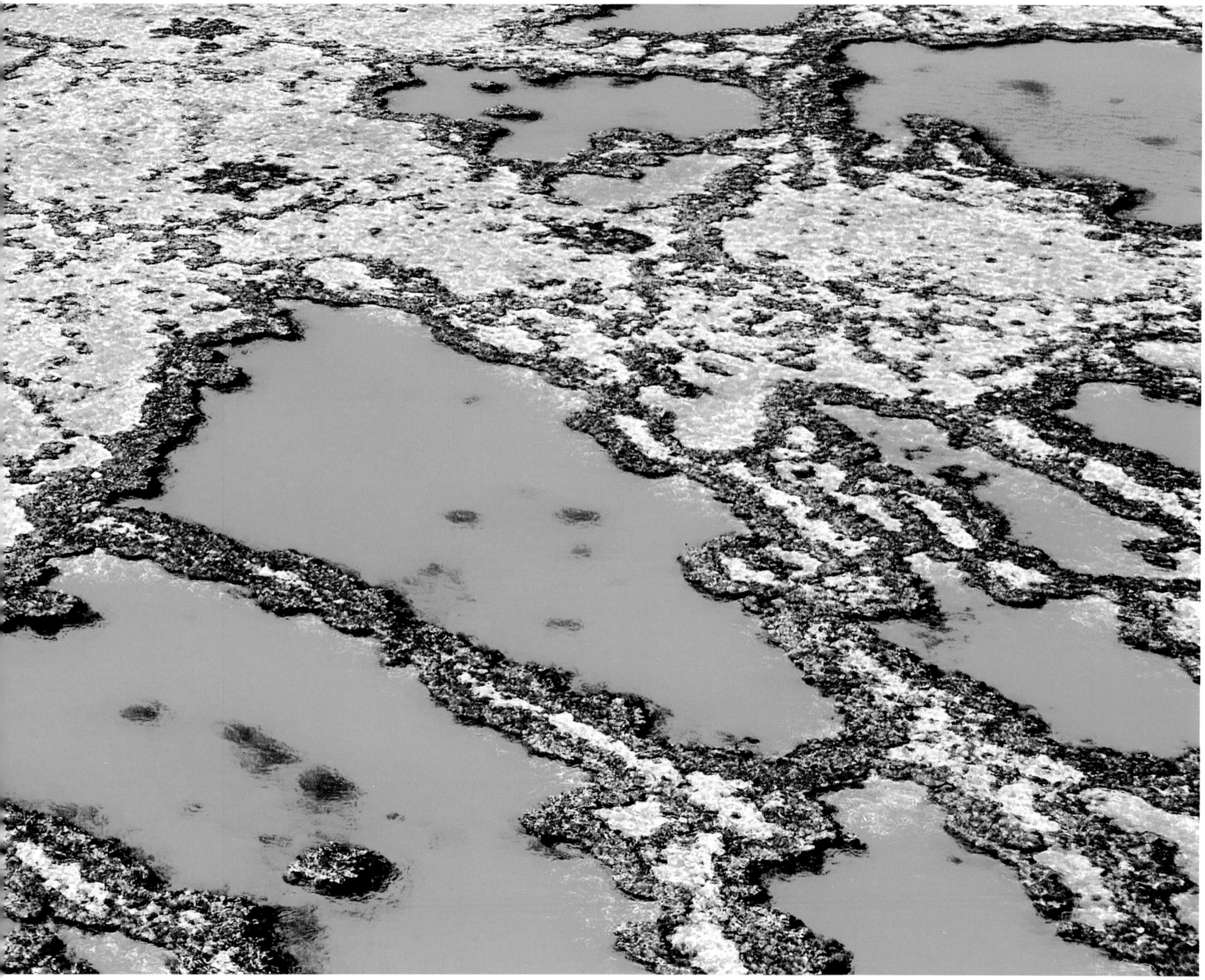

Fossilized reef
Some of Earth's earliest reefs were made during the Cambrian period by animals called archaeocyathid sponges. Their rocky colonies were built from a series of hollow cones.

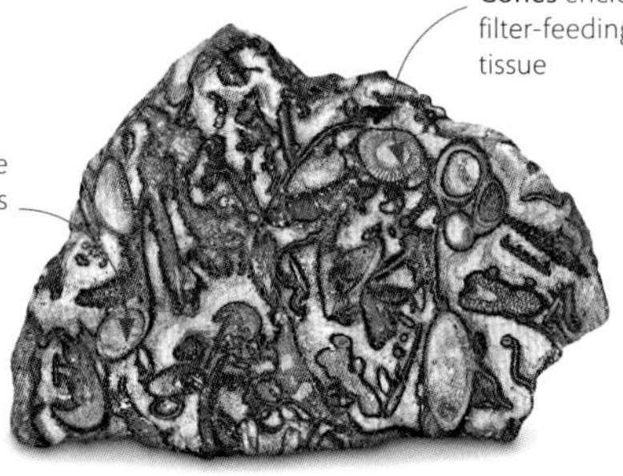

REEF BUILDERS, PAST AND PRESENT

Modern reef-forming corals have been around since the time of the dinosaurs, but earlier reefs were built mainly by different kinds of organisms. The first were made from microbial films or sponges. Most of these were later replaced by hornlike rugose corals and then by modern stony corals.

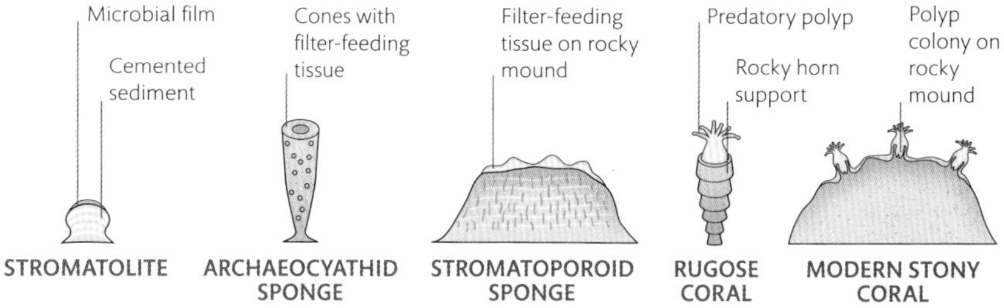

THE DEVELOPING FISH

Vertebrates have a skeleton made of cartilage and bone. Their fishlike invertebrate ancestors, from more than 500 million years ago, had neither—but their body had a rubbery rod, called a notochord, which ran through their back and helped to strengthen their side-to-side swimming movements. Over time, the notochord was reinforced with cartilage and bone that also enclosed and protected the brain and spinal cord, as well as bracing the gills. The gill braces, or arches, kept the gills open for absorbing oxygen. Later, the front gill arches evolved into hinged jaws, which helped manipulate and bite food.

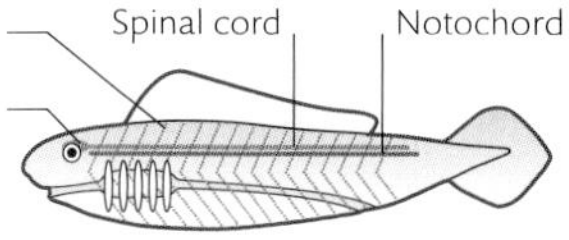

1 JAWLESS PROTO-FISH WITH NO CARTILAGE OR BONE

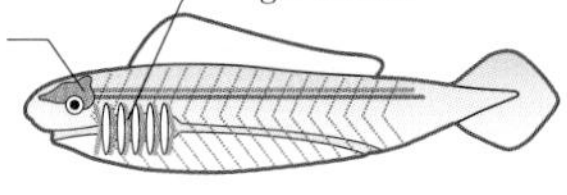

2 JAWLESS FISH WITH SIMPLE CARTILAGINOUS SKELETON

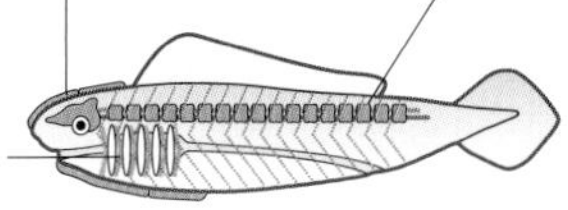

3 JAWLESS FISH WITH COMPLEX CARTILAGINOUS SKELETON

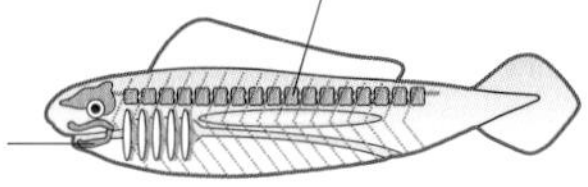

4 JAWED FISH WITH BONY SKELETON

Jawless fish
The earliest fish—such as ostracoderms—lacked jaws and probably grubbed around near the ocean floor. This *Zenaspis* from the early Devonian lived in shallow seas and river mouths.

Giant predator
This fossil of *Dunkleosteus*, a giant armor-plated fish from the late Devonian, shows that by around 400 million years ago, backboned animals had evolved into the biggest biting species of the time—something that made them top predators of increasingly complex food chains. *Dunkleosteus* belongs to a group of fish, called placoderms, that were the earliest vertebrates with jaws.

age of fish

As living things filled Earth's waters, animals evolved to be bigger, faster, and stronger, especially among the first backboned animals (vertebrates)—the ocean's fish. Their skeletons supported growing bodies and in some groups became protective armor. Fish had descended from small soft-bodied invertebrates that used gills to filter nourishment, but now sucked larger morsels of food into their mouths with their muscular throat. This left their gills to focus on extracting oxygen from the water. By the Devonian period, around 400 million years ago, the oceans were filled with numerous groups of fish. In some, the gills' supporting rods had evolved into another key innovation: biting jaws.

Fleshy finned jawed fish
Tristichopterus was a late Devonian fish more closely related to living jawed vertebrates than placoderms. Some of its later cousins evolved walking limbs and became land vertebrates.

Hinged neck joint allowed head to pivot upward, helping to widen mouth before biting
Jaw's jagged edge provided cutting surface similar to the teeth that evolved in later jawed fish, like sharks
Joint between upper and lower jaw was worked by strong muscles that could deliver powerful bite force

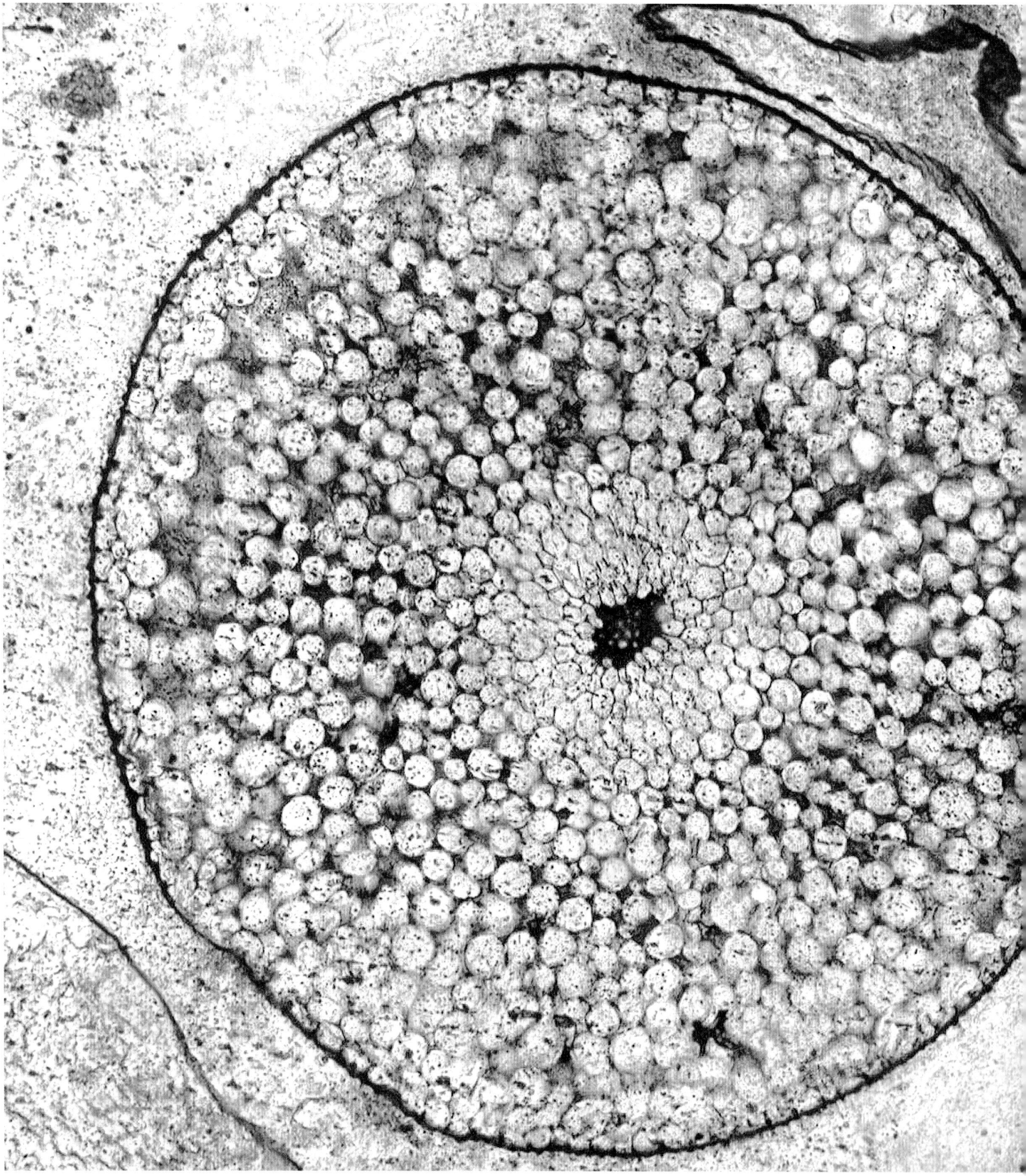

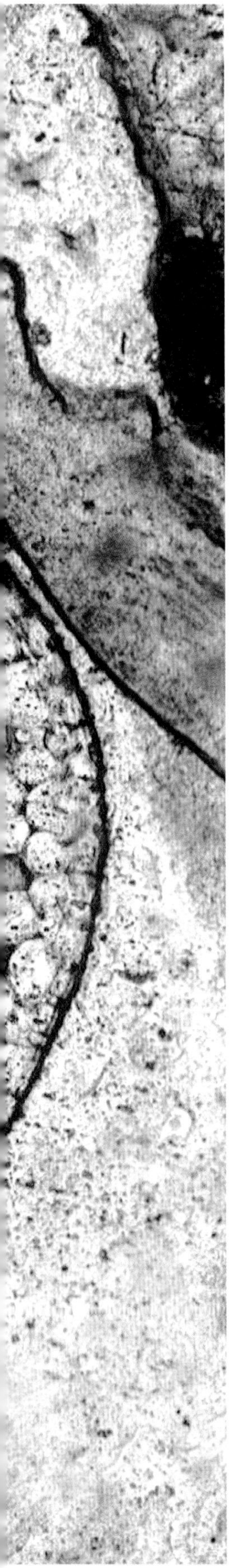

Pioneering vegetation
One of the earliest land plant pioneers, this fossilized *Thursophyton* had branching stems as long as a human hand. Ground-hugging plants like these provided shelter and food for the earliest land animals—paving the way for the first complex food chains on land.

invading land

Land and air were inhospitable to early life, which evolved in water. Because living bodies are at least 70 percent water, any organism that ventures onto land must prevent water loss by evaporation. The first to do so—bacteria—probably colonized rock more than 3 billion years ago. When larger life forms, such as plants and animals, invaded land, they needed strengthening tissues to enable them to stand upright. The first land plants—descended from aquatic algae—appeared more than 400 million years ago and had waxy, waterproof shoots and leaves, stiff internal pipework, and roots providing anchorage.

Plant plumbing
This fossilized cross section of an early mosslike plant, *Rhynia major*, shows a pipe—called a xylem vessel—running through the center of a pithy stem. Just like in modern land plants, this pipe carried water from the roots and strengthened the upright stem.

MULTI-LAYERED COMMUNITY

The evolution of branching, erect land vegetation meant that plants, such as *Cooksonia*, could absorb more light energy for photosynthesis and disperse their reproductive spores on the wind. However, *Cooksonia* contributed only to the lowest level in new, multi-layered land communities. Taller pioneers called nematophytes (including *Germanophyton*, *Mosellophyton*, and *Prototaxites*) towered above them. Only known from the fossil record, they may have been related to non-photosynthetic fungi rather than plants.

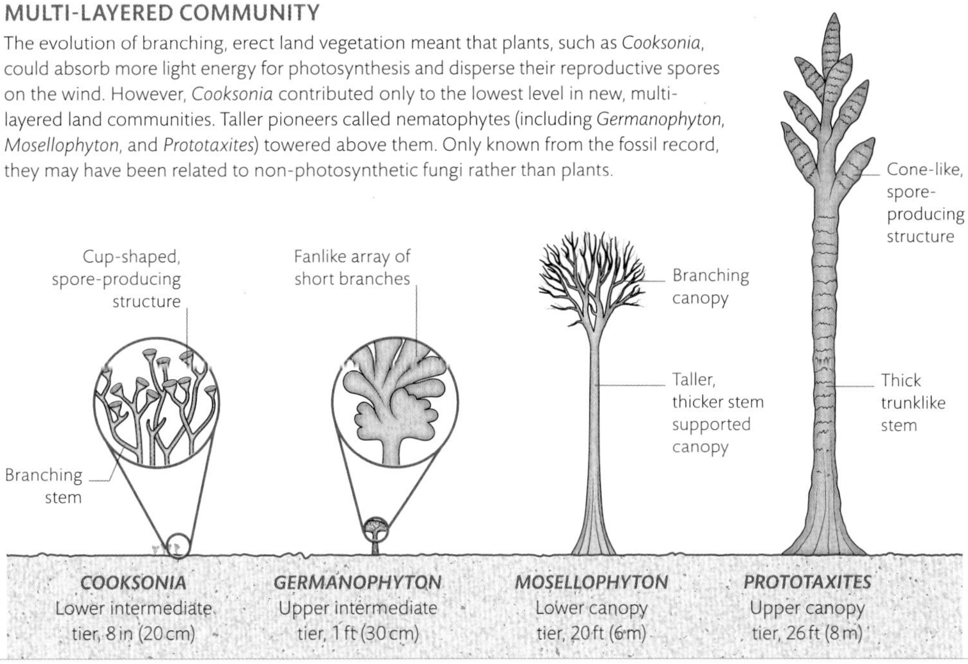

Primary succession in action
Anak Krakatau, a volcanic island off the Indonesian islands of Java and Sumatra, first appeared in 1927. Repeated eruptions have limited the growth of vegetation, but as seen in this satellite image, plant life is gradually becoming established.

succession

The sequence of organisms that colonize newly available habitats is predictable, and is known as succession. In areas where no life existed before, such as in the path of lava flows or retreating glaciers, succession is considered primary, whereas after disturbances that destroy vegetation but not soil, such as wildfires, succession is secondary. In both types of succession, early colonizers (pioneer species) tend to be fast-growing, rapidly reproducing species that capitalize on the wealth of new nutrients and resources. Later arrivals are slower growing but hardier, ensuring their survival. Early species may facilitate the arrival of later species by modifying the environment, or they may inhibit invaders until they are outcompeted.

THE STAGES OF PRIMARY SUCCESSION

Primary succession occurs when completely new habitats arise, as on a cooled lava flow. Barren rock is broken down by microorganisms, lichens, and algae, forming soil, which supports rapidly growing plants. More complex ecosystems form as generations of decaying plants build up the soil, allowing other plants to grow. Eventually, a relatively stable and complex ecosystem may form. This is called the climax community.

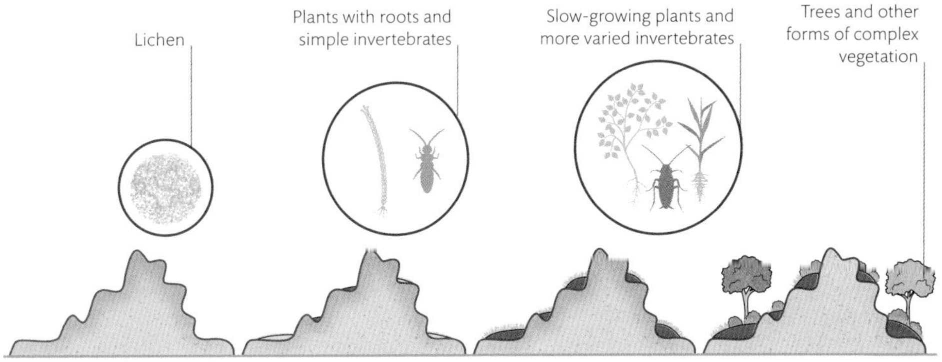

Bare rock
Living organisms that colonize bare rock include microbes, algae, and lichens.

Thin soil
Fast-growing vascular plants set down roots in thin soil. Populations of invertebrates are present.

Developing community
Soil builds up and slower growing plant species appear, forming a more complex ecosystem.

Climax community
Mature, complex vegetation established, forming a stable ecosystem that supports a range of animal life.

RECONSTRUCTION OF A CLADOXYLOPSID FOREST

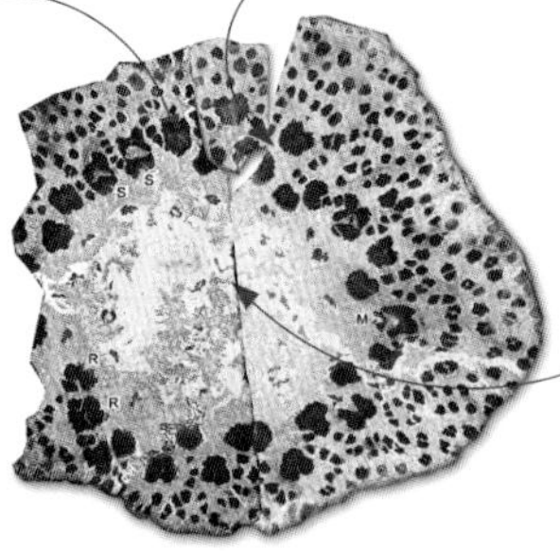

FOSSILIZED CROSS SECTION OF A CLADOXYLOPSID TRUNK

Devonian forests
Fossils show that the first forests grew nearly 400 million years ago in the Devonian period. They were formed by distant cousins of ferns called cladoxylopsids, which were among the first plants to develop thick stems supported by rigid water-carrying pipes.

forests form

Plants that expose more of their foliage to the sun's energy-giving rays generate, by their photosynthesis, more nourishment for growth. The first land plants sprawled on the ground—but as they competed for light, some evolved taller stems that allowed them to escape shadows cast by their neighbors. In the race to grow upward to the light, they strengthened their stems with wood and evolved into the first trees. Today trees are the biggest single—non-colonial—organisms on the planet. And, growing together, they make up the most complex land habitats of all: forests, whose foliage can reach higher than a 15-story building.

Today's rainforests
In the warm, moist tropics, conditions in the forest canopy stay humid and wet under regular rainfall, meaning that smaller plants—called epiphytes—can set root and grow on branches of trees. Here, beyond the shaded ground, a huge diversity of species—including bromeliads, orchids, and ferns—can thrive, providing a rich habitat for arboreal rainforest animals such as insects, spiders, frogs, and birds.

FOREST LAYERING

Trees cast shadows and create dark, humid microclimates close to the ground. Trees and shrubs, and the vines and epiphytes they carry, develop into layers, especially in rainforests where a huge number of species are adapted to different conditions. Undergrowth and understory trees are most tolerant of shade and many even complete their life cycles there, whereas canopy and emergent trees need full exposure to the sun to set seed.

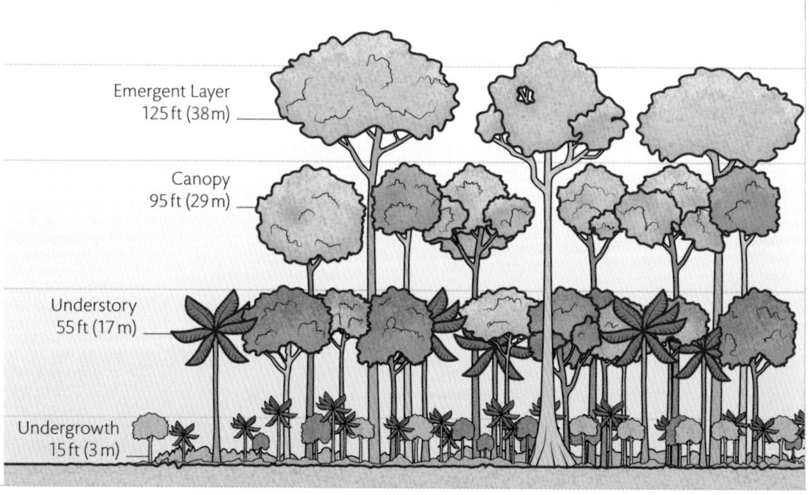

LAYERS OF A RAINFOREST

fossilization

The process that transforms formerly living organisms into long-lasting remains, fossilization usually involves the replacement of original tissues with more durable minerals. This can happen in many different ways depending on the environment in which the animal or plant was buried, and the process can take millions of years to replace the minerals in a skeleton. The resulting remains can withstand the destructive forces of burial and erosion and retain their original shapes and features of the tissues, even down to the cellular structures, providing a wealth of information for today's scientists.

Fossil formation
The two main varieties of fossils are body fossils (parts of the life form itself) and trace fossils (evidence of its behavior). Most body fossils, such as bones or teeth, have their tissues replaced with minerals, in a process called permineralization. Trace fossils preserve signs of a creature's activity, such as footprints or burrows.

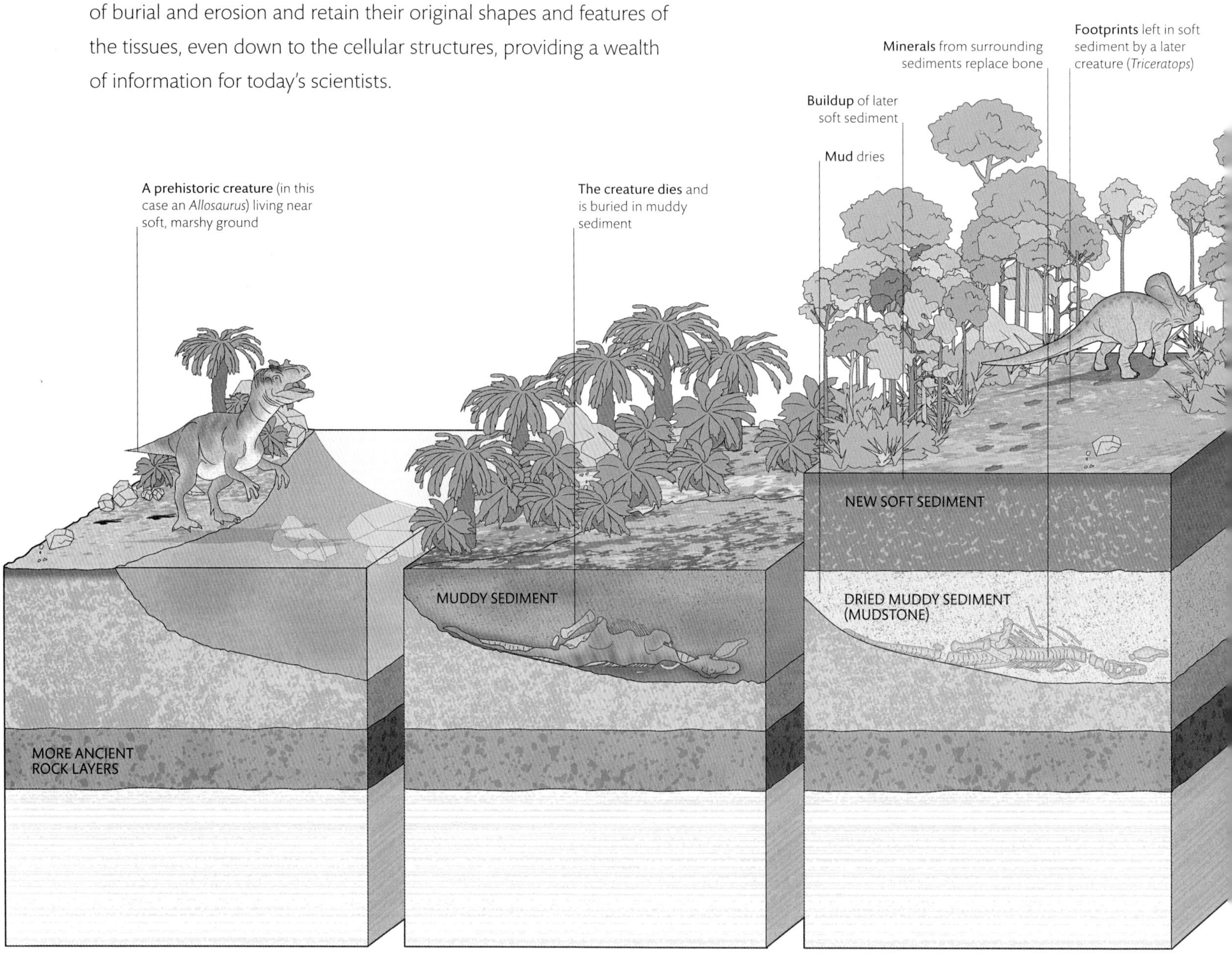

Ancient life
Creatures most likely to be preserved as fossils lived in places where there was a buildup of sediment in which they were buried after death.

Death and burial
If, after death, the creature's remains are protected from scavenging and further decay by overlying sediments for long periods of time, they may be fossilized.

Sediment buildup
Over time, soil and sediment build up and begin to turn to rock. During this process, the organic remains within the sediment are replaced by minerals.

Preserved in amber
Organisms can be buried and preserved in amber (hardened tree resin). Amber prevents the decay of the delicate structures and provides a window into the evolution of many small plants and animals that were too fragile to be fossilized in rock.

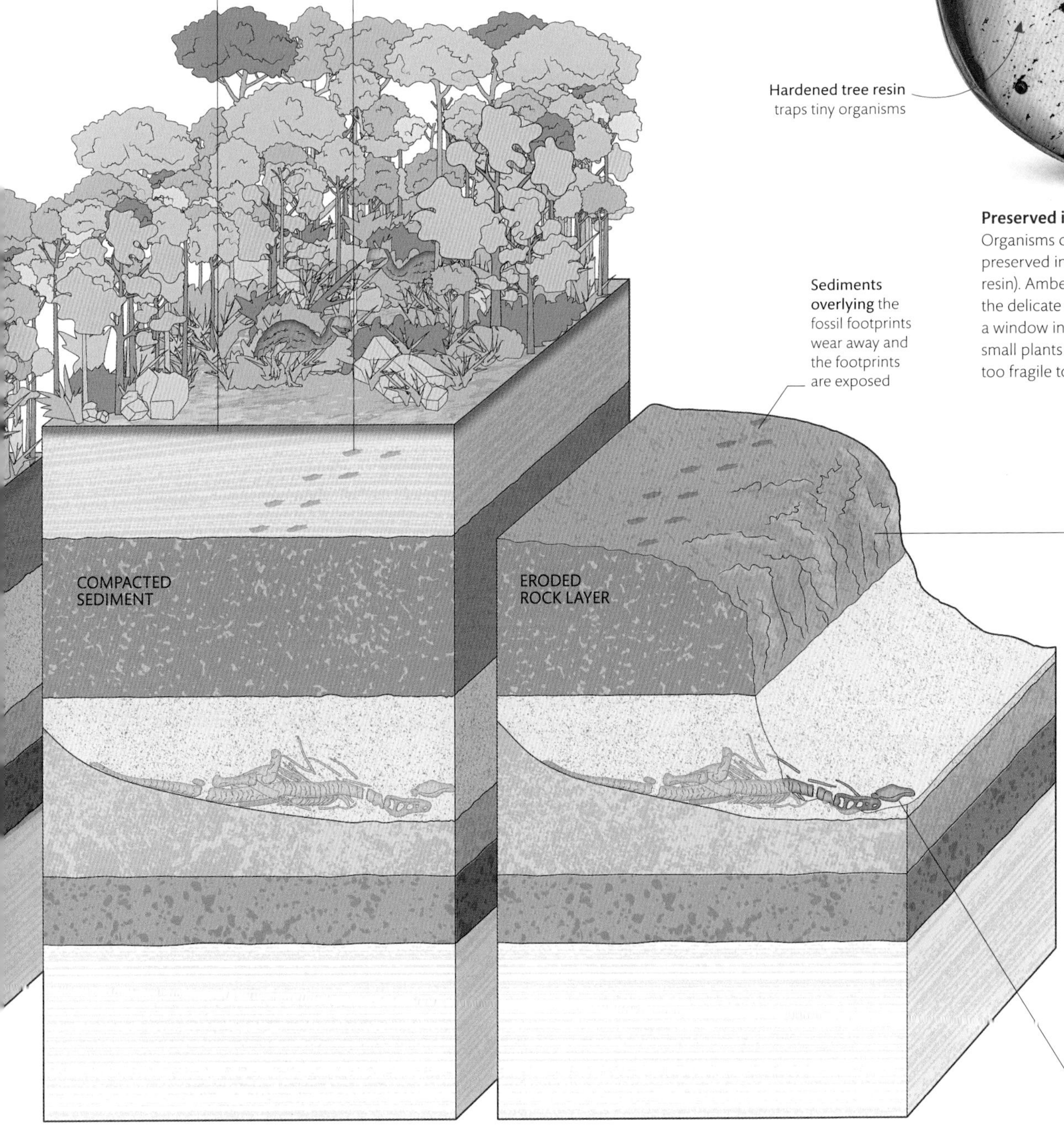

Later fossils
Fossils of later life forms or fossil footprints may be preserved in newer rock layers. Footprints are formed as impressions or casts, depending on the type of rock.

Exposure and discovery
Fossils are usually found when they are exposed at the surface, in locations where erosion has removed overlying layers of sediment and rock.

coal forms

Black seams of coal are testament to the intimate link between the living biosphere and the rocks beneath it. All living things are part of a cycle of matter that persists long after they die and decay. Most are destined to rot completely into minerals that nourish new growth in other organisms. But under specific circumstances, decomposition slows almost to a halt. By 360 million years ago, plants had evolved a woody material called lignin that helped them grow into tall forests. However, decomposers such as fungi had not yet evolved the ability to break the lignin down. As a result, the woody trunks became compacted in the ground and—over many thousands of years—turned into carbon-rich coal.

Fossilized tree bark
Coal-forming swamp forests were composed mainly of spore-producing trees, such as this *Lepidodendron*—a relative of today's ferns.

Seam of coal

Coal is a type of sedimentary rock made almost entirely from black carbon, and most was deposited at least 100 million years before the time of the dinosaurs, during a period appropriately called the Carboniferous. Seams—like this one seen above ground along a beach in New South Wales, Australia—are the organic remains of Carboniferous swamp forests.

LONG CARBON CYCLE

Coal forms when dead vegetation becomes compacted. It is rotted down by bacteria into peat. This is buried and warmed under accumulated sediment, squeezing water and gases from the remains, producing first lignite and then coal. The process concentrates carbon that can only then be liberated by combustion into carbon dioxide (CO_2); faster, human-caused forms of combustion raisesCO_2 levels.

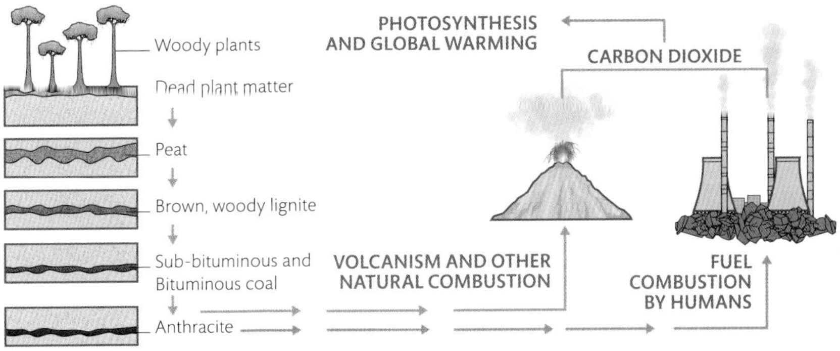

oxygen-fueled giants

Once microbes flooded the air with oxygen more than 2 billion years ago (see p.252), atmospheric levels of oxygen remained fairly steady. But during the Carboniferous, oxygen levels nearly doubled. That this happened at a time when coal seams were building up is no coincidence. As carbon was locked away underground, less of it combined chemically with oxygen, so oxygen gas built up in the air. Such a highly charged atmosphere meant more wildfires and the evolution of huge arthropods, such as millipedes the length of cars.

Fire consumes oxygen and emits ash, but also releases water vapor and carbon dioxide

Spontaneous forest fires
Prehistoric rocks show that forest blazes were common in the Carboniferous. Today, some plants found in hot, dry climates are fire resistant, while others actually need the fires for germination.

ATMOSPHERIC OXYGEN

Photosynthesis generates oxygen, but other processes consume it, such as fire, and respiration by living things. A balance between the two has helped stabilize oxygen levels for much of prehistory. But a peak happened in the Carboniferous when woody trees evolved that initially resisted attack by decomposers. Less decomposer respiration meant that less oxygen was consumed, so levels rose.

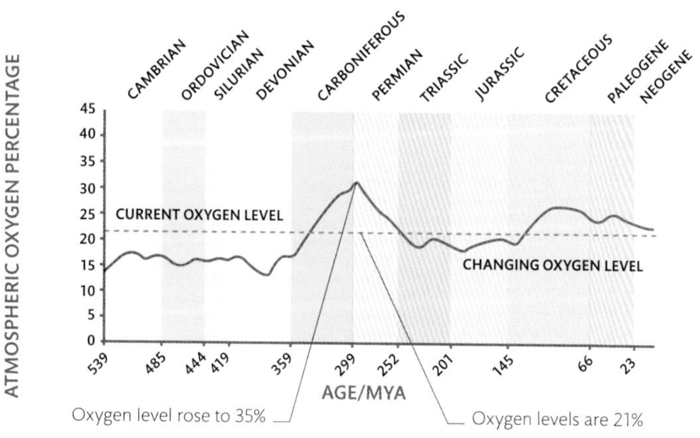

Largest flying insect
Insects rely on oxygen seeping directly into their tissues through perforations in their body wall. The high oxygen levels of the Carboniferous could diffuse deeper into tissues, so could satisfy the demands of bigger bodies. The result was giants like this dragonfly-like insect, *Meganeura*, which had a 28-in (70-cm) wingspan.

walking on land

Once algae and plants had colonized the land from the oceans where life had begun, animals followed—the vegetation providing shelter and food. The first animals to venture out of water probably did so fleetingly, perhaps at night, just as some marine snails do today. Prehistoric footprints indicate that shrimplike animals made the move from water to land more than half a billion years ago. Their armor-plated exoskeleton doubtless helped shield them from the sun's drying rays. But it took millions of years of evolution to turn the tentative pioneers into permanent air-breathing insects and spiders. By the time of the coal-producing swamp forests (see pp.270–271), backboned animals (vertebrates) had shifted landward, and a fleshy finned fish had evolved into primitive walking tetrapods.

Walking vertebrate
Living between 295 and 272 million years ago, *Seymouria*, an animal measuring 2 ft (60 cm) in length that resembled a giant modern day salamander, had limbs strong enough to lift its body off the ground. Like all land-living backboned animals, it evolved from a group of fish whose thick fins had bony supports that would become the digits of walking feet.

Reinvasion of the land
Today many groups of fish continue to evolve ways of conquering the land. Descended from oceanic gobies, gold-spotted mudskippers (*Periophthalmus chrysospilos*) have forward-positioned pectoral (chest) fins that are strong enough to haul them across mudflats.

MULTIPLE LAND INVASIONS

During the history of their evolution, animals have conquered the land several times—in independent groups. By comparing DNA of living animals to date their ancestry, and studying fossil records, scientists estimate when different groups of land animals developed from their aquatic ancestors. The oldest land animals—jointed-legged arthropods (including insects, crustaceans, and spiders), and nematode worms—had already filled the prehistoric forests by the time the first four-legged vertebrates, or tetrapods, walked on land. All of these pioneers started out being partly aquatic, but eventually evolved the means to stay out of water. Snails, slugs, and earthworms turned landward much later—around the time of the dinosaurs.

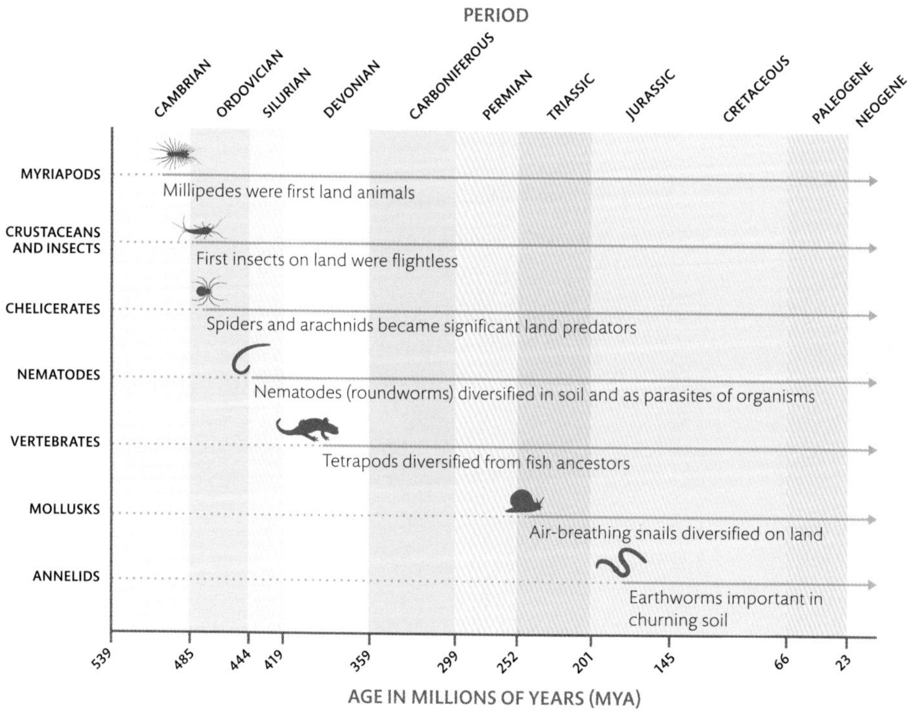

The Permian's great survivor
A distant reptilian predecessor of mammals, *Lystrosaurus* accounts for up to 95 percent of vertebrate fossils deposited after this event, and a burrowing habit may have protected it.

Evidence of extinction
The dark layers visible in Siberia's Putorana Mountains built up as lava flooded the land from volcanic fissure eruptions, in which Earth's crust fractured over a wide area. The lava solidified in layers to form a vast plateau, of which this is a small part. This enormous volcanic event came as close to eliminating complex life as at any time in Earth's history.

the great dying

Earth has been described as the "Goldilocks planet"—for most of its existence, conditions have been just right for living things to thrive and evolve. But several events have caused colossal loss of species, or mass extinctions. One of these had an extraterrestrial cause—an asteroid strike that killed the dinosaurs (see p.295)—but most events were triggered by Earth's own violent geology. The biggest happened 250 million years ago at the end of the Permian—before the dinosaurs. The climate change that resulted from huge volcanic eruptions wiped out more than three-quarters of all species. But there are always winners as well as losers, and from the survivors evolved new kinds of organisms that took over the biosphere.

PERMIAN ERUPTION

The end of the Permian is marked by an event in which a huge plume of molten rock rose up beneath Siberia, causing massive outflows of lava, which solidified into plateaus of dark, fine-grained rock called traps. The initial event resulted in global warming from trapped greenhouse gases, acid rain, and ash fallout, which washed into the oceans.

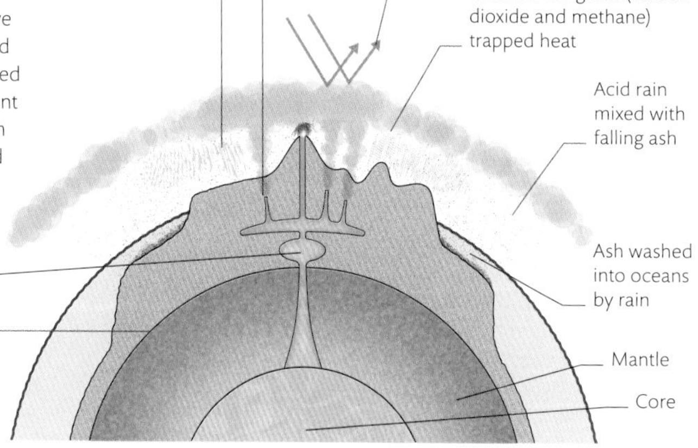

EFFECTS OF SIBERIAN TRAP ERUPTION

Blind trilobite had eyeless cephalon

CONOCORYPHE

Central axis between distinct pleural lobes

ELRATHIA

Long, crescent-shaped eyes

OLENELLUS

Body was up to 20 in (50 cm) long, from head to tail

PARADOXIDES

Rise of the trilobites

Trilobites originated early in the Paleozoic era as part of the Cambrian explosion about 539–520 million years ago (see pp.256–257), and their durable exoskeletons form a major component of its fossil record. Trilobites were very successful and reached a peak in their diversity toward the end of the Cambrian (around 485 million years ago), and into the Ordovician.

Median face spine

MOROCCONITES

Spiny tail, or pygidium, ornamentation

ZLICHOVASPIS

Large, slightly raised eye

HUNTONIATONIA

Pair of recurving horns

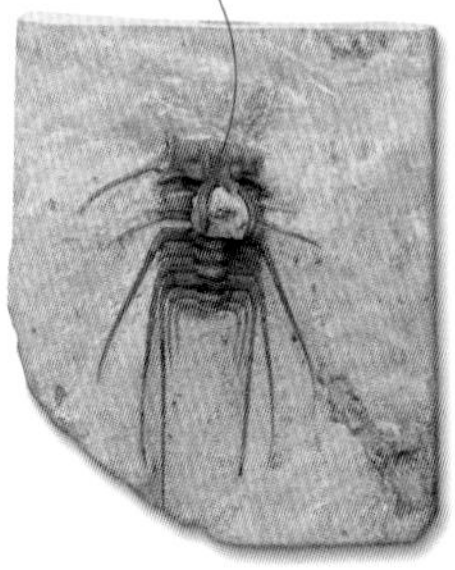

DICRANURUS

Decline of the trilobites

Throughout the Silurian and Devonian periods that followed the late Ordovician mass extinction, the number of trilobite families continued to decrease, although many groups were still lavishly ornamented. Only one order of trilobites, the Proetida, survived past the Devonian, and finally went extinct in the Permian mass extinction (see pp.276–277).

trilobite diversity

Trilobites were marine arthropods (animals with an exoskeleton; segmented body; and paired, jointed appendages), part of the group that includes today's crustaceans, arachnids, and insects. The trilobite body was divided into three segments: a cephalon (head), thorax, and pygidium (tail). The mouth opened on the underside, below the cephalon. However, the name trilobite, meaning "three-lobed," is due to the thorax being divided into three parts—a central axial lobe, and a pleural lobe on each side, each pleural lobe bearing legs with gills. With this body plan, trilobites diversified in the oceans and lived for 270 million years, leaving a rich fossil record of more than 20,000 species.

Three-pronged trident-like horn extended from front of cephalon

ONNIA

TRIARTHRUS

SELENOPELTIS

ENCRINURUS

Turnover of groups
During the Ordovician period, new groups of trilobites with distinctive body plans continued to develop, but the number of families steadily decreased. During the late Ordovician mass extinction, about 445 million years ago, many of the groups originating in the Cambrian went extinct, but were replaced by those that had evolved in the Ordovician.

Trident trilobite
Although trilobites declined during the Devonian, some, such as this *Walliserops*, were magnificent. The function of its distinctive cephalon horn is unclear—researchers believe it may have served in sexual display, for battling rivals, or as a signal of a healthy individual—all functions served by deer antlers today.

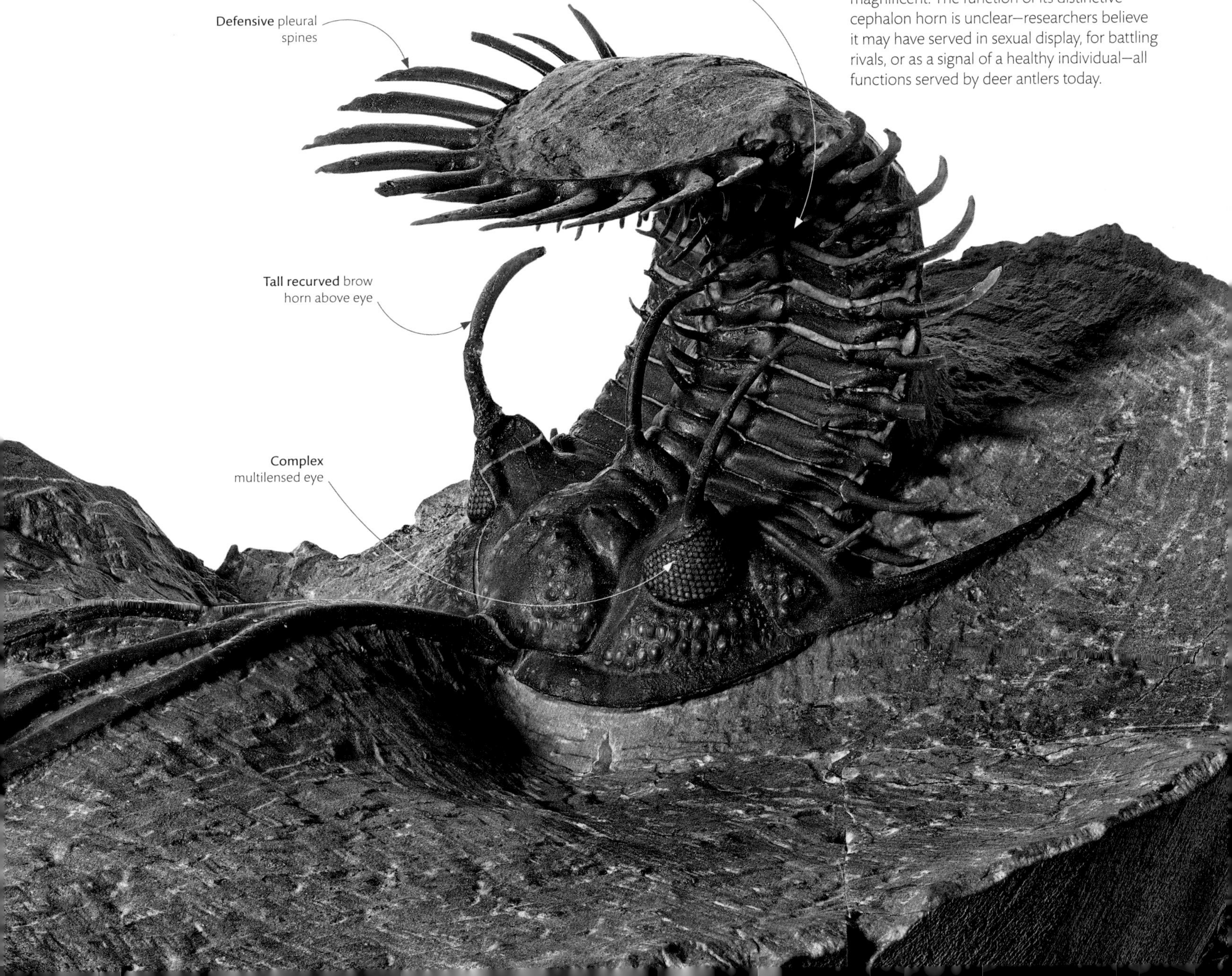

surviving drought

All organisms need water, but many have evolved ways of living in the driest habitats. As continents shift positions, land turns to desert for different reasons and places can be so far inland that rainfall scarcely reaches them. Around 250 million years ago—just before the rise of the dinosaurs—most of Earth's land was amassed into a supercontinent called Pangaea with a vast desert interior, and global temperatures peaked. Amphibians that laid eggs without shells in water declined. But reptiles, whose hard-shelled eggs developed on land, could thrive in the driest regions.

Dry interior of Mongolia
In the center of continental Asia, the Gobi Desert is in a "rain shadow" created by the Tibetan Plateau, which prevents moisture-laden winds from reaching it. With its extremes of temperature—bitterly cold winters and hot, dry summers—vegetation is sparse and only specialized desert animals survive.

Hard-shelled egg
The evolution of shelled eggs—such as this 66-million-year-old fossil—was critical in allowing reptiles to breed independently of water bodies.

ADAPTING TO ARIDITY

The aridity, or dryness, of a habitat is determined by the relative levels of rainfall and evaporation. A subhumid desert loses no more than twice the volume of its annual rainfall. But the driest desert can evaporate 200 times more water than it receives, so living things require extreme adaptations to extract it from their surroundings.

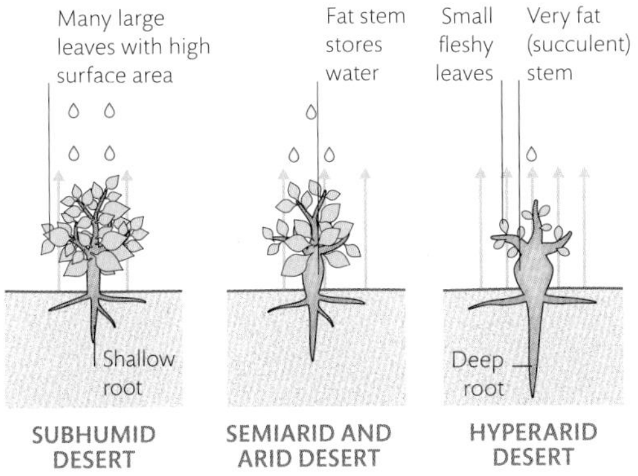

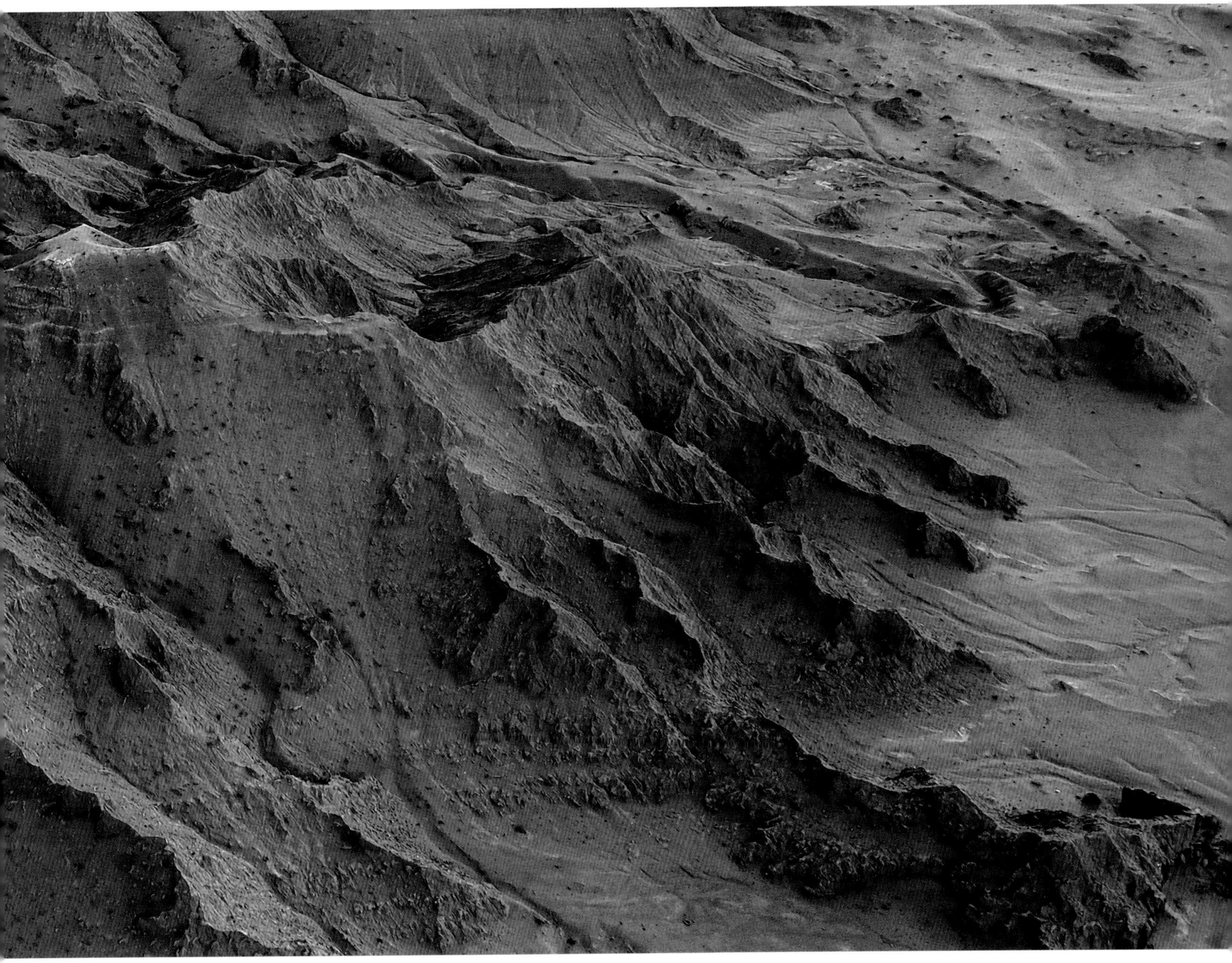

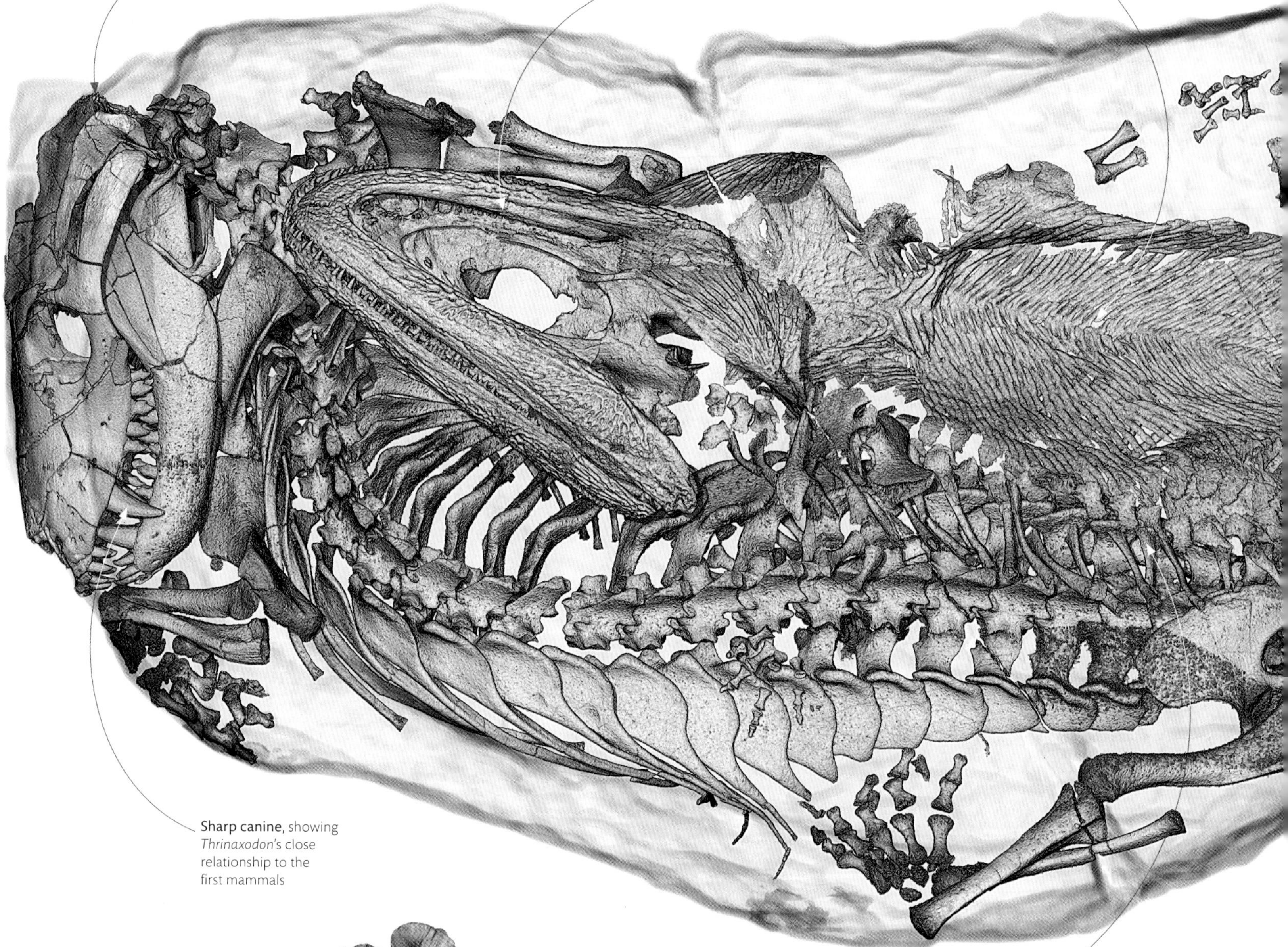

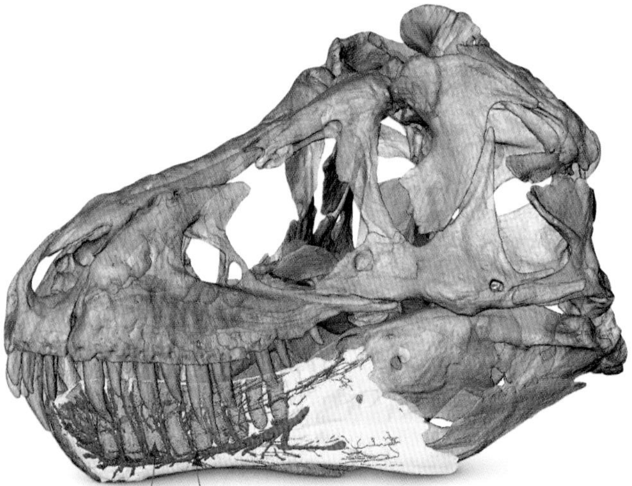

Revealing nerve pathways
This CT scan of the skull and lower jaw of a *Tyrannosaurus rex* reveals a dense network of nerves and blood vessels, providing evidence of a highly sensitive snout.

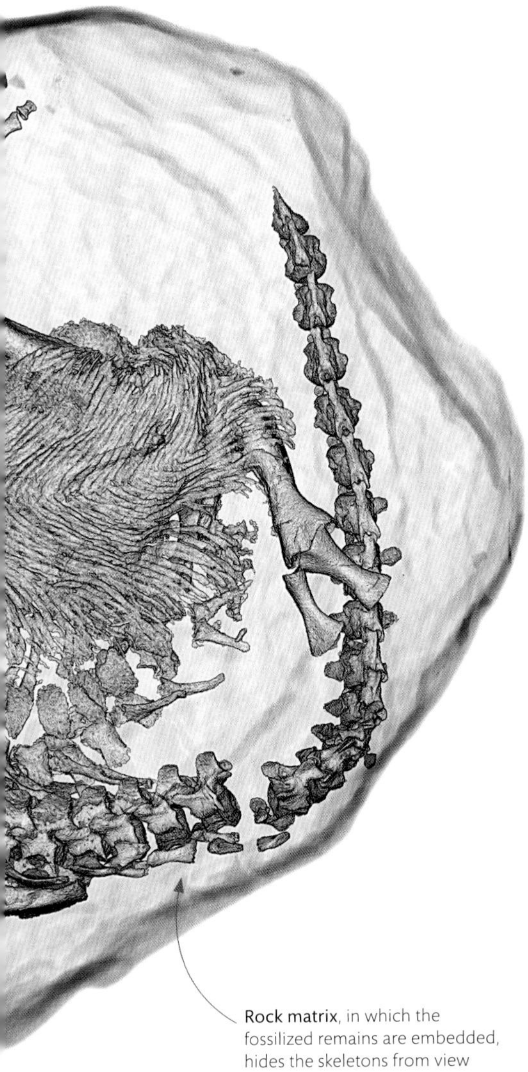

Looking inside a dinosaur
The internal structures of the fossilized skull of an *Arcovenator*, a theropod dinosaur from France, are viewed in great detail using a CT scanner.

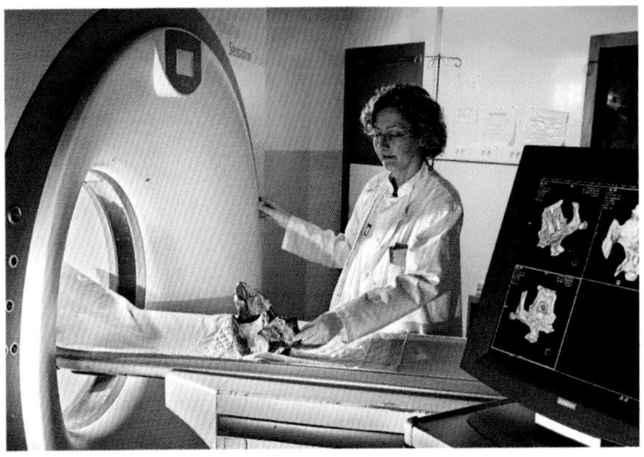

history of Earth science

scanning fossils

The external features of bones and teeth reveal only some aspects of the biology of long-extinct animals. To investigate other characteristics, scientists must look inside the skeleton, where soft tissues once existed. In the past, this could only be done by destroying parts of the fossil. Modern scanning methods avoid this problem, and also enable us to investigate fossils that cannot be removed from the rock in which they are embedded.

Rock matrix, in which the fossilized remains are embedded, hides the skeletons from view with the naked eye

Strange bedfellows
Thrinaxodon (grayish pink), a type of extinct mammal-like animal called a cynodont, and a *Broomistega* (dark gray), an amphibian, were buried together in a Triassic burrow, where the *Thrinaxodon* may have been lying dormant to avoid a summer drought. This amazing double fossil was revealed only when scientists used a high-powered synchrotron particle accelerator to scan rock that they suspected of containing a fossilized cynodont burrow.

Since the 1970s, X-ray technologies have been used by scientists to look inside fossils. High-powered X-rays penetrate the rock containing the fossil and show the different densities of the internal structures. The advent of more powerful and accessible computing in the 1980s and 1990s led to wider use of CT (computerized tomography) scanning by palaeontologists. In this technique, hundreds of X-ray images from different angles are aligned by the computer, creating a three-dimensional rendition of the interior of the fossil.

CT scanning techniques have steadily improved, allowing scientists to digitally reconstruct the internal anatomy of ancient organisms with unprecedented detail. Researchers have now visualized the brains of long-extinct animals, including dinosaurs, marine reptiles, and early mammals. By comparing the sizes of different parts of the brain, digital models can be used to understand the importance to the creature of senses such as eyesight or smell.

CT scanning also shows other structures, such as the delicate chambers of the inner ear, nerve networks, and even daily growth layers in teeth. Today, new and more powerful imaging tools—for example, the synchrotron particle accelerator—reveal ever-more detail, even down to microscopic levels.

> “Almost all the traditional problems associated with recovering fossil data from rocks can be overcome with modern 3D imaging.”

JOHN CUNNINGHAM *ET AL.* *A VIRTUAL WORLD OF PALEONTOLOGY* 2014

SOURCES OF HEAT

All animals generate some heat by their metabolism. Cold-blooded (endothermic) reptiles rely on the sun for heat, but warm-blooded (ectothermic) animals, including mammals, can produce so much heat that they can keep their body temperature high, even when surroundings are cold. They do it by burning specialized brown fat. Moreover, they can also control their body temperature day and night via a gland in their brain called the hypothalamus.

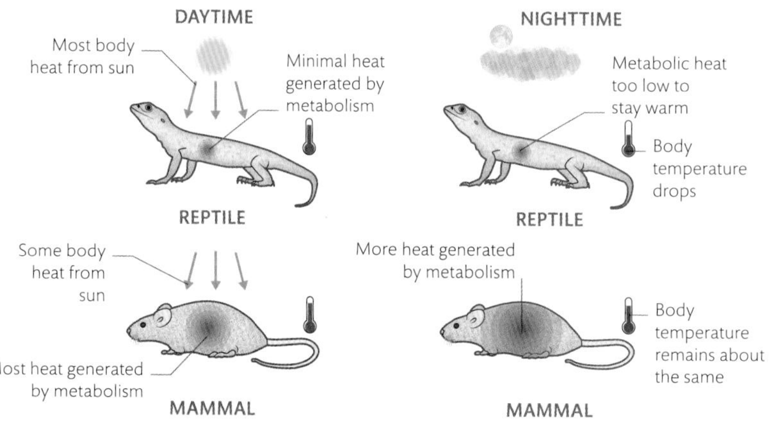

Furry fossil
The rock around the fossilized bones of this shrew-size *Eomaia*, meaning "dawn mother," show the impression of a coat of fur: clear evidence that this was a mammal. *Eomaia* lived 125 million years ago in China—during the Cretaceous peak of the dinosaurs—and was an ancient "sister" of living marsupial and placental mammals (see pp.298–299), but too early to be placed in either of these groups.

Dark halo is a carbon impression of the hairy coat that covered this animal's body

Tail vertebrae were elongated—entire tail was twice as long as the rest of the spine in a complete specimen

Hedgehogs inhabit grasslands, hedgerows, and woodland, searching for insects, beetles, and worms

Predatory mammal
Most living mammals, like this European hedgehog (*Erinaceus europaeus*), are nocturnal—suggesting that night foraging was adopted early in the group's evolution. Daytime habits emerged in later groups, such as squirrels and monkeys.

becoming warm-blooded

While cold-blooded reptiles walked the land and were warmed by the sun's rays, some of their descendants found different ways of surviving. These were the mammals, or warm-blooded animals. They adopted a more frenetic life: their sharp-toothed fossils point to them chasing scurrying insect prey, and their skeletons show evidence that they matured quickly. Some fossils even have impressions of a novel innovation: hairy skin. Their diet, rich in fat and protein, fueled a body that stayed warm, even during cold nights, which helped them avoid competition with reptiles that were mainly active by day. Insulated by a furry coat, their descendants could survive in permanently cold habitats, such as Arctic tundra, that were inhospitable to their cold-blooded cousins.

Reconstruction of *Eomaia*
Long, clawed digits and a long tail that could have helped with balance suggest that *Eomaia* was an agile small animal that might have climbed trees.

Thick, furry coat helped trap body heat

The Red Deer River Valley, incised like a wound into the prairies of Alberta, Canada, is home to some of the richest dinosaur fossil beds in the world. The striped hills provide a record of life in the late Cretaceous (78–66 million years ago) along the shores of an ancient intercontinental sea. The valley was carved by the drainage of glacial lakes after the last ice age (11,700 years ago), which exposed the soft Cretaceous sediments.

spotlight Red Deer River Valley

This valley is home to numerous First Nations peoples, who were the first fossil collectors here. An area called the badlands produces the *iniskim*—buffalo-shaped stones sacred to the Niitsitapi (Blackfoot), formed from ammonite fossils.

In the 1880s geological surveys recognized the incredibly rich geological resources of the valley, including thick coal seams, as well as an abundance of dinosaur fossils. Some areas in the valley, like Dinosaur Provincial Park or the Horsethief Canyon, are so rich in fossils that paleontologists walk over some fossil skeletons to reach the most scientifically important examples. The fossils found here reveal some of the best known dinosaur ecosystems, spanning 12 million years in time, and include familiar groups like the iconic tyrannosaurs, duck-billed hadrosaurs, frilled ceratopsians, and fearsome dromaeosaurs. Living alongside these dinosaurs in the lush subtropical climate was a diverse array of fish, lizards, turtles, crocodiles, and early mammals.

Fossil is curled into the "death pose," common among dinosaur skeletons found in the valley

Painted valley
The colors of the hills in the Horsethief Canyon are caused by variations in rock type and reflect different environments. The mudstones (brown), sandstones (white), ironstones (orange), and coal seams (black) are home to millions of fossils, giving paleontologists a clear picture of the ecosystems of the Late Cretaceous.

THE ORIGINS OF POWERED FLIGHT

The first animals that took to the skies were insects. Their wings evolved as flap-like extensions of their exoskeleton. Backboned fliers adapted their forelimbs as wings. Birds then used a fringe of stiff flight feathers as their wing blade. Other vertebrates used a stretch of skin instead—between the fingers of bats, or between finger and ankle of pterosaurs.

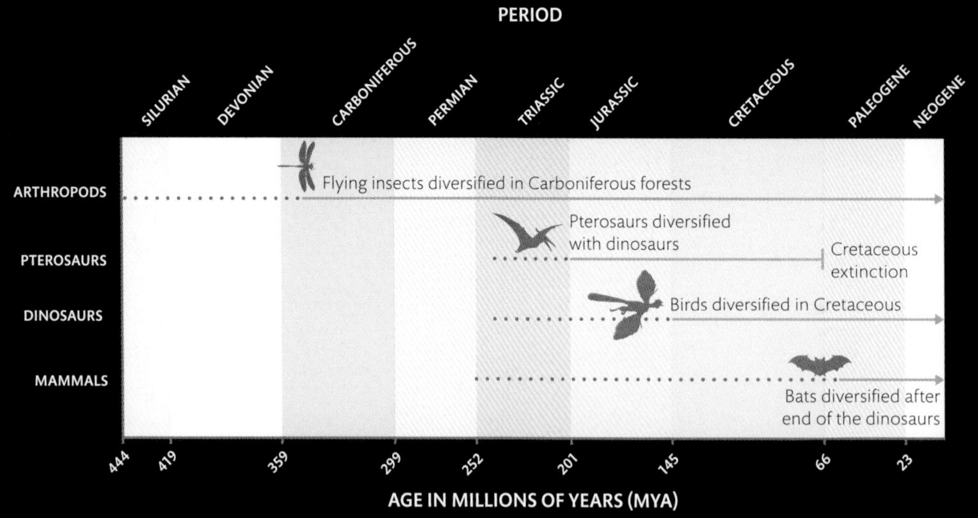

taking to the air

Many living things break free from gravity and rise into the atmosphere, which becomes an aerial highway for crossing distances quickly, largely unhindered by friction, as well as a potential source of airborne food. Plants and fungi scatter seeds, pollen, and spores, while flying animals evolved wings—airfoils that are sufficiently broad to generate lift, but thin enough to keep them lightweight. Many animals—such as flying squirrels—use static wings just to glide (see p.299). But other animals use flapping wings to generate forward thrust—which allows them to exploit the air to a much greater extent; a few birds today can stay airborne for months at a time.

Jurassic dragonfly
At the time of the first flying birds, insects like this fossilized *Cymatophlebia* dragonfly had been airborne for 200 million years.

Wings were formed from the exoskeleton, quite separately from their limbs

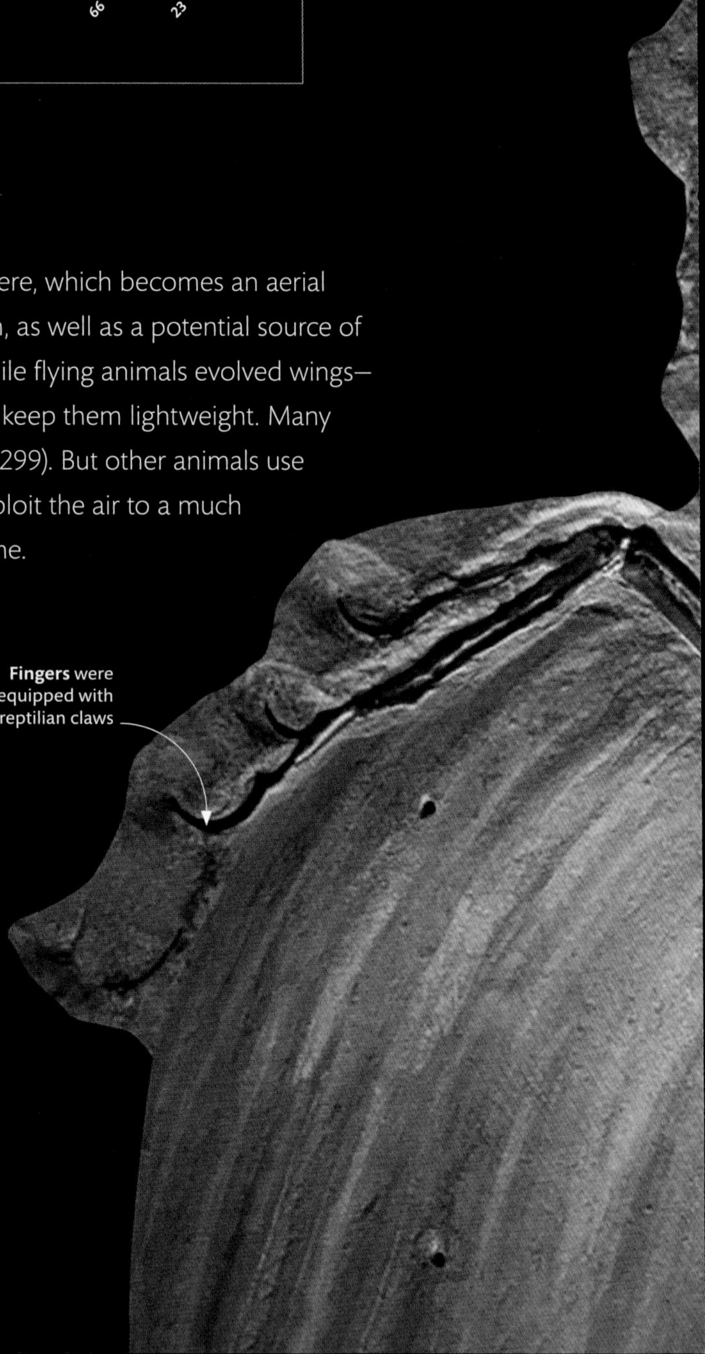

Fingers were equipped with reptilian claws

Earliest birdlike animal

All flying animals evolved from land-bound ancestors. Birds descended from two-legged predatory dinosaurs. But *Archaeopteryx*, one of the oldest flying dinosaurs, dating back 150 million years, shows transitional features. It had reptilian teeth and tail, but avian, or birdlike, beak and feathers. Scientists are not certain if it flapped its wings or was a glider.

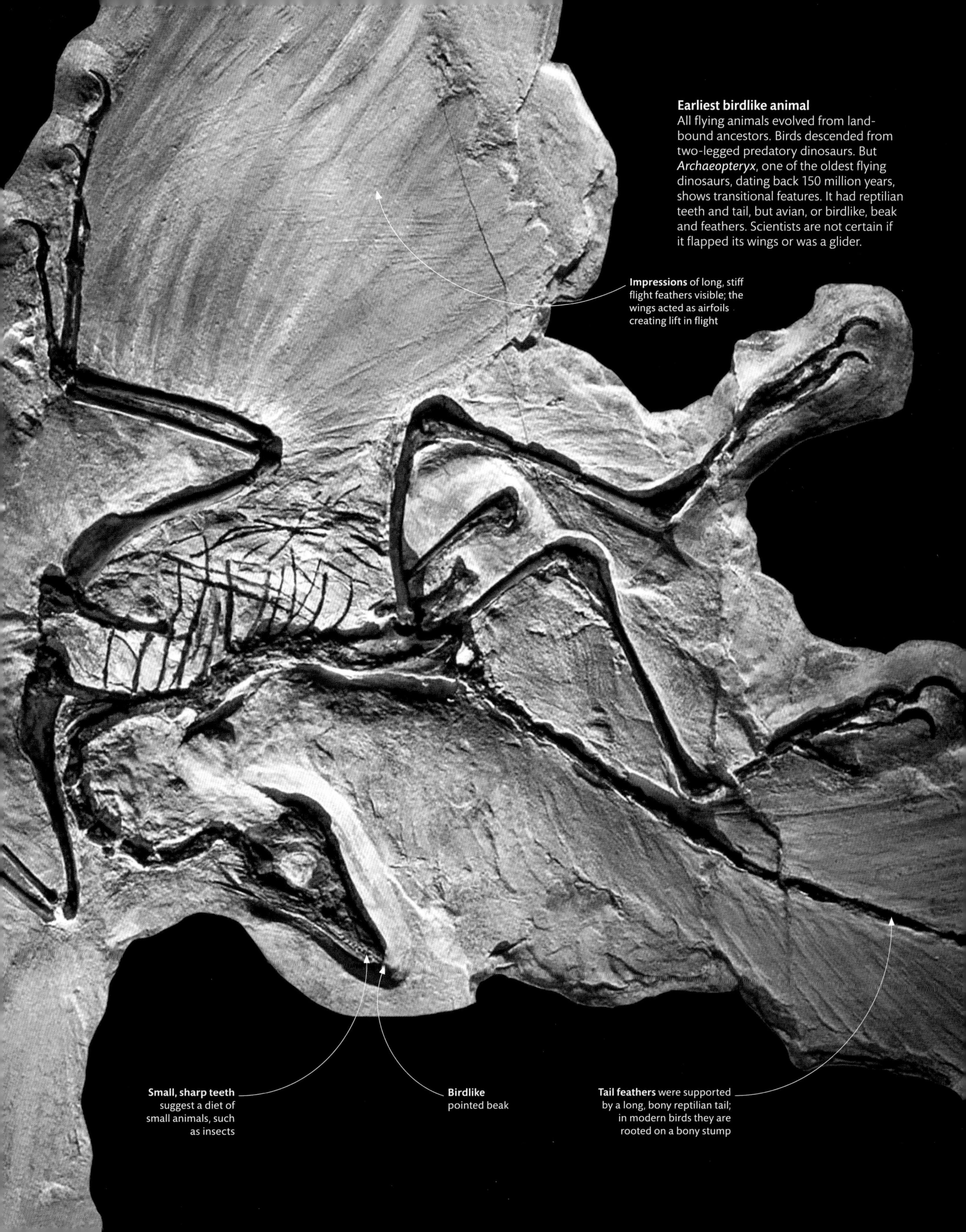

giants of the land

Dinosaurs lived in the Mesozoic era, c. 252–66 million years ago, and included the largest animals that ever lived on land. Animals grow big by converting the food they eat into flesh, and giants have huge appetites; dinosaurs probably managed this by combining appetite with efficient physiology. Their breathing relied on a system of supplementary air sacs that helped their lungs extract more oxygen from air. In today's birds—descendants of the dinosaurs (see p.288)—air sacs lighten the body, which helps with flight. In dinosaurs, the air sacs eased muscular effort, which helped with walking. An oxygen-enriched fast metabolism also meant speedy growth: estimates suggest that these giants could have gained 2 tons in body weight each year.

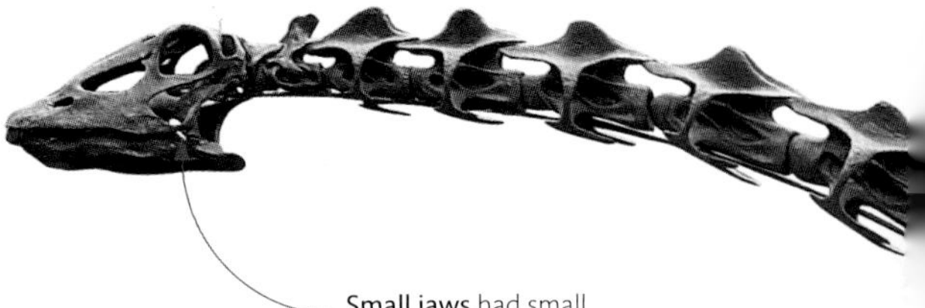

Small jaws had small peg-like teeth that could crop leaves, but not chew them

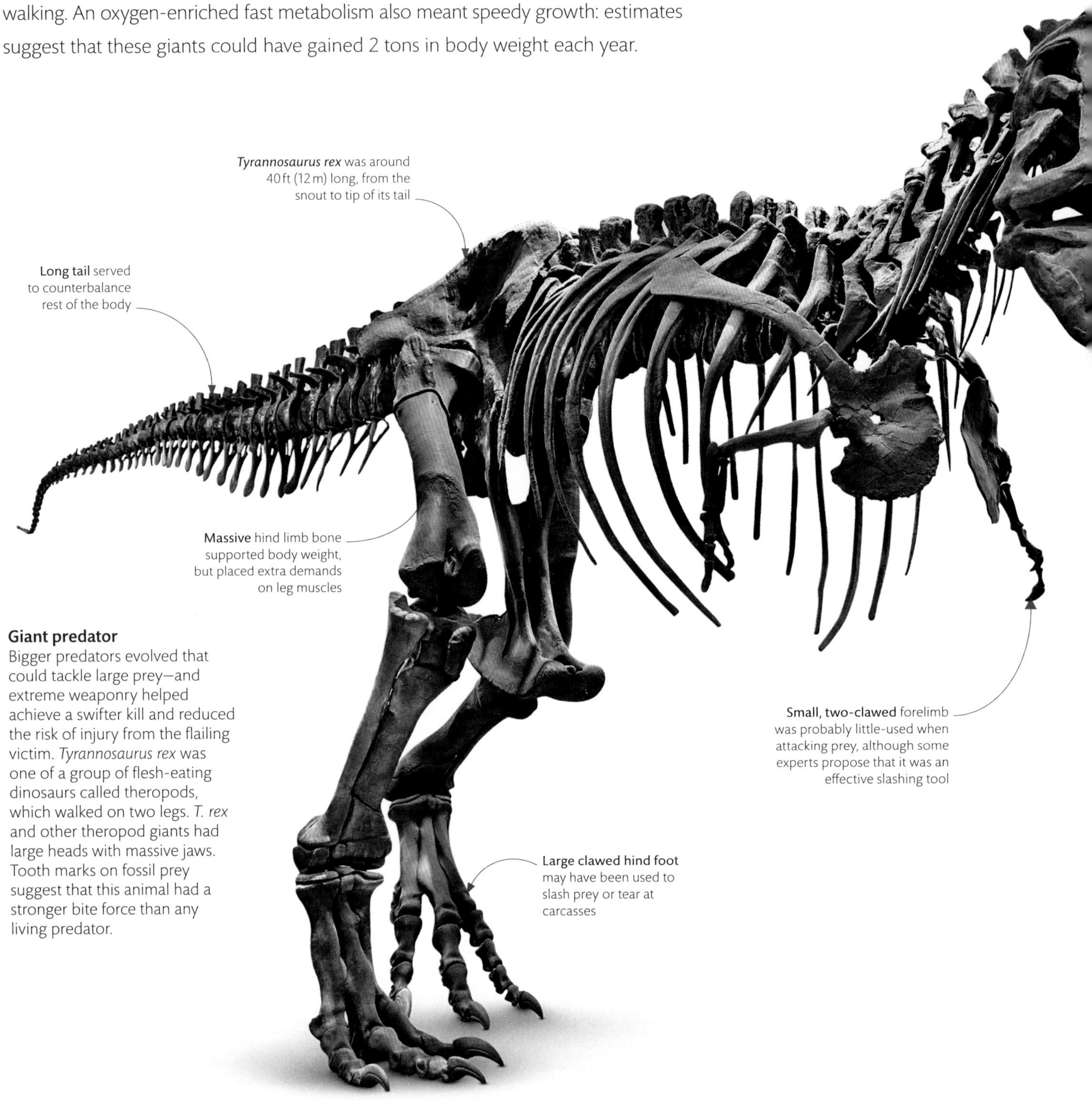

Tyrannosaurus rex was around 40 ft (12 m) long, from the snout to tip of its tail

Long tail served to counterbalance rest of the body

Massive hind limb bone supported body weight, but placed extra demands on leg muscles

Small, two-clawed forelimb was probably little-used when attacking prey, although some experts propose that it was an effective slashing tool

Large clawed hind foot may have been used to slash prey or tear at carcasses

Giant predator
Bigger predators evolved that could tackle large prey—and extreme weaponry helped achieve a swifter kill and reduced the risk of injury from the flailing victim. *Tyrannosaurus rex* was one of a group of flesh-eating dinosaurs called theropods, which walked on two legs. *T. rex* and other theropod giants had large heads with massive jaws. Tooth marks on fossil prey suggest that this animal had a stronger bite force than any living predator.

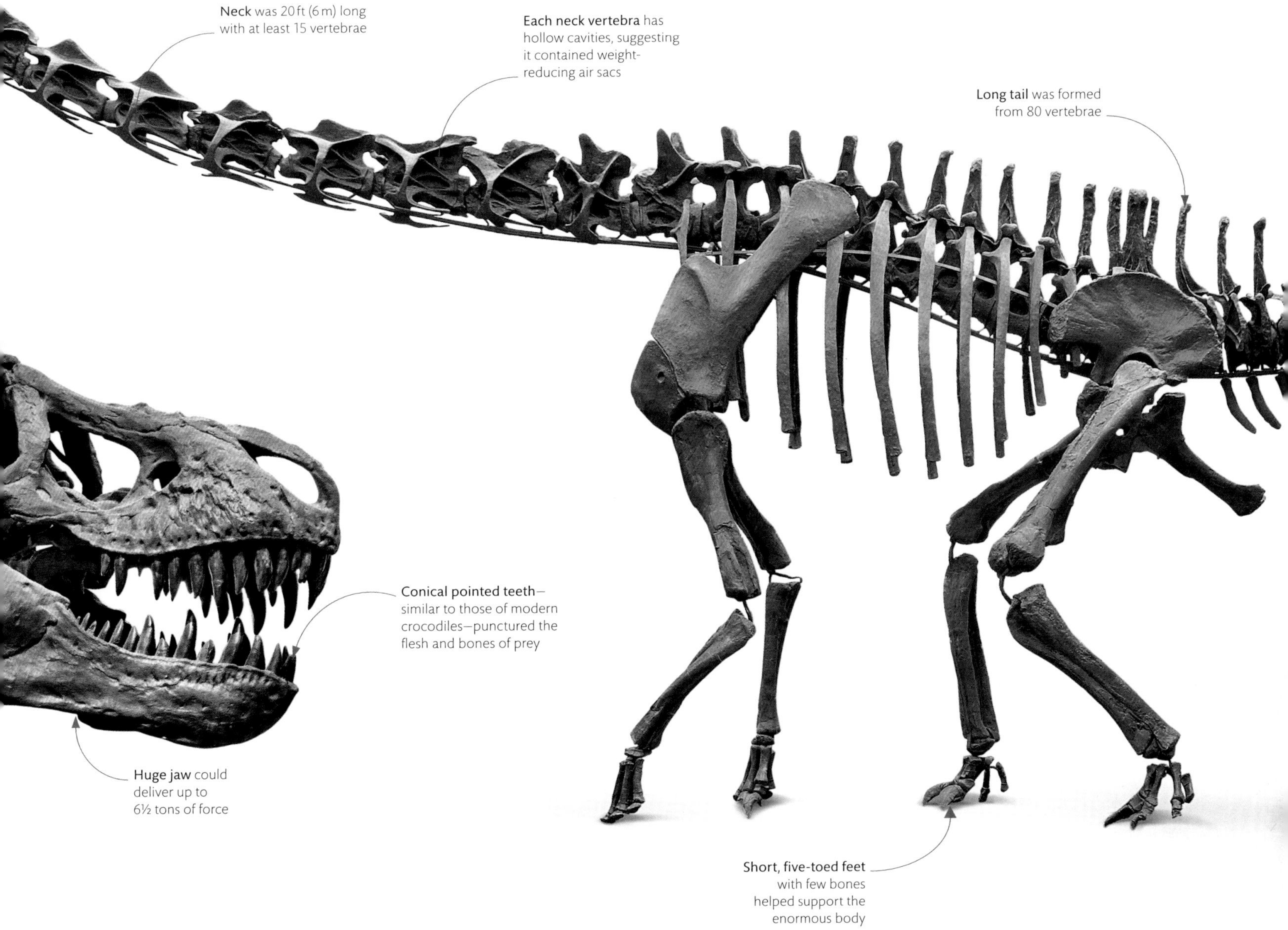

LIFESTYLE FROM FOOTPRINTS

Fossilized trackways can be used to estimate the walking speed of dinosaurs. By measuring the distance between successive prints of forefeet and hind feet, scientists can estimate the length of an animal's stride. When this is combined with the shoulder-to-hip length, known from excavated skeletons, scientists can also calculate the dinosaur's likely speed. The tracks of a large sauropod called *Titanopodus* suggest a slow, ponderous gait—less than 3 mph (5 km/h)—which is consistent with that of a large herbivore that spent a long time digesting food.

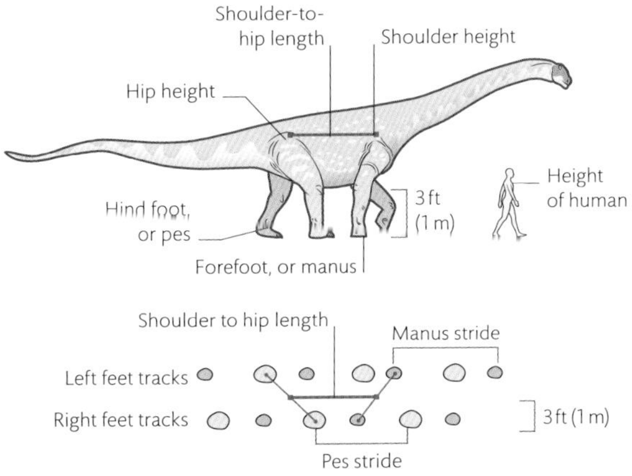

CALCULATING *TITANOPODUS* STRIDE LENGTH

Giant herbivore

The biggest dinosaurs known were herbivorous, longnecked sauropods, such as *Diplodocus*, which averaged 90 ft (27 m) in length. Their teeth indicate that they swallowed mouthfuls of leaves whole, then relied on an enormous gut to digest food over a long time. Their long neck could sweep their head through a wide arc, enabling them to browse a wide area while remaining in one place.

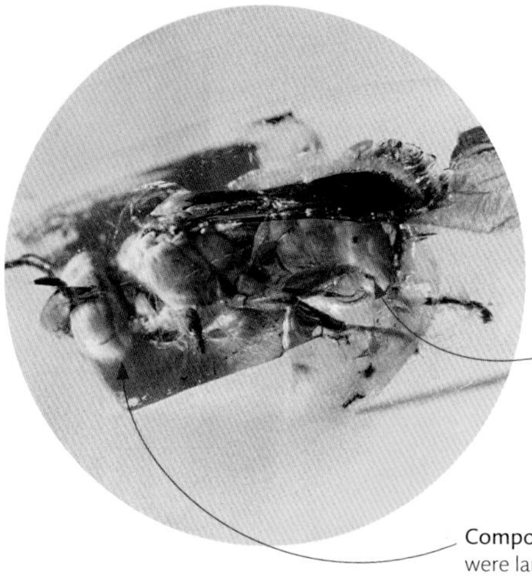

Fossilized bee
The groups of insects that rose to prominence after the demise of the dinosaurs included bees that diversified alongside flowering plants. This bee, from around 50 million years ago, is embedded in Baltic amber—the solidified sticky resin of trees that trapped many small prehistoric animals, including flying insects.

Hair-fringed cavity on hind leg was used to carry pollen

Compound eyes were larger and likely had more color receptors than those of beetles

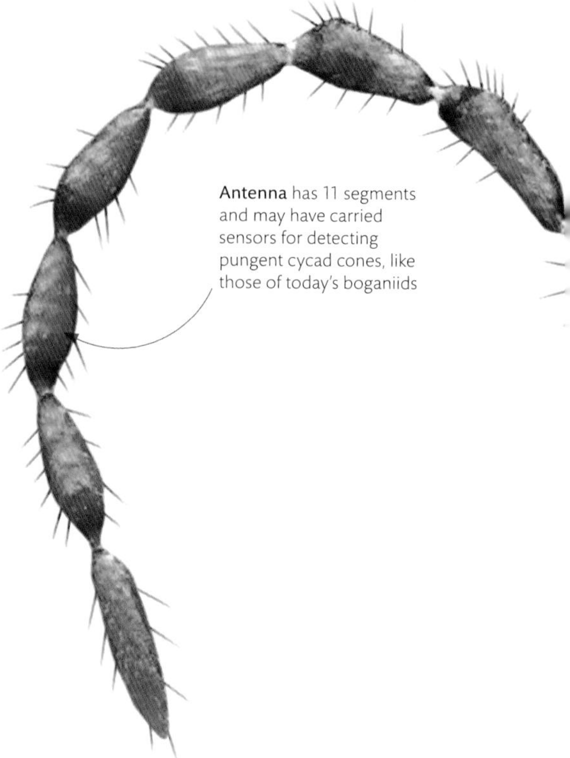

Antenna has 11 segments and may have carried sensors for detecting pungent cycad cones, like those of today's boganiids

THE EARLY POLLINATORS

The Cretaceous (145–66 million years ago) is best known as the final period of the dinosaurs' reign. But it was also a time of major change in vegetation and insects: cone-bearing seed plants, such as cycads, declined as flowering plants diversified, and with that, their pollinators changed, too. The cone-bearers had relied on beetles and flies, and the first flowers were probably pollinated by them, too, but the showy, sweet-scented flowers later attracted hymenopteran insects, such as bees, and supplied nectar for their specially adapted lapping mouthparts.

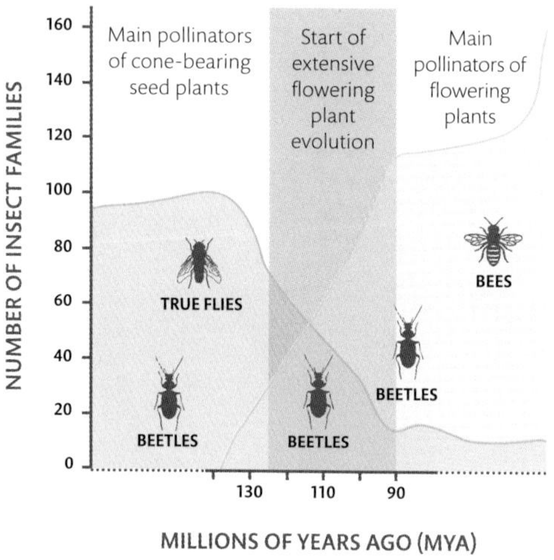

CRETACEOUS POLLINATORS

THE FIRST FLOWERS

By comparing flowers from a range of living plant families with fossils, biologists have reconstructed the most likely appearance of an early Cretaceous flower. The robust, thickly petaled bloom was similar to a magnolia. Like magnolias—and the cycads that dominated before—it was probably pollinated by beetles with poor color vision and chewing mouthparts. Pollen, nectar, and parts of the flower may also have been food sources for the insects.

RECONSTRUCTION OF AN ANCESTRAL FLOWER

Cycad pollinator
This *Cretoparacucujus* beetle from around 100 million years ago is so well-preserved in fossilized amber that it can be confidently classified in the family of boganiid beetles, which still pollinate cycads today. And, tellingly, cycad pollen grains were found in the same piece of amber. Cycads produce a large amount of pollen—some is sacrificed to feed the beetles, and some is carried by beetles to fertilize the plant's neighbors.

insect pollinators

Plants have had varying fortunes with the animal life around them. Herbivores drove many plants to make unpalatable chemicals, while animals responded by evolving ways to detoxify them. But during the time of the dinosaurs, insects evolved a mutually beneficial relationship with seed plants: insects carried pollen for the plant's sexual reproduction, while plants rewarded the carriers with a meal. It began with cone-bearing cycads—distant cousins of wind-pollinated conifers (see p.309)—but the arrangement was so successful that it changed the face of the living planet. A new kind of seed plant evolved, with flowers instead of cones, and Earth blossomed.

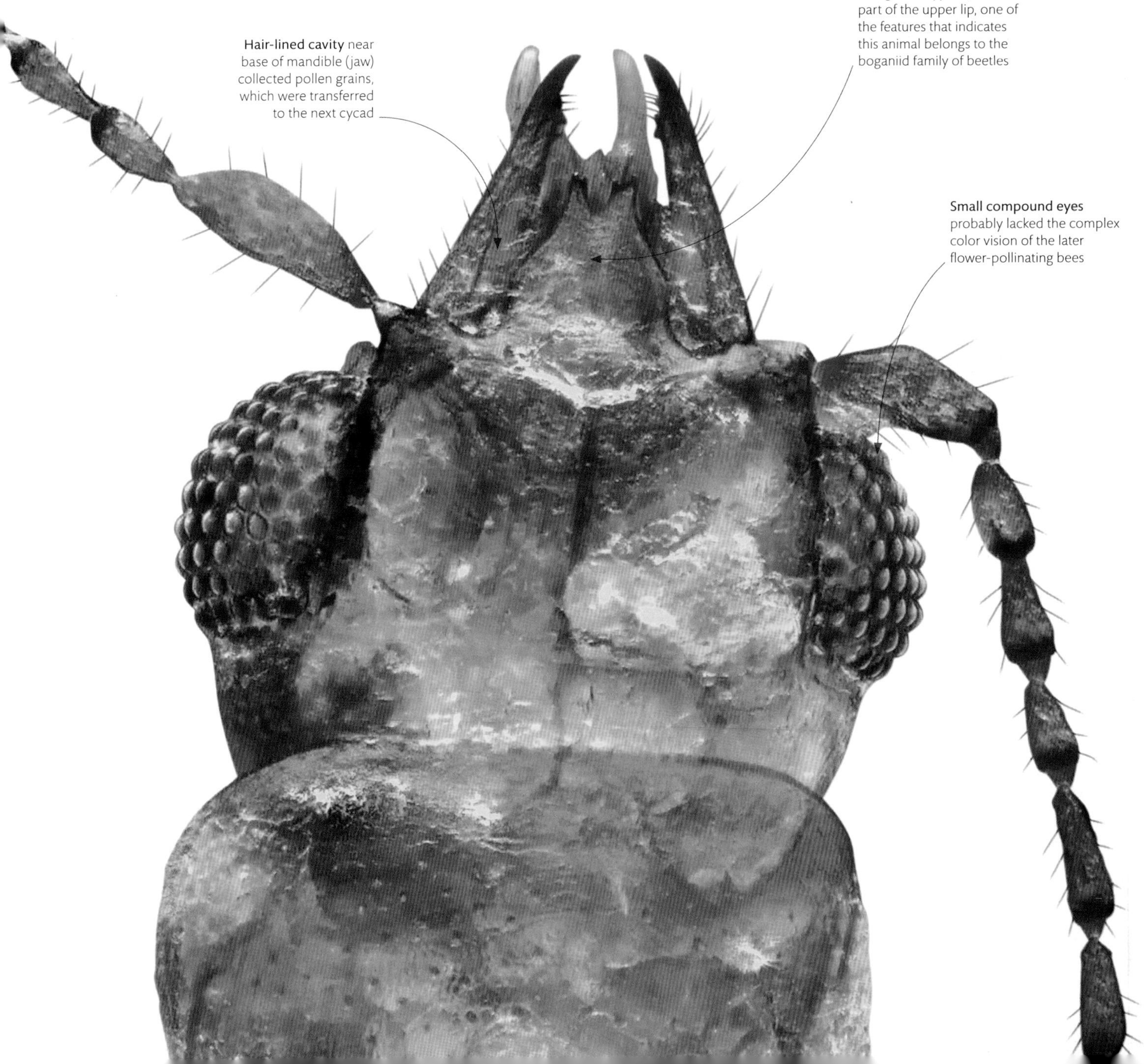

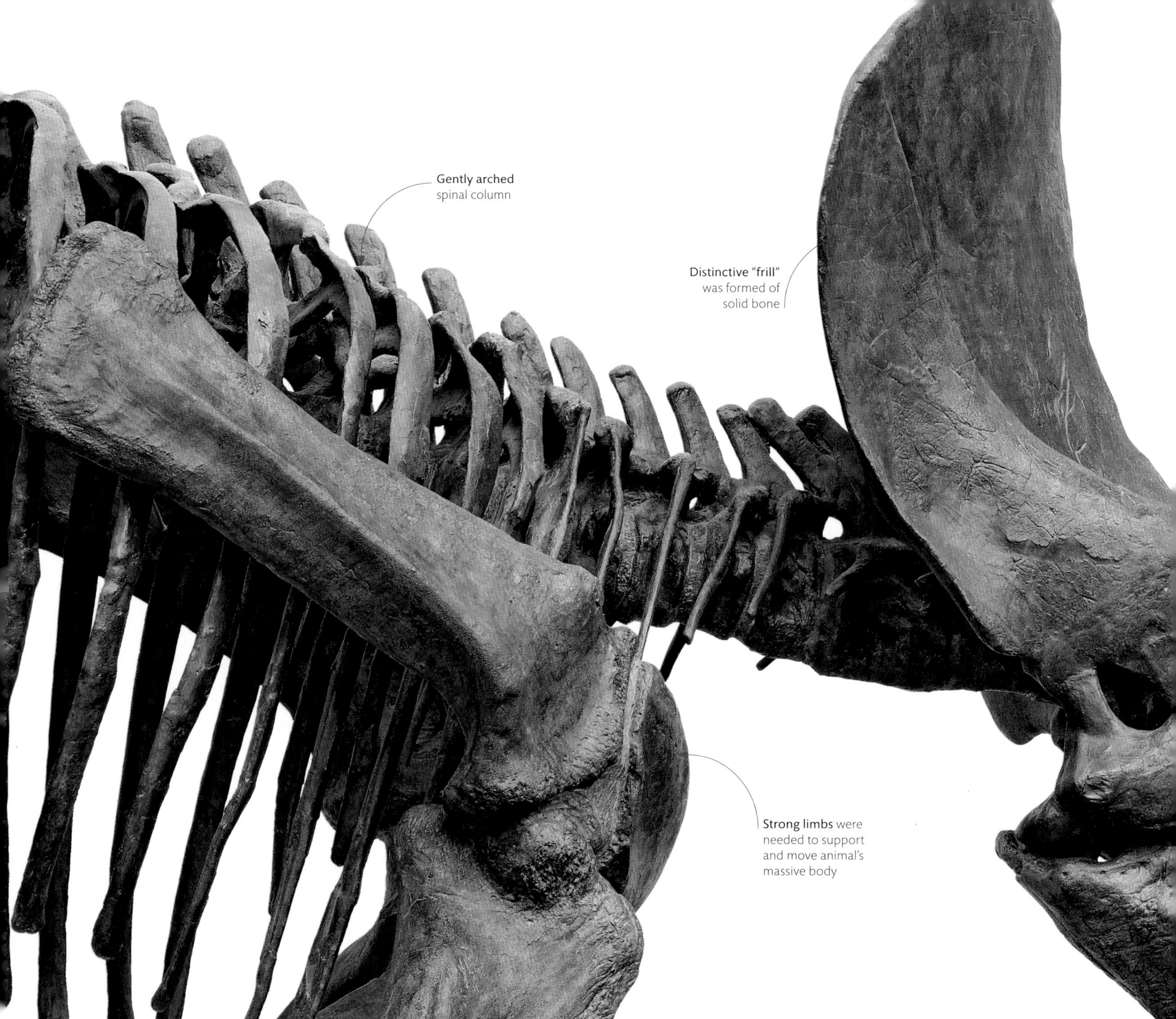

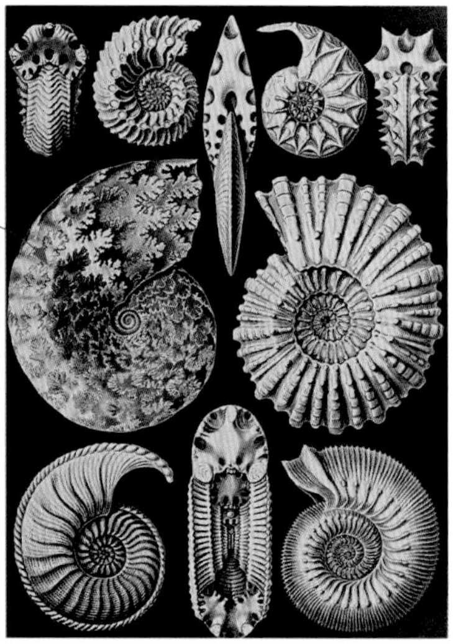

Fossilized spiral shell once enclosed a soft tentacled body

Lost mollusks
The end-of-Cretaceous extinction was as devastating for marine life as it was to land species—and many groups that lived alongside the dinosaurs disappeared, too. Among them were ammonites, the spiral-shelled relatives of modern squid and octopus.

The last dinosaurs

By the end of the Cretaceous period Earth was rich with dinosaurs, such as the three-horned *Triceratops*. But their long, successful reign was already being threatened by massive volcanic eruptions in Asia—the so-called Deccan Traps—that spewed toxic fumes. The asteroid impact was their coup de grâce: an abrupt break in the fossil record suggests that within millennia, it had killed them all. *Triceratops* had an array of effective defensive and feeding adaptations, but these did not prepare it for this catastrophe.

ASTEROID EVIDENCE

In 1978, geophysicists uncovered anomalies in rocks that suggested the Gulf of Mexico had been struck by a massive asteroid. The impact left a crater 125 miles (200 km) in width centered near Chicxulub Pueblo at the northern tip of what is now the Yucatán peninsula. A thin band of iridium, an element common in asteroids, was then discovered in end-of-Cretaceous rocks all around the world—showing how the smashed remains of the asteroid were scattered far and wide, having a truly global effect.

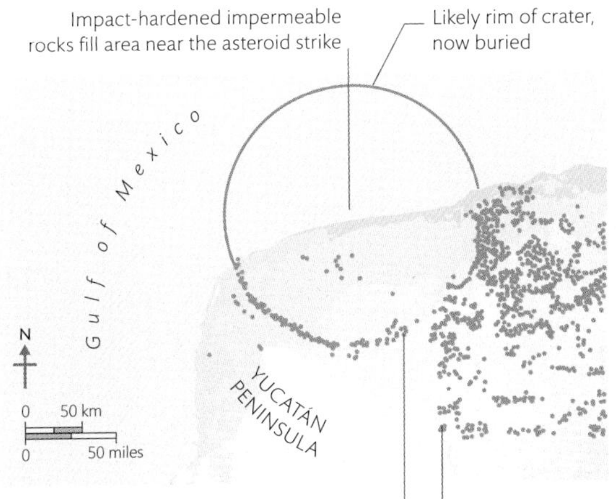

THE CHICXULUB CRATER

death of the dinosaurs

In the aftermath of its creation, Earth was bombarded with asteroids. This coincided with the period of life's single-celled origins, but life's birthplace on the ocean floor possibly helped shield it from harm. Since then, most asteroids striking the planet have been too small to have a lasting influence on the biosphere, so life has been free to evolve into extraordinary complexity. But around 66 million years ago Earth was hit by an asteroid that changed the course of evolution. It scattered dust and soot into the atmosphere and plunged Earth into a dark winter that lasted for thousands of years, driving three-quarters of plant and animal species to extinction—including the giant dinosaurs.

Insects
More than 20,000 insect fossils have been recovered from the Messel Pit, representing all major groups of insects living today. Delicate features such as wings, antennae, and eyes have been preserved. In some specimens the light-reflecting nanostructures of their exoskeletons are still evident, resulting not only in iridescent fossils, but also allowing scientists to reconstruct the vibrant colors of these ancient insects.

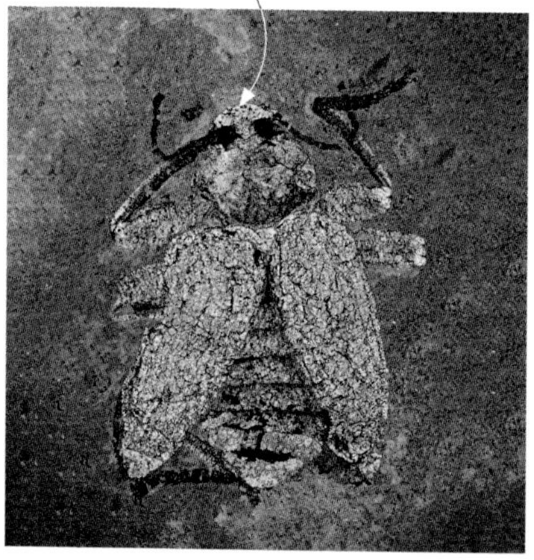

GROUND BEETLE
Ceropria messelense

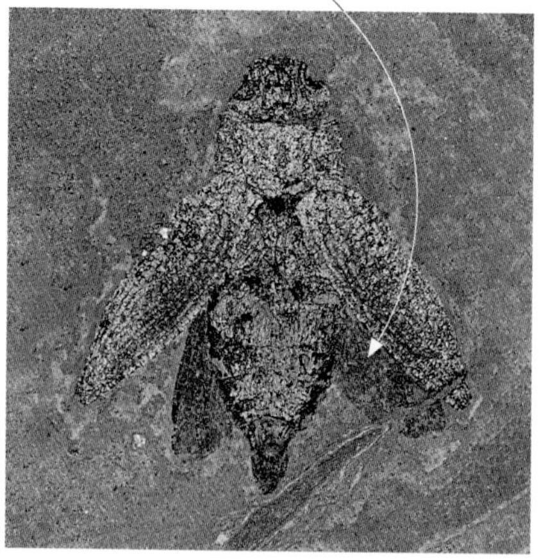

JEWEL BEETLE
Psiloptera weigelti

Cold-blooded vertebrates
The warm, greenhouse climate of the Eocene supported a rich diversity of fish, amphibians, and reptiles. The Messel Lake and surrounding subtropical forests were home to more than 30 species of reptiles, including turtles, lizards, snakes, and crocodiles. Some of the skeletons had been "frozen" in their final moments; there are fossilized pairs of turtles mating and even snakes with their last meals in their stomachs.

FISH
Amphiperca multiformis

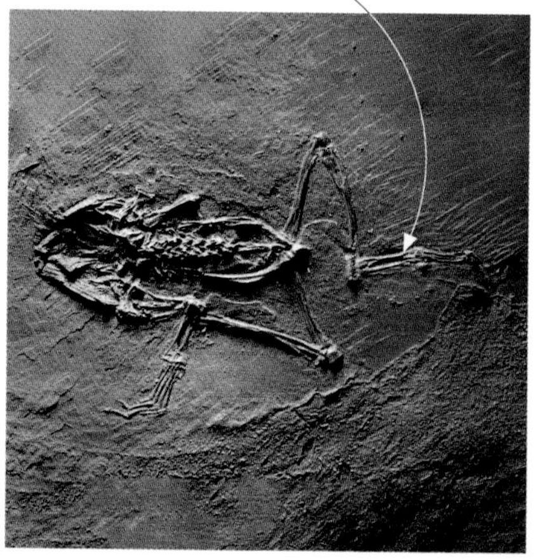

FROG
Eopelobates wagneri

Warm-blooded vertebrates
An incredible range of mammals and birds have been found in the Messel Pit. Early representatives of major animal groups living today show that ecosystems of the Eocene were beginning to resemble modern communities. Hoofed proto-horses mingled with early tapirs, primitive rodents, and tree-living primates. Bats shared the skies with a true aviary: more than 70 species of birds are known from the area.

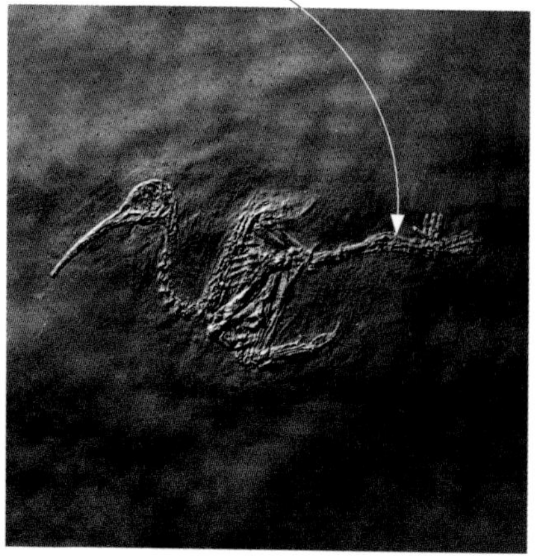

IBIS
Rhynchaeites messelense

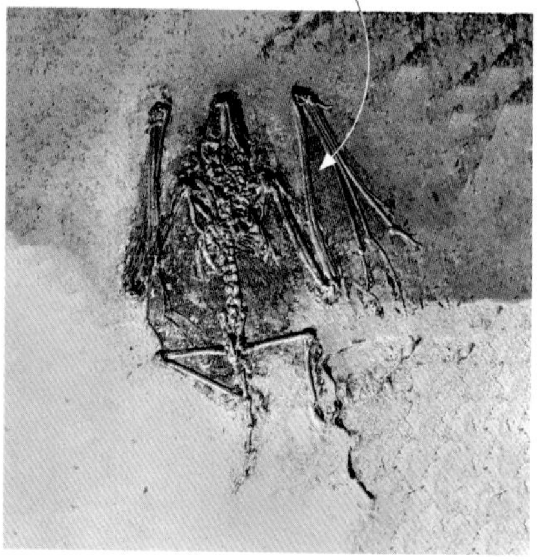

BAT
Palaeochiropteryx tupaiodon

fossilized community

Lagerstätten, from the German word meaning "storage places," are sedimentary deposits worldwide in which fossils are exceptionally preserved. One of these, the remains of an ancient volcanic lake in Germany called the Messel Pit, provides an unparalleled window into an Eocene ecosystem of around 47 million years ago; more than 53,000 plant and animal fossils have been found. The lack of oxygen and currents at the bottom of the lake ensured preservation, but scientists debate why so many specimens were fossilized. One explanation is that toxic gases were periodically released from the lake, killing animals in the surrounding subtropical forest. Alternatively, algal blooms may have poisoned animals that drank the water.

Large ovipositor (organ used to deposit eggs) is prominent

PARASITIC WASP
Xanthopimpla messelensis

Domed carapace compacted by sediment

TURTLE
Allaeochelys crassesculptata

Keeled scales help animal move through vegetation

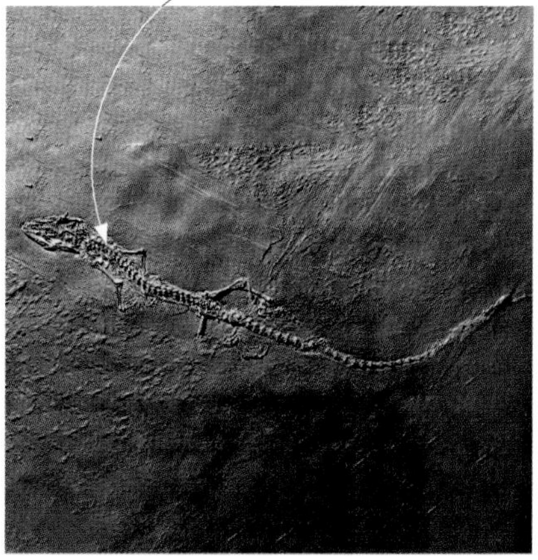

MONITOR LIZARD
Paranecrosaurus feisti

Remains of powerful trunk muscles that constrict prey

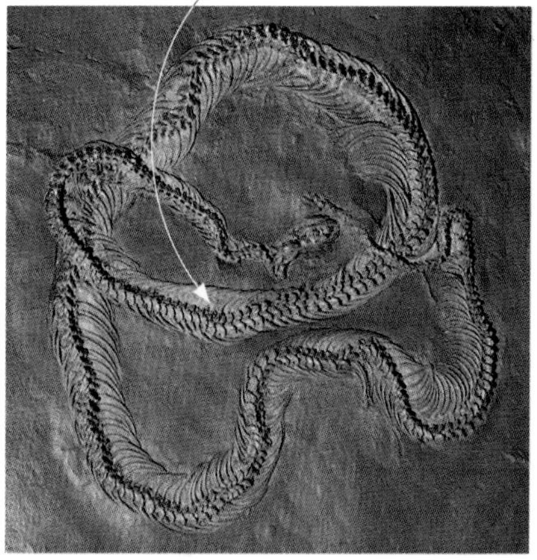

SNAKE
Eoconstrictor fischeri

Jaw musculature similar to that of a guinea pig or porcupine

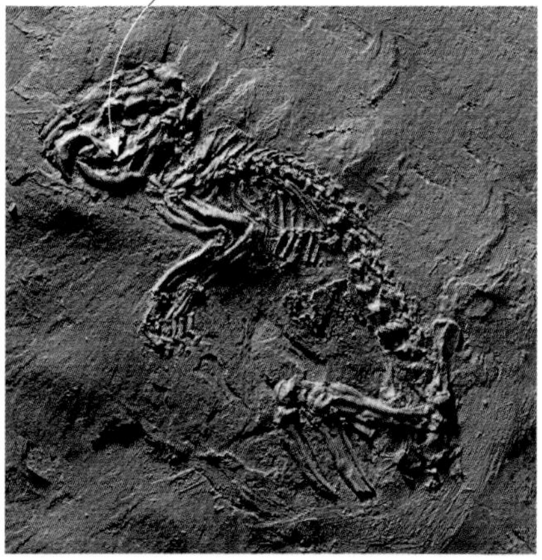

RODENT
Masillamys beegeri

Opposable toes well-suited to climbing

PRIMATE
Darwinius massillae

Rhinoceros-like teeth supports ancestral links to both tapirs and rhinos

TAPIR-LIKE HOOFED ANIMAL
Hyrachyus minimus

the age of mammals

As organisms evolve, they adapt to play out different roles, or niches, within the biosphere. Over 100 million years ago—when dinosaurs still dominated the land—mammals occupied the roles of small nocturnal hunters. After dinosaurs were wiped out (see pp.294–295), live-bearing mammals rose up to fill their niches—diversifying, in the absence of competition, into new kinds of herbivores and carnivores. They diverged into two main groups—marsupials, which nurtured their young in a pouch, and placentals, which nourished young for longer inside the womb. But some evolutionary trajectories converged—including predatory marsupials in Australia that came to resemble placental big cats in other parts of the world.

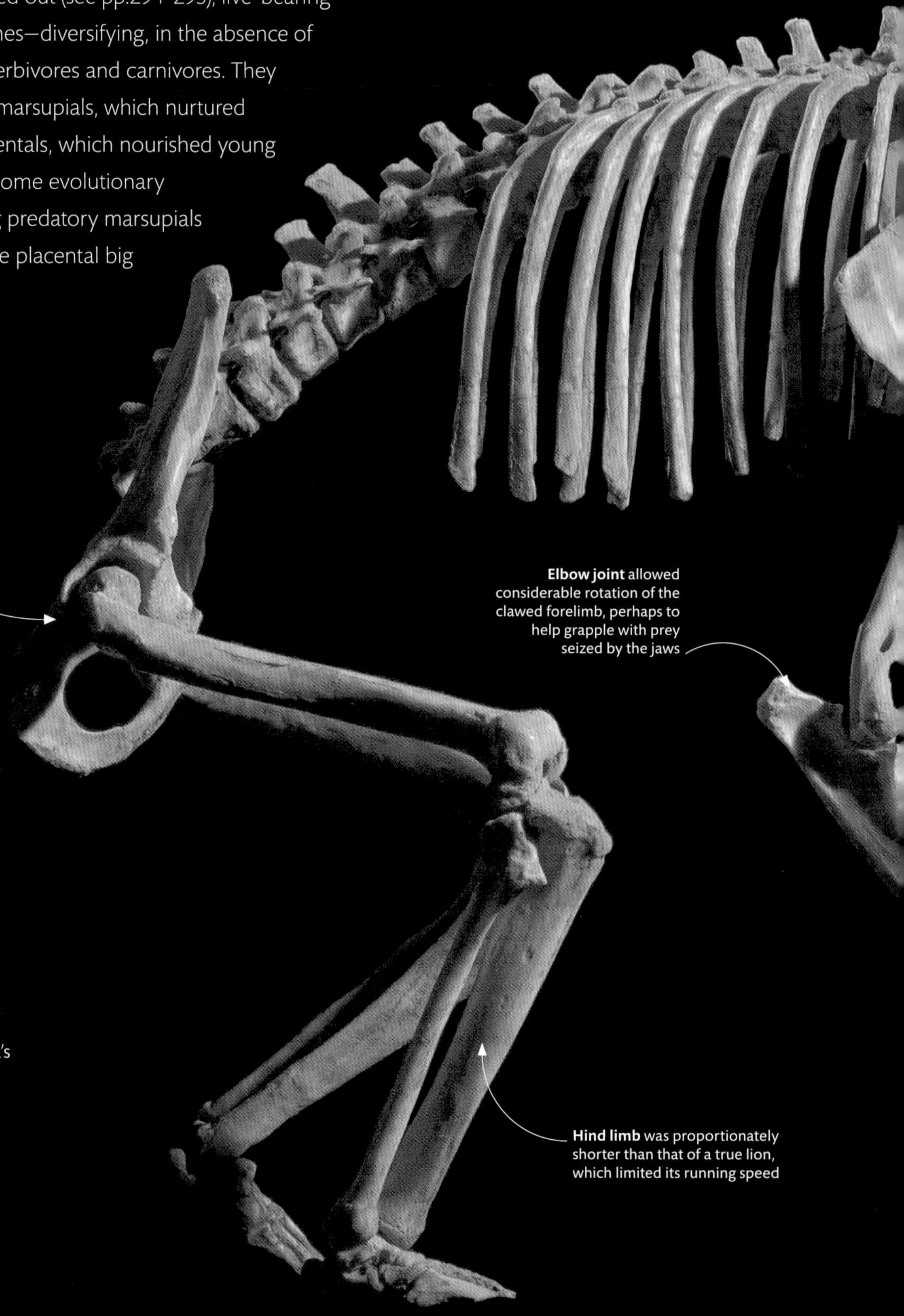

Marsupial lion skeleton
Living around 2 million years ago, *Thylacoleo carniflex* was one of Australia's top predators. It had the cutting teeth and sharp claws of the big cats, but—as a marsupial—was more closely related to modern kangaroos and wombats. It occupied an ecological niche similar to that of a true cat and may have climbed trees—perhaps to hide its kills, in parallel with the African-Asian leopards of today.

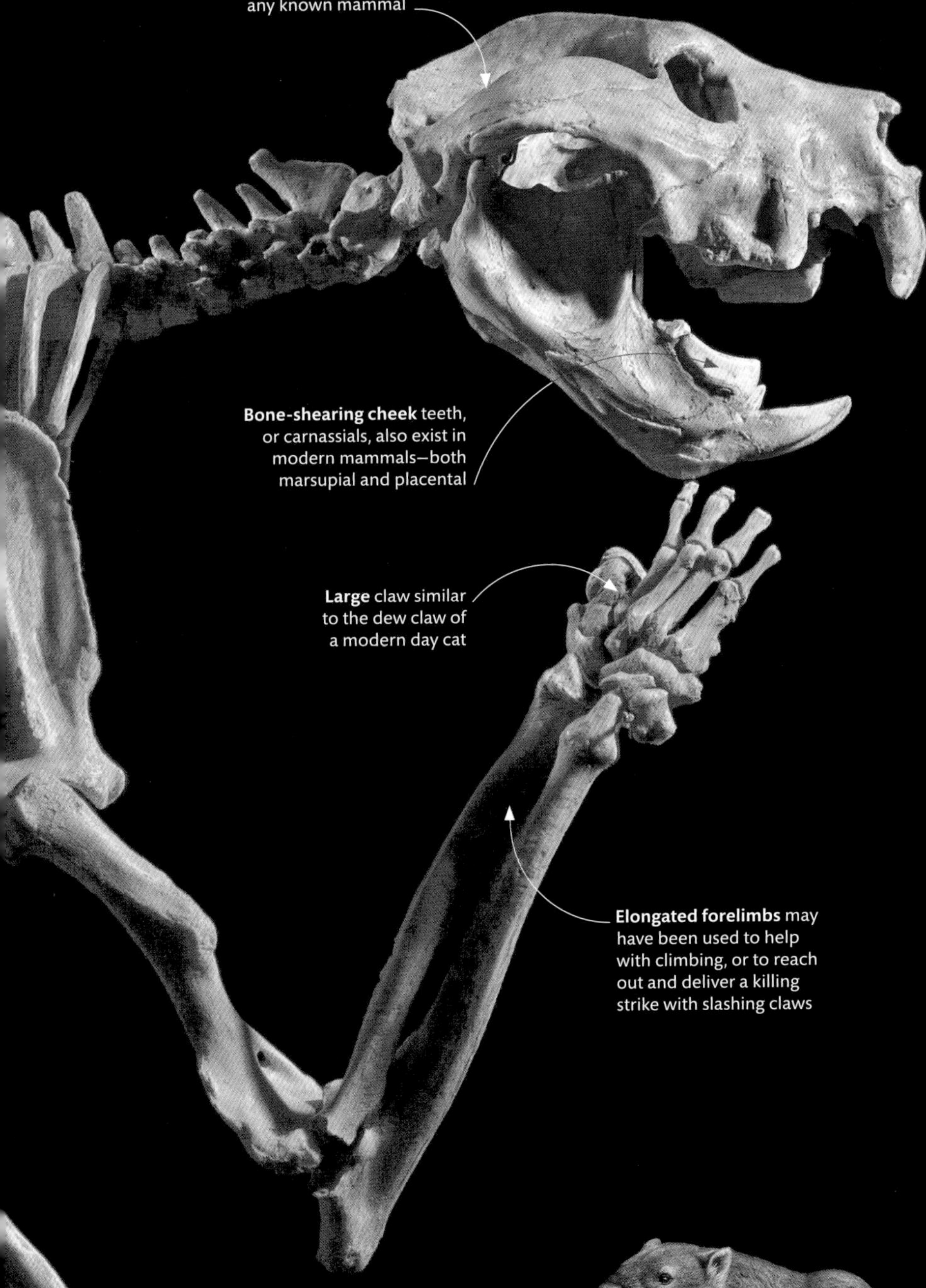

EVOLUTION CONVERGES

Although mammals diverged into placentals and marsupials, after 60 million years of independent evolution, there are many examples of convergence. This happens when unrelated species fill similar niches in different parts of the world, leading to similar adaptations. There are tree-dwellers that evolved parachute-like skin membranes to glide between branches, and ground-living insectivores that lap ants and termites with a long tongue. Marsupial and placental moles show striking convergence: a voracious appetite for burrowing prey, spade-like feet, and reduced vision for an underground life.

Glides in Australian forests

SUGAR GLIDER
Petaurus breviceps

North American forest dweller

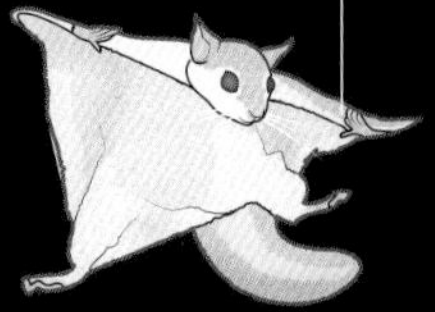

FLYING SQUIRREL
Pteromyini

Eats termites in Australian woodland

NUMBAT
Myrmecobius fasciatus

Eats ants and termites in South American grassland

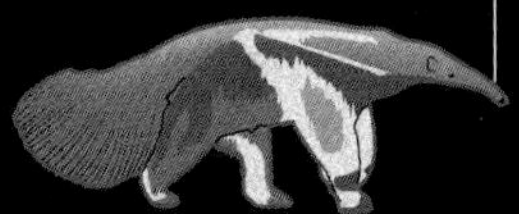

GIANT ANTEATER
Myrmecophaga tridactyla

Digs in Australian sand dunes

MARSUPIAL MOLE
Notoryctes typhlops

European woodland digger

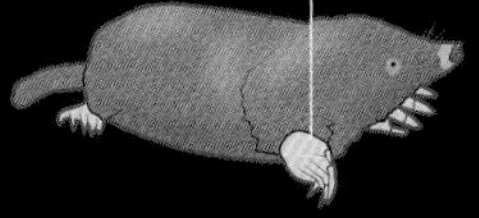

EUROPEAN MOLE
Talpa europaea

MARSUPIAL MAMMALS | **PLACENTAL MAMMALS**

Ambush predator
The long, powerful forelimbs, each equipped with a huge hooked claw, and shorter hind legs suggest that *Thylacoleo* probably relied more on ambush than speed to bring down its prey.

Grazers on the move
Blue wildebeest (*Connochaetes taurinus*) are dependent upon the African savanna for food, as grass makes up 90 percent of their diet. Here, in the Masai-Mara, Kenya, they undertake annual migrations in their thousands following seasonal rains that generate the best grazing.

Temperate-zone grazer
On the steppes of Central Asia, the wild Przewalski's horses (*Equus przewalskii*) search for the best grazing in summer, building fat reserves that help sustain them through winter when grass growth stops.

FROM BROWSERS TO GRAZERS

Mammals that graze on low grass descended from browsers (those that eat leaves, shoots, and fruit of taller, woody vegetation) like the *Hyracotherium* and *Mesohippus*. More specialized grazers—such as *Merychippus* and modern horses, asses, and zebras—have higher-crowned, more deeply rooted cheek teeth with self-sharpening ridges. They clip the vegetation with their front incisors, before the cheek teeth grind grass blades as the jaw moves from side to side.

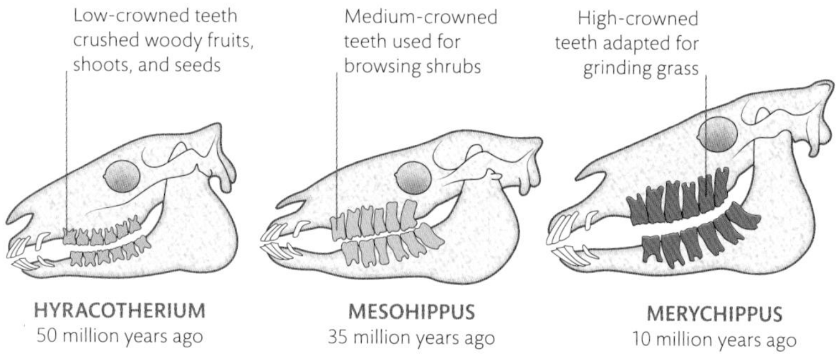

grasslands

No single group of plants has had as much bearing upon the nature of the modern biosphere's habitats as grasses. Grasses defend themselves with shards of silica, which makes them tough to eat, and their leaves grow from the base of their blade—so they can quickly regenerate after being "attacked" by any grazing herbivore. These adaptations have also helped drive their success—especially in the wake of the asteroid collision that sent the dinosaurs and so much other life to extinction (see pp.294–295). In the "new" post-asteroid biosphere when forests waned, the grasses grew so well that they created an entirely different kind of habitat: the open, mainly treeless steppes and savannahs of a new world dominated by large mammals (see pp.312–313).

Waller's terror bird
Standing around 5 ft (1.5 m) tall, *Titanis walleri* belonged to a group of flightless, predatory birds that evolved in South America, but migrated northward when the two American continents joined. In North America it became one of the most formidable predators—successfully preying on native mammals that perhaps had no fear of birds.

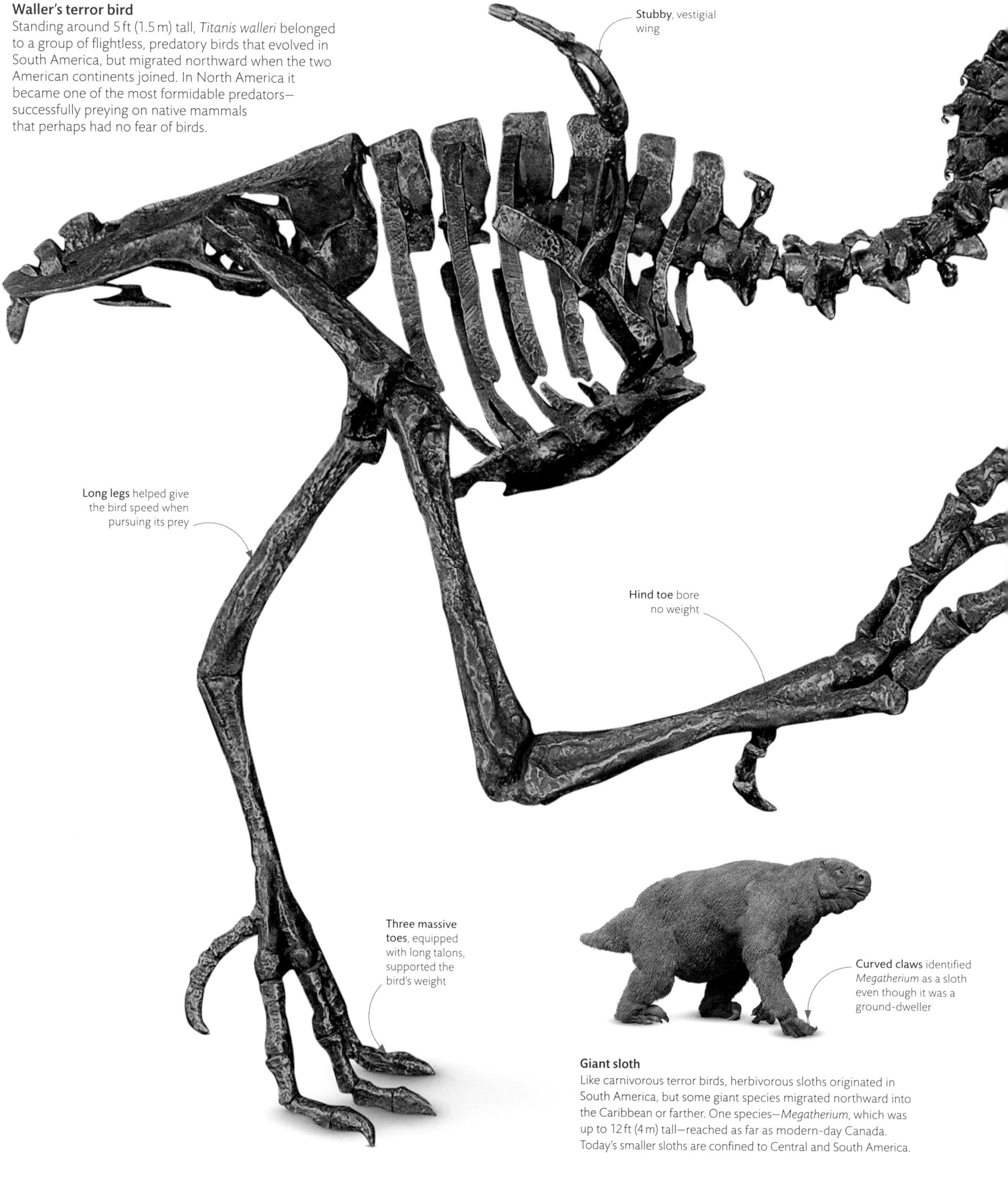

Giant sloth
Like carnivorous terror birds, herbivorous sloths originated in South America, but some giant species migrated northward into the Caribbean or farther. One species—*Megatherium*, which was up to 12 ft (4 m) tall—reached as far as modern-day Canada. Today's smaller sloths are confined to Central and South America.

life on the move

Earth's changing geography plays a pivotal role in the evolution of its biosphere. Over millions of years, as continents drifted and islands rose, a dynamic Earth helped shape animal distributions—such as kangaroos in Australia, sloths in South America, and flightless birds on oceanic islands. When continents split apart and collide, their faunas follow suit. A dramatic meeting of different worlds happened several million years ago, when North and South America came together, and animals living on different continents came face-to-face with counterparts that had been separate since the time of the dinosaurs.

THE GREAT AMERICAN INTERCHANGE

Groups of mammals evolved in isolation on North or South America when the continents were separate. Camels, for instance, evolved in North America, while sloths originated in South America. Both were part of a mammal interchange, first via Caribbean islands, and then across the isthmus that joined the landmasses. Sloths migrated north, and today's guanacos—descendants of early camels—live as immigrants in South America.

KEY

Camelids
Xenarthrans (sloths, anteaters, and armadillos)

ISOLATION ON ISLANDS

Islands form as new volcanic lands erupt from the seafloor or as fragments split away from drifting continents, but new life can evolve in isolation on both. Volcanic islands must be colonized over seas, but without predators or strong competitors, their pioneers can evolve in radical new ways. The large, flightless dodos on Mauritius descended from ancestral pigeons that flew from Asia, but like many island species, evolution left them vulnerable to outsiders. Dodos failed to survive the onslaught of human hunters and went extinct around 1662.

FLIGHTLESS DODO
Raphus cucullatus

THE MAMMOTH STEPPE

This vast expanse of grassland stretching from Europe across northern Asia to Canada was historically the largest single expanse of biome by area, and provided grazing for ice-age herbivores. With so much water locked up in ice, this continental habitat was drier than today's tundra. Today the steppe has largely been replaced by boreal forest (see pp.308–309), but patches remain in Central Asia.

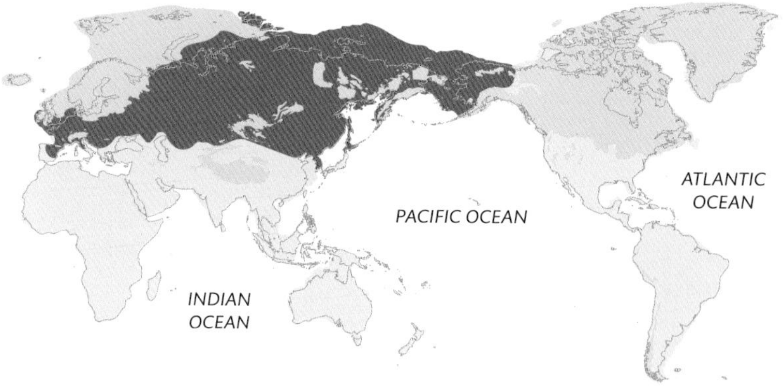

MAMMOTH STEPPE DURING THE PLEISTOCENE ICE AGE

surviving ice

Cold is an enemy of life: it slows metabolism, while freezing kills tissues. The coldest places today are at the poles, where the sun's rays are too feeble to provide much warmth, so life must adapt. But throughout prehistory, tilts in planetary orbit or shifting continents have resulted in parts of Earth being covered with ice. The most recent ice age began just over 2.5 million years ago in the Pleistocene, and its coldest glacial periods (see p.243) saw much of North America and Eurasia under an Arctic ice sheet. It turned the land beyond into freezing steppe, where cold-resistant large mammals grazed where few other species could.

Tusks found on both males and females were longer and more curved than those of today's elephants

Ice-age rhinoceros
In an example of convergence (see p.299), the woolly rhinoceros (*Coelodonta antiquitatis*) evolved a large body size and shaggy coat of long hair that helped trap body heat—just like the mammoth's.

Shoulder hump housed fat, muscle, and well-developed neck ligament that supported head and horn

Ice-age mammoth

Descended from tropical elephants, the woolly mammoth (*Mammuthus primigenius*) is one of the best-known animals to thrive during the Pleistocene Ice Age. Remains of mammoths found preserved in sub-zero temperatures show that, in life, they had a cold-adapted metabolism and were insulated by a subcutaneous fat layer and a thick coat. They also relied more on grazing grass than does any living elephant.

DEEP FREEZE

Any soil or rock that stays frozen for at least two consecutive years is classed as permafrost. Continuous permafrost lies beneath most of the land inside the Arctic Circle. Above the permafrost lies an active layer of soil that thaws in summer, in which shallow-rooted plants can grow. Farther south, the permafrost breaks up—so named discontinuous permafrost—so trees can grow and tundra gives way to coniferous forest (see pp.308–309).

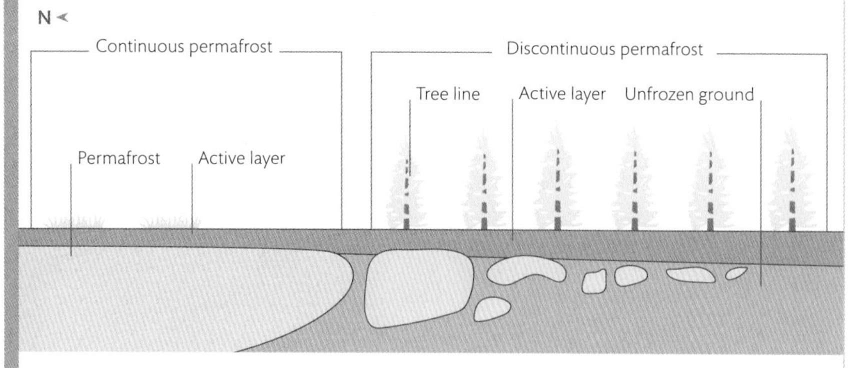

SOIL PROFILE OF PERMAFROST IN SUMMER

permafrost

At times, Earth's climate has been so warm that even the poles were unfrozen (see pp.242–243). But today the Arctic and Antarctic are both capped by polar ice sheets that wax and wane with the seasons. Around the Arctic circle, however, even in summer, the ground remains frozen beneath a certain depth. The ice blocks the downward growth of roots and restricts the upward growth of vegetation. This is the world of the Arctic tundra—a place dominated by low-growing plants, such as sedges and mosses—where few woody trees and shrubs can survive.

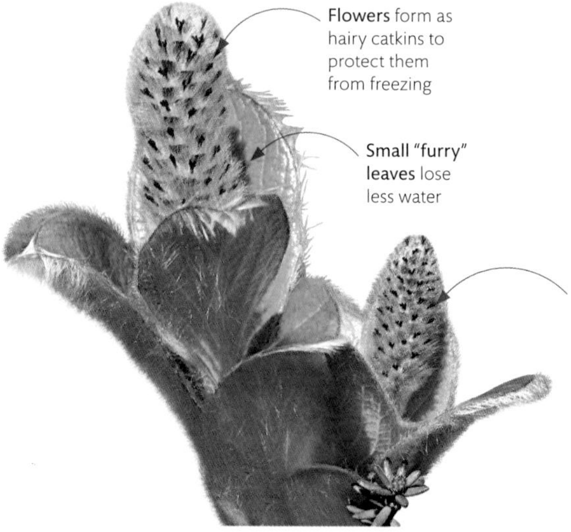

Tundra polygons
Although very little rain falls in the tundra, the low temperatures slow its evaporation, so the land stays wet and permafrost stops water draining away. Water-filled cracks in the ground freeze with wedges of ice, which then thaw during the summer, and the meltwater collects in huge polygonal pools, as seen in this northern Alaskan landscape.

Northern shrub
No woody plant lives farther north than the Arctic willow (*Salix arctica*). Despite being a relative of the willow trees of temperate latitudes, this plant is tiny, and it hugs the ground. Its buds are insulated from the bitter, icy winds by woolly hairs.

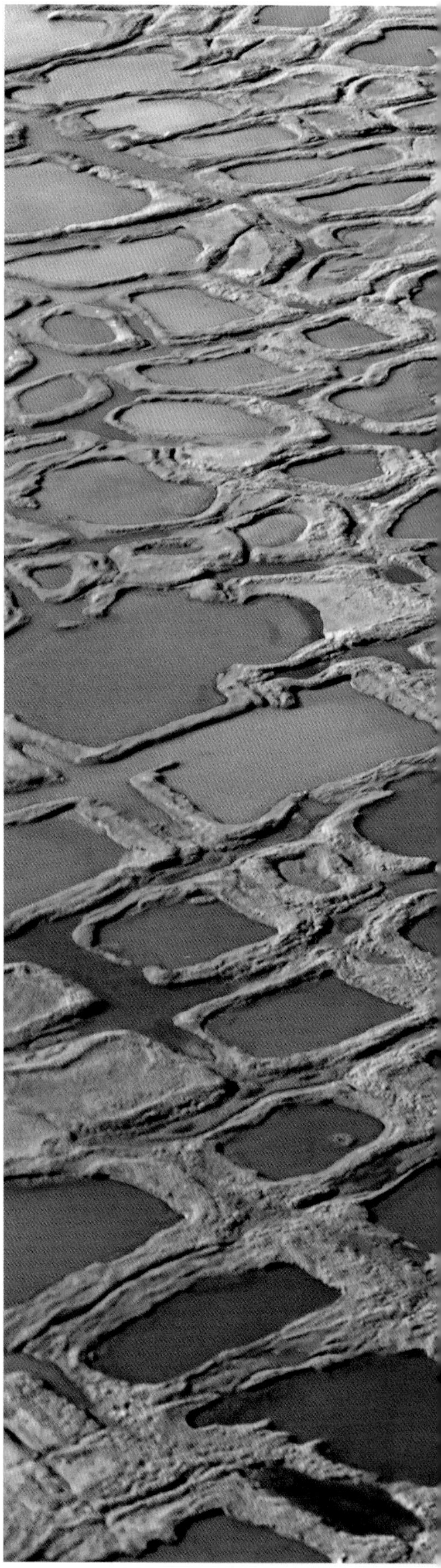

boreal forest

Also known as taiga, boreal forest is the biosphere's newest and most extensive natural habitat. Around 12,000 years ago, as the last glaciation (see pp.304–305) was receding, the polar icecaps shrank, uncovering more land in North America and Eurasia. Ice that had long laid buried in soil—the permafrost (see pp.306–307)—melted, and the dry steppes gave way to damper forest. Boreal forest encircles the globe between tundra to its north and broadleaved forest and grasslands farther south. But the climate here, close to the Arctic Circle where the sun stays low above the horizon, is still very cold and the coniferous trees that grow here must be able to survive these conditions.

Forests of the north
Cold-adapted forests, such as this one in Finland, are dominated by needle-leaved, cone-bearing conifer trees. These conifers include evergreen spruces and pines and deciduous larches, species that are hardy enough to survive temperatures that remain below freezing for most of the year.

Circumpolar animals
The boreal forests that stretch across North America and Eurasia provide similar kinds of habitats, so many species of animals, such as the moose (*Alces alces*), range across the entire region.

CONIFER ADAPTATIONS

Foliage of conifers is quite different from that of broadleaved, flowering plants. Their leaves are shaped like needles or reduced to small scales, which reduces their surface area, limiting water loss by evaporation, so the trees can survive when so much ground water is frozen. These adaptations help conifers to grow in tropical, arid parts of the world, too. Conifer tree sap also contains resins that work as antifreeze, and the broader leaves have a waterproof waxy coating.

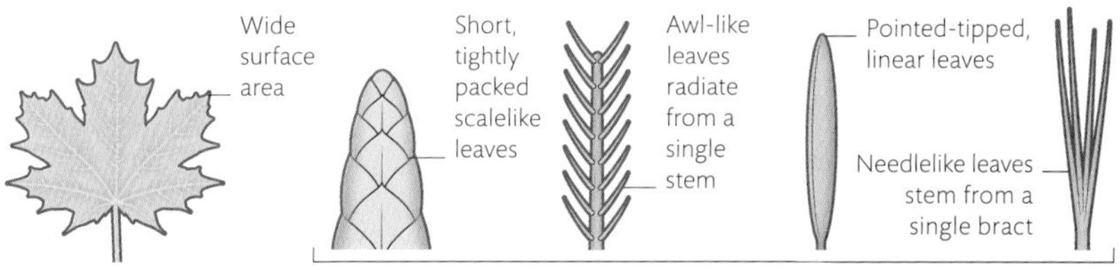

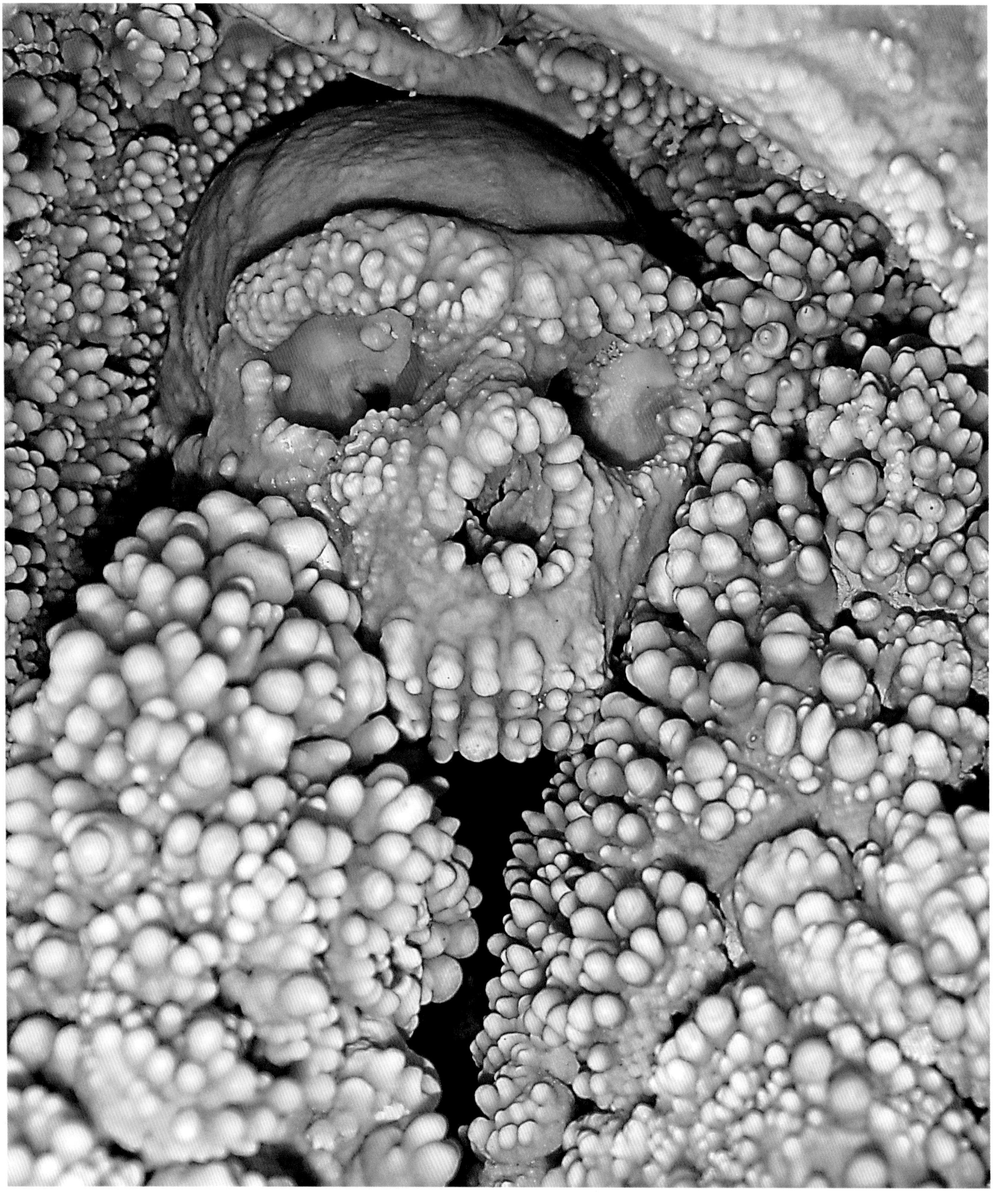

Altamura Man
The fossilized remains of Altamura Man (left), discovered in Italy in 1993, have provided a rich source of Neanderthal DNA. Analysis of DNA from Neanderthal remains shows that they interbred with modern humans—an insight that has revolutionized paleoanthropology.

> “The challenge for the molecular paleontologist is to discover chemical traces of ancient life.”
>
> DEREK E. G. BRIGGS AND ROGER E. SUMMONS 2014

history of Earth science

molecular evidence from fossils

SINOSAUROPTERYX FOSSIL

Exceptionally preserved fossils—including those of "soft" parts that are not usually mineralized—have long provided insight into the shapes and features of extinct organisms. Modern analytical methods are now showing that these spectacular fossils of prehistoric life may preserve more than meets the eye: cellular details, molecular signals, and possibly even fragments of DNA.

The potential preservation in fossils of biological molecules (biomolecules), including proteins, carbohydrates, and genetic material, was first recognized in the 1950s. In the 1970s and 1980s, analysis of biomolecules progressed, and with advances in DNA sequencing, the first DNA from an extinct animal was extracted from the 19th-century remains of the extinct zebralike quagga. By refining these approaches, scientists working with remains that have been preserved, but not fossilized, have since been able to identify the genetic makeup of many extinct animals, such as human relatives, ice-age horses, and even million-year-old woolly mammoths.

In the mid-2000s, the recovery of original soft tissues from *Tyrannosaurus rex* spurred debate about whether DNA might be preserved in fossils. Remnants of soft tissue have now been found in a wide range of dinosaur fossils. However, scientists still debate the nature of these preserved tissues, and whether it is possible to distinguish between original tissues and those contaminated by the presence of microorganisms or other genetic material.

Further studies have shown that many different kinds of biomolecules, especially proteins, can be transformed during fossilization into stable forms that provide information about the organisms when they were alive. Transformed biomolecules have now been used to study, for example, the relationships of some of the earliest life forms, the colors of dinosaurs, and even to identify fossilized blood in the stomach of an Eocene mosquito (56–34 million years ago). Ongoing advances in these techniques continue to push the boundaries of the study and understanding of ancient biomolecules and life forms.

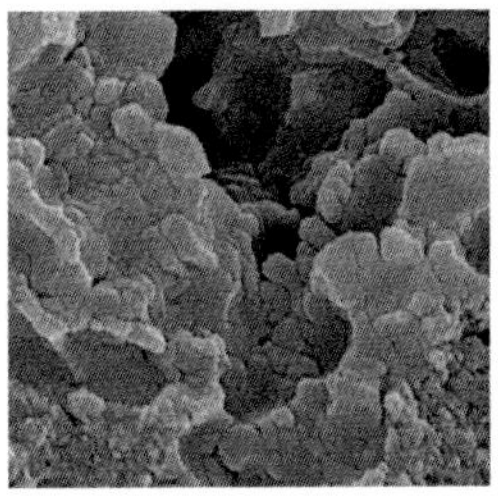

REDDISH MELANOSOMES

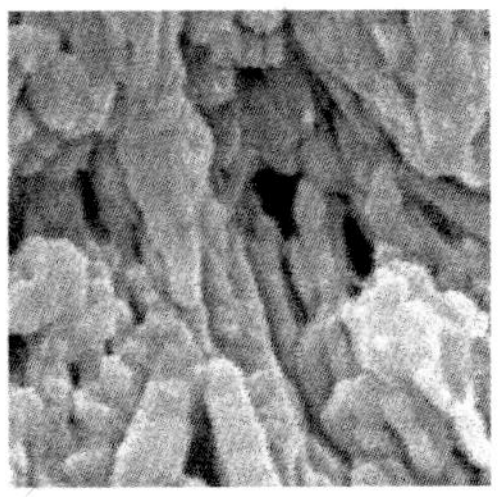

BLACK-GRAY MELANOSOMES

Ancient melanosomes
Under the electron microscope (left), the melanosomes (microscopic sacs of pigment within cells) of fossilized feathers can be seen. Red melanosomes tend to be spherical, while black ones are sausage-shaped. The fossilized remains of *Sinosauropteryx* (above left) indicate that the creature had feathers with dark and pale stripes.

megafauna

In the last 60 million years, mammals became the new big beasts on Earth—from huge ground sloths in America (see p.303) to giant marsupials in Australia. Like the giant reptiles before them (see pp.290–291), they followed a trend that may have been driven by the benefits of outsizing competitors and predators. In a world that had been cooling since the peak of the dinosaurs, large warm-blooded mammals were good at maintaining body temperature. Today's megafauna is greatly reduced—most giant mammals became extinct within the last 50,000 years, a timing that coincides with the dispersal of humans, suggesting that these formidable hunters were at least partly to blame.

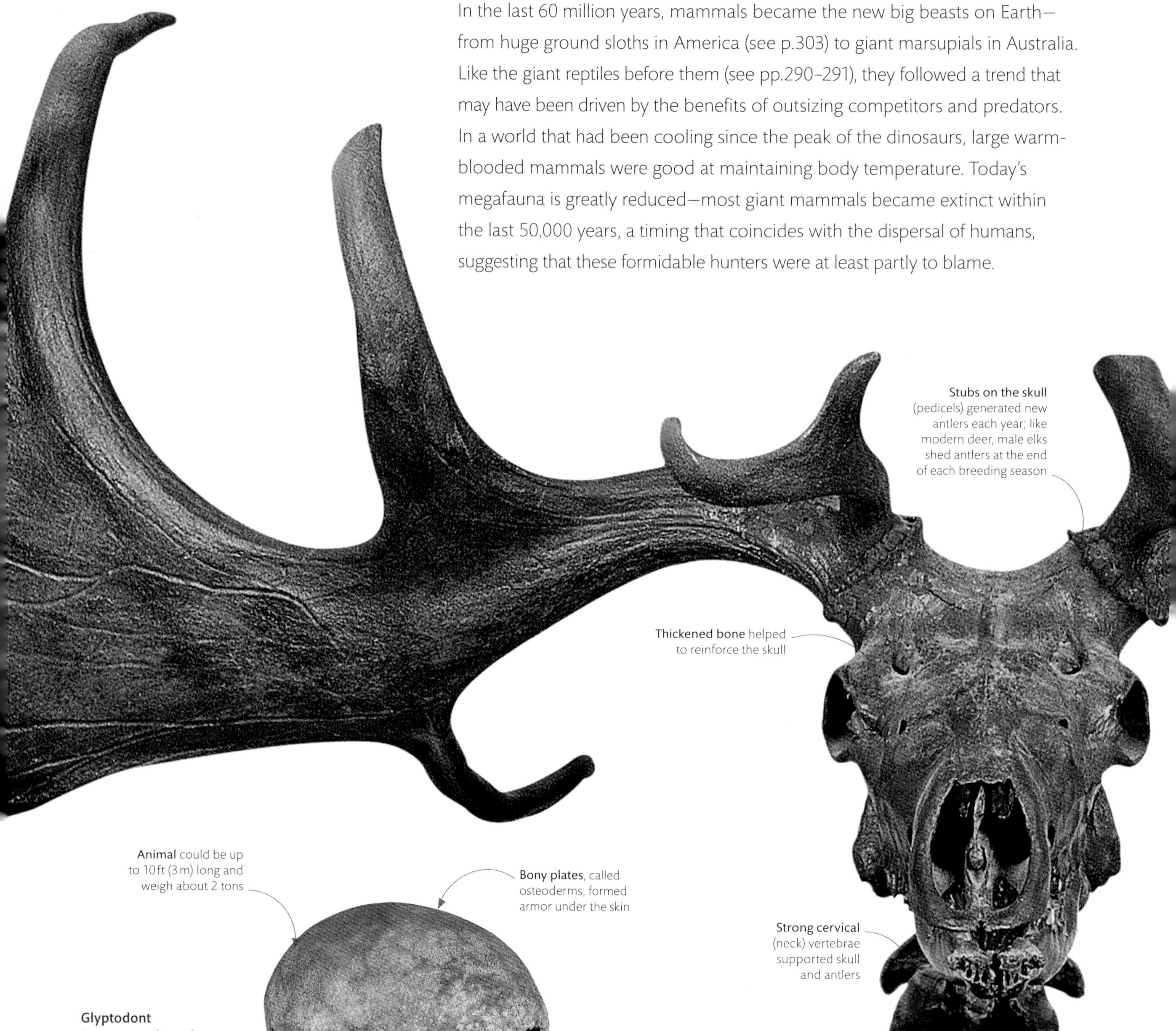

Glyptodont
Extinct megafauna from South America, *Glyptodonts* were giant armadillos that grew to the size of a small car and survived by browsing on vegetation. The armadillos of today are smaller than a dog and eat mainly eat insects.

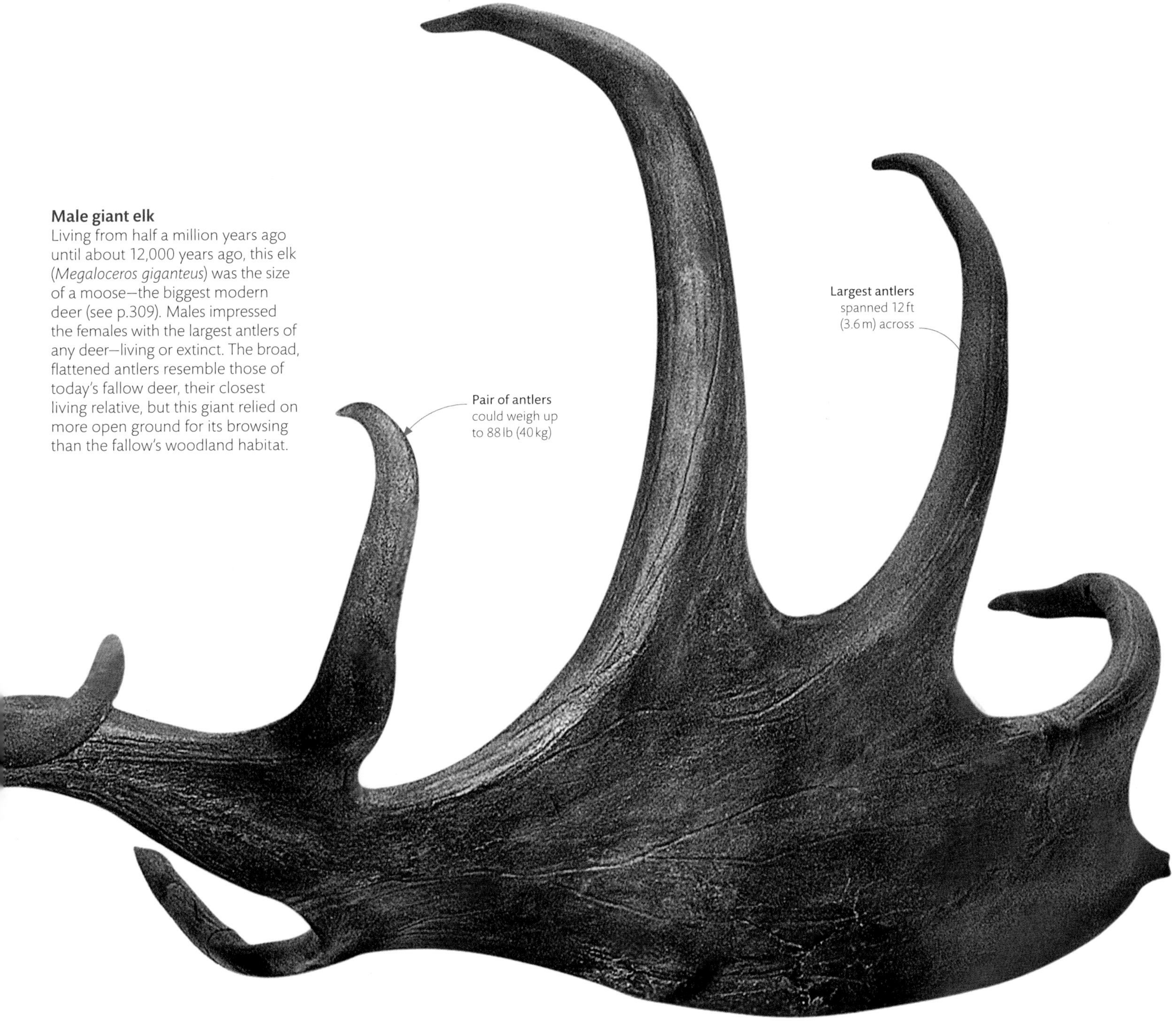

Male giant elk
Living from half a million years ago until about 12,000 years ago, this elk (*Megaloceros giganteus*) was the size of a moose—the biggest modern deer (see p.309). Males impressed the females with the largest antlers of any deer—living or extinct. The broad, flattened antlers resemble those of today's fallow deer, their closest living relative, but this giant relied on more open ground for its browsing than the fallow's woodland habitat.

MEGAFAUNA DIMENSIONS

The heaviest megafauna, all weighing several tons, include now extinct mammoths, as well as living giants such as giraffes and bison, all of which are herbivores: processing tough plant food requires big guts. But the carnivores that hunted them also became huge. Megafauna species numbers reached their peak in the Pleistocene epoch—about 2.5 million years ago—a time that corresponded with the last ice age (see pp.304–305).

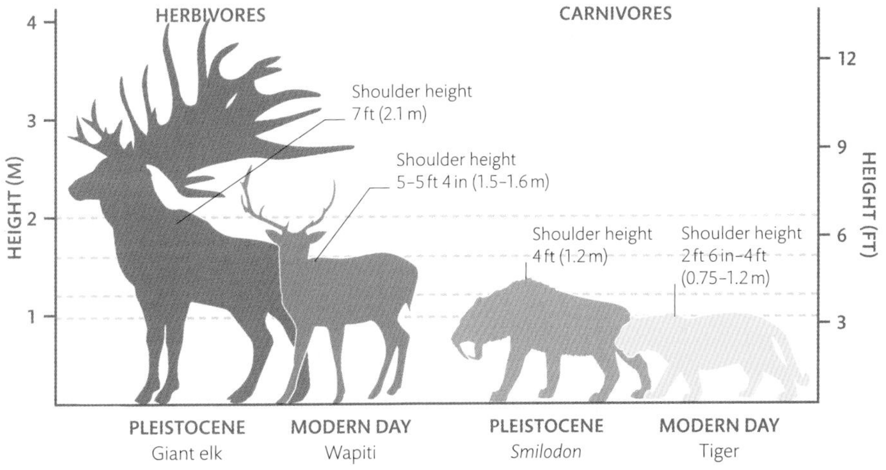

Rising eerily from the beach along Cardigan Bay on the west coast of Wales, the submerged forest of Borth is a reminder of a bygone age. Stretching for 4 miles (7 km) between the towns of Borth and Ynyslas, these preserved trunks of ash, oak, birch, and pine are entombed in a layer of peat some 6,000 years old, which protected them as they were submerged by rising seas. Today, they are sporadically exhumed by storms, which scrape

spotlight Forest of Borth

away overlying sand and leave the twisted roots exposed at low tide. Sightings of the prehistoric forest are becoming more commonplace, and more and more of the forest is exposed as warming oceans produce more frequent and powerful storms.

Locals link the trees to the legend of Cantre'r Gwaelod, or the Lowland Hundred—a sunken ancient kingdom. As the story goes, the civilization was inundated after a prince—either forgetful, lustful, or inebriated—left open a sluice gate, allowing the sea to wash over the low-lying area. Scholars are less eager to jump to conclusions. The ephemeral exposure of the trees does make them challenging to study, but the reefs imagined as city walls are more likely to be glacial moraines. Nonetheless, the area was clearly once inhabited as stone hearths, wooden walkways, and stone tools have been found among the stumps.

Similar submerged forests are found around the coasts of Great Britain, which was connected to mainland Europe during the last glacial period (around 115,000–11,700 years ago, see pp.304–305). Only when the glaciers receded did sea level rise, flooding the once-fertile land bridge and isolating the island.

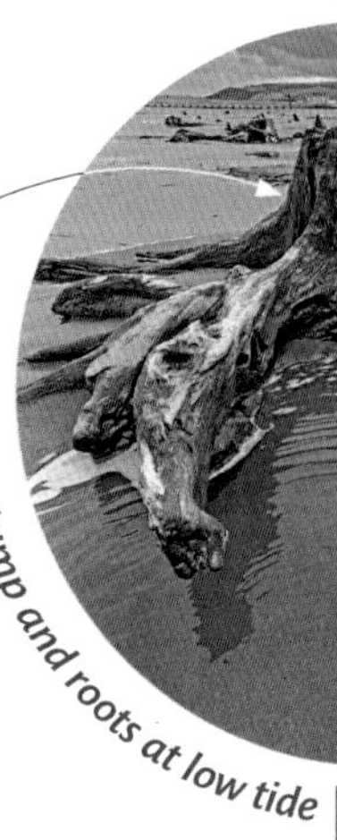

Algae-covered tree remains are exposed as overlying sand is washed away

Tree stump and roots at low tide

Landscape of the forest
In 2014, a major storm scoured the shoreline of Ceredigion County and, as the waves receded, the stumps of an ancient forest once buried in peat were exposed. Research shows that the trees stopped growing 4,000–6,000 years ago. The remains are visible at very low tides; they could begin to rot if exposed for long periods, but seawater also has a preservative effect.

shaped by humans

One species, *Homo sapiens*, has had a far greater impact on the rest of the biosphere than any other single kind of living thing. Its entire fossil record is less than half a million years old, but its impact on the environment will last far longer. So great are the effects of humankind that scientists have proposed a new geological epoch—the Anthropocene—to mark its significance. By 10,000 years ago, humans had colonized much of the world, altering landscapes with their agriculture and hunting large animals to extinction. Since then, a new age of industry and technology is polluting the world and changing the climate, causing global warming.

Humanity's mark
Less than one-quarter of Earth's land surface remains as wilderness—largely untouched by humanity. These giraffes in Nairobi National Park, Kenya, stand in a 39 square mile (100 square km) area of protected land—but one that is fenced in on three sides and has Kenya's capital city, Nairobi, looming in the distance.

Discarded cup is a plastic substitute for a natural shell

Plastic crisis
This plastic cup may be a temporary home for a hermit crab (*Coenobita* sp.), but plastics resist decomposition and are ultimately part of a pollution crisis in the ground and oceans that is threatening the natural world.

THE GRADUAL ONSET OF THE ANTHROPOCENE

Like all species, humans use resources and generate waste, but they also modify the environment extensively through agriculture, transportation, urbanization, and fuel combustion. Forests and other habitats make way for fields and cities, invasive species are introduced, and technology and industry generate pollutants. No single event defines the Anthropocene—experts differ on when it began.

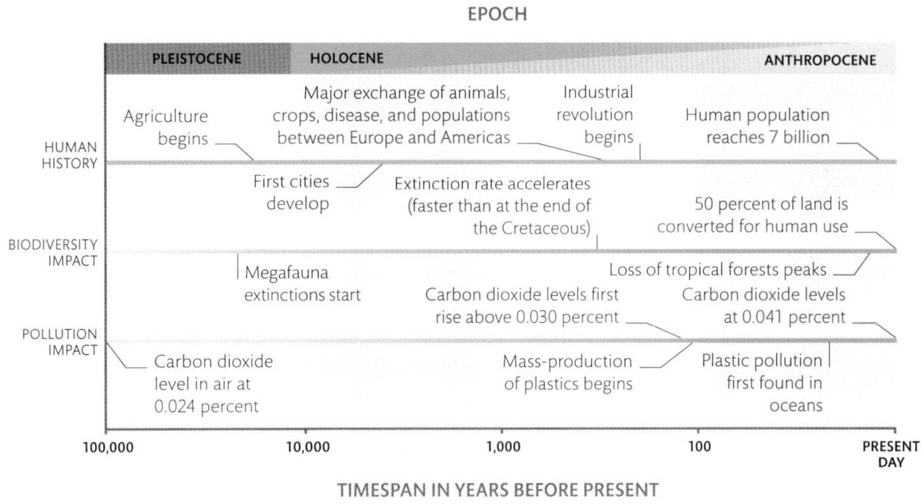

spotlight Danum Valley

Although the Anthropocene (see p.317) has been bad news for nature, pockets of pristine, diverse wildlife habitat do survive. While the tropical rainforest on the island of Borneo is under threat (more than 50 percent of old-growth forest has been lost since 1973), nestled on the northeastern coast of the island is the Danum Valley Conservation Area—more than 150 square miles (400 square km) of uninhabited rainforest. One of the last expanses of undisturbed old-growth forest left in Southeast Asia, it is dominated by dipterocarp trees, some of the tallest flowering trees in the world. There are 270 dipterocarp species on Borneo, of which 155 exist only on the island, and most, if not all, are found in the Danum Valley. The area is home to unique plant species, like the insect-eating pitcher plants, the world's largest flower—the corpse flower *Rafflesia*—and the tallest tropical tree,

yellow meranti (*Shorea faguetiana*). Inhabiting the forest are populations of critically endangered animals such as Bornean orangutans (*Pongo pygmaeus*), Bornean pygmy elephants (*Elephas maximus borneensis*), and the clouded leopard (*Neofelis nebulosa*). The valley is a sanctuary for more than 124 mammal and 340 bird species vulnerable to hunting or habitat fragmentation.

Biodiversity research in designated conservation areas such as Danum Valley—protected from hunting and logging—demonstrates that conservation efforts work and can successfully guide management policies. As a result, the rate of deforestation has recently decreased, and protected areas have expanded.

Heart of Danum Valley
Morning mist hangs over the rainforest of the Danum Valley in Sabah, Borneo. The dipterocarp trees seen here produce distinctive winged seeds once in a decade, and do so en masse, flooding the jungle with billions of seeds. Many endemic plant and animal species survive here away from the threat of deforestation and hunting.

directory

minerals

A mineral is a naturally occurring, usually inorganic, chemical compound with a definite composition and crystalline structure (see pp.36–37). Minerals are divided into groups depending on their chemistry. They not only form the bulk of Earth's solid matter but also contribute to the composition of animal shells and skeletons.

German mineralogist Georgius Agricola is generally called the Father of Mineralogy. His *De Natura Fossilium* (*On Natural Fossils*), published in 1546, described and classified minerals, and remained the standard reference work for hundreds of years. In 1837, American geologist, mineralogist, and zoologist James Dwight Dana published *System of Mineralogy*, which featured the classification system referred to as "Dana's System of Mineralogy." Many museums (and the following pages in this book) still use Dana's system to curate their display of minerals into systematic collections. Over the years, scientists have expanded Dana's system as their understanding of minerals has increased.

There are about 5,000 minerals, classified into 78 recognized mineral classes. A few, such as quartz and feldspars, are important as rock-forming minerals. Others, such as borates, molybdates, chromates, tungstates, sulfosalts, vanadates, and arsenates are mainly valued for their uses in the agricultural, chemical, and health industries. Ore-forming minerals are mined to obtain metals—for example, chalcopyrite is an ore of copper.

Using Dana's classification, minerals are sorted into groups ranging from those with a simple chemistry, such as the native elements, to more complex minerals. Each mineral has a chemical formula, in which letters signify the different elements and numbers show their amounts. For example, silicates range from a simple tetrahedron with the formula SiO_4—in which one silicon atom is bonded to four oxygen atoms—to complex frameworks in which several tetrahedrons are joined together.

Vesuvianite crystals
Classified as a sorosilicate, vesuvianite is named after Mount Vesuvius in Italy, where the mineral was first discovered in 1795. It forms tetragonal crystals.

Precious stones
This colored print from German mineralogist Adolph Kenngott's "Mineralogy" section in a late 19th-century natural history book shows flint (hornstein, illustration number 1), conglomerate (5), chrysoprase (8), agates (such as 3 and 4), and opal (7, 19, and 20).

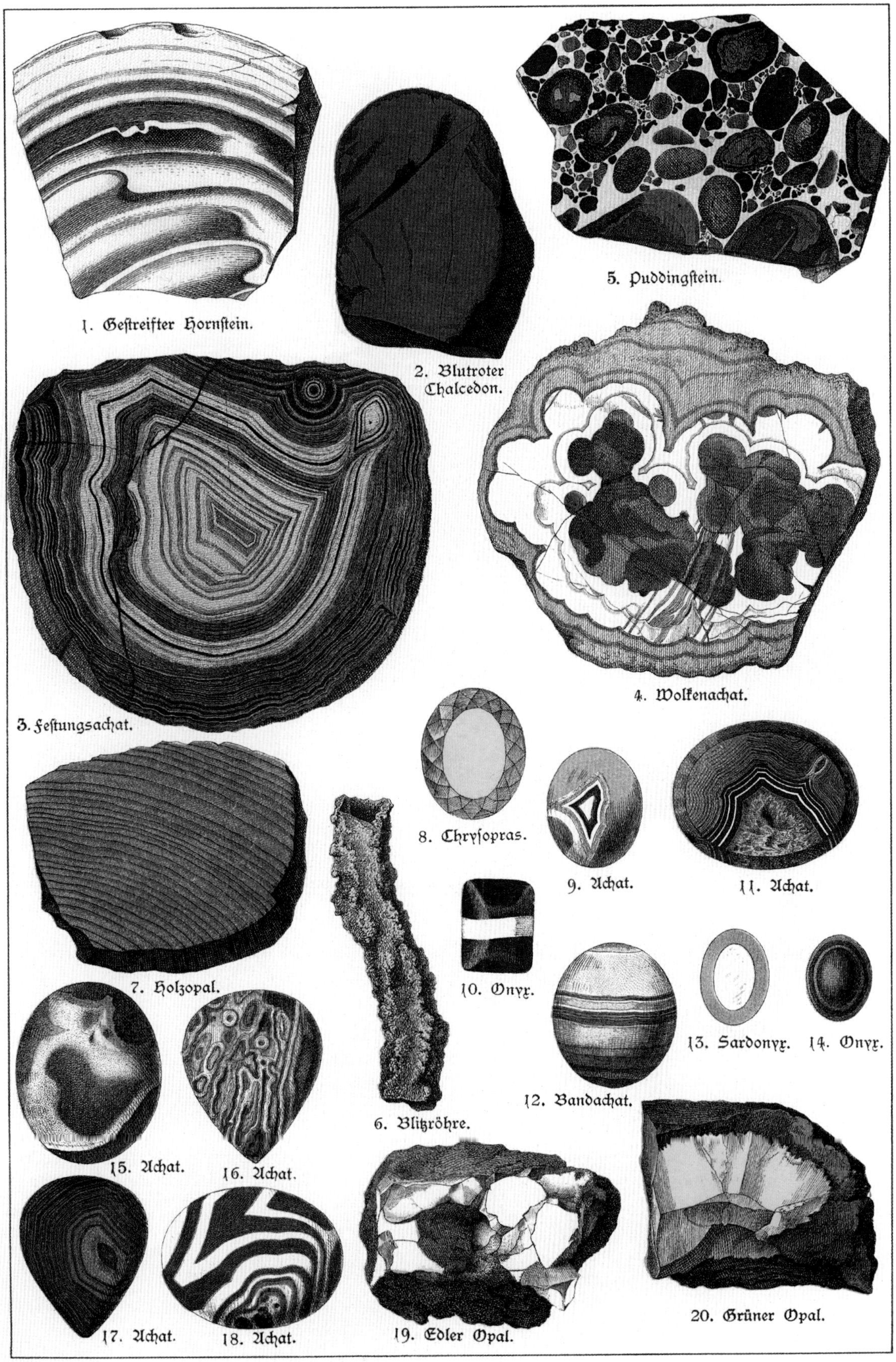
1. Gestreifter Hornstein.
2. Blutroter Chalcedon.
5. Puddingstein.
3. Festungsachat.
4. Wolkenachat.
8. Chrysopras.
9. Achat.
11. Achat.
7. Holzopal.
10. Onyx.
13. Sardonyx.
14. Onyx.
12. Bandachat.
6. Blitzröhre.
15. Achat.
16. Achat.
17. Achat.
18. Achat.
19. Edler Opal.
20. Grüner Opal.

native elements

Native elements are chemical elements that occur in nature in a pure form, uncombined with any other silver, copper, or platinum group—platinum, palladium, rhodium, ruthenium, osmium, and iridium. A few intermetallic elements—naturally occurring alloys, such as electrum made of gold and silver—metalloids such as arsenic; and nonmetals such as sulfur, graphite, and diamond are also native elements.

Gold has the chemical symbol Au, derived from the Latin *aurum*, meaning "gold." In its native form, gold is a bright golden yellow. It is dense, soft, malleable, and ductile. It is also one of the least reactive chemical elements. Gold can be found embedded in the rock in which it has formed, or as nuggets or fine grains if it has been transported by water. It can be extracted by panning—sifting sediments in a pan to separate the gold—from river and beach gravels.

Silver has the chemical symbol Ag, derived from the Latin *argentum*, meaning "shiny," although it tarnishes when exposed to air. It is often found with copper. Silver is used in photography, electronics, and chemistry, and it is the best known conductor of heat and electricity.

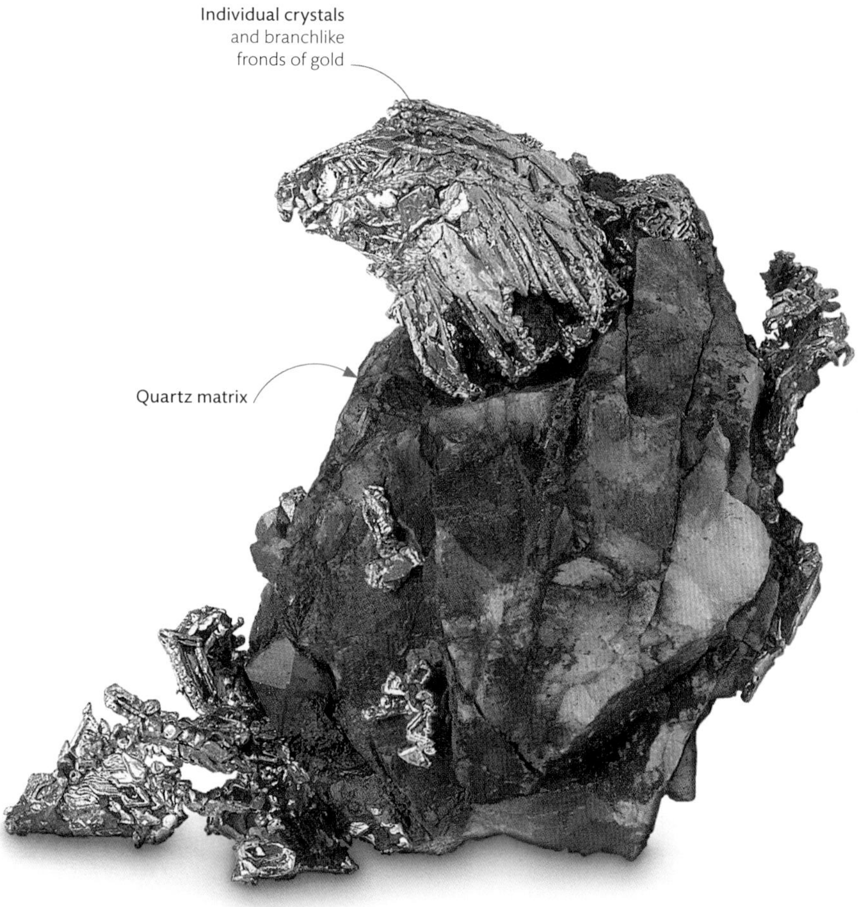

Rare crystal
In this specimen, crystals of native gold have formed in a quartz matrix. Such large crystals are rare, as gold mostly occurs as small grains or weathered nuggets.

Silver

Copper

Platinum

Graphite

nesosilicates

Silicates are the most common minerals on Earth and make up about 90 percent of the planet's crust. They are classified by their structure. Nesosilicates (formerly called orthosilicates) are the simplest form of silicates, consisting of isolated silicate tetrahedrons, each with one silicon atom bonded to four oxygen atoms (see p.50), combined with other elements. The negatively charged silicate tetrahedron (an anion) forms bonds with neighboring positively charged ions (cations) of other elements only at the points of the tetrahedron. The single tetrahedrons are held together by interspersed cations, such as iron and magnesium.

Nesosilicates include the minerals topaz, howlite, kyanite, sillimanite, zircon, staurolite, and the olivine and garnet groups. They have a compact habit and are dense. Nesosilicates are generally hard—7 or more on Mohs scale of hardness—and have a high refractive index.

Minerals in the olivine group range in composition from the magnesium-rich forsterite at one end to the iron-rich fayalite at the other end of what is called an isomorphous series—they have the same structure but different cations. The garnet group also forms with a range of compositions.

Hard gem
Clear crystals of topaz have a hardness of 8 on Mohs scale, and may be cut and fashioned into gemstones.

Kyanite

Zircon (brown crystal)

Olivine

Garnet

sorosilicates

Formerly known as disilicates, sorosilicates form double tetrahedrons—they have two silicate tetrahedrons (see p.50) linked together. One tetrahedron, or SiO_4 unit, is connected to the other at one of its points, and the remaining three points are available to join with neighbouring cations (positively charged ions) that are not silicon. Silicates with pairs of tetrahedrons are rare. They include minerals such as hemimorphite and vesuvianite, and the minerals in the epidote group. The arrangement of the tetrahedrons results in dense, moderately hard minerals (6–7 on Mohs scale, see pp.42–43), often with distinct cleavage.

Vesuvianite, also known as idocrase, is a hydrous calcium magnesium aluminum silicate. It forms as stubby tetragonal prismatic crystals and rarely occurs in the form of aggregates. The crystal faces often have fine striations (grooves) parallel to their length. The abundant rock-forming mineral epidote is hydrous calcium aluminum iron silicate. Cations of calcium, aluminum, and iron join the SiO_4 tetrahedrons, along with hydroxyl (OH). Epidote forms as prisms, often with finely striated faces.

Blue mass
Hemimorphite often occurs in altered zinc deposits. This specimen from Sardinia has formed as a lumpy blue crust with rounded protuberances on a rock matrix.

Vesuvianite (dark crystal)

Epidote

Piemontite

Zoisite var. tanzanite

inosilicates

Inosilicates (formerly called metasilicates) are formed from chains of silicate tetrahedrons. The chains result in minerals with mainly acicular (needlelike) habit that are mostly green, greenish gray, or white. Inosilicates with single and double chains form the major rock-forming mineral groups pyroxenes and amphiboles, respectively.

Pyroxenes are the most common inosilicates. They have two planes of cleavage—planes along which a mineral splits preferentially—that are at 90 degrees to each other. This group is further divided into orthopyroxenes and clinopyroxenes. Orthopyroxenes, such as enstatite, contain little or no calcium (when compared to clinopyroxenes) and belong to the orthorhombic crystal system. Clinopyroxenes, such as diopside, augite, jadeite, and spodumene, belong to the monoclinic crystal system. Jade refers to two different minerals—the clinopyroxene jadeite jade and the amphibole nephrite jade.

Amphiboles are a group of inosilicates that have cleavage planes at 120 degrees to each other. They are hydrous silicates and have the hydroxyl (OH) group as an essential part of their general formula. Amphiboles, such as asbestos, are often fibrous.

Green crystal
Dark green, bottle-green, and light green variants of the pyroxene mineral diopside are common, although colorless and violet crystals also occur.

Enstatite

Spodumene var. kunzite (pink crystals)

Asbestos

Nephrite jade

cyclosilicates

Cyclosilicates are made up of rings of silicate tetrahedrons. The rings may have three, four, or six elements, and these determine the crystal symmetry of the mineral. The crystals are characterized by their prismatic habit and may be trigonal (three-sided), tetragonal (four-sided), or hexagonal (six-sided). The most common cyclosilicates have hexagonal symmetry and groups of six rings, as seen in beryl. The rings give beryl moderate density and a hardness of 7.5–8 on Mohs scale.

Cyclosilicates occur in a wide range of colors. Dioptase, a rare copper mineral, is bright green. Pure beryl, called goshenite, is colorless. Colored beryl gets its color from trace elements—very small amounts of other elements. Chromium and vanadium make emerald green; iron and titanium give aquamarine its blue color; manganese makes morganite pink; and iron makes heliodor yellow. Iolite, also called cordierite, is pleochroic—it appears gray, violet, or yellowish blue when viewed from different directions. Benitoite, also pleochroic, occurs as blue, pink, or colorless crystals. Tourmaline is a complex borosilicate and is found in many colors. Some tourmaline crystals have two or more colors, for example, watermelon tourmaline.

Crystal cluster
The most common gemstone variety of beryl—aquamarine—typically forms large, clear crystals. This specimen shows a cluster of pale blue hexagonal prismatic crystals.

Cordierite

Benitoite

Dioptase

Watermelon tourmaline

phyllosilicates

In phyllosilicates, the silicate tetrahedrons are joined together to form parallel sheets held together by positively charged ions (cations). Each tetrahedron shares three of its four oxygen ions with neighboring tetrahedrons to produce a hexagonal sheet. The remaining oxygen ion forms a weak bond between two sheets. Phyllosilicate minerals often occur as thin lamellae or flakes that split relatively easily along planes of weakness called cleavage planes. Examples include the micas, talcs, and clays. These minerals generally have low density and a hardness of 2.5–3 on Mohs scale.

Micas have monoclinic crystal symmetry and are good electrical and heat insulators. They are divided into light- and dark-colored varieties, and those that contain lithium, such as lepidolite. The color is determined by the cation that joins with the tetrahedrons. Light-colored muscovite is rich in aluminum, while dark-colored biotite is rich in iron and magnesium.

Talc has a monoclinic crystal symmetry and is a poor conductor of heat. It feels slippery and can form pale, compact masses, called steatite or soapstone, with a hardness of only 1 on Mohs scale.

Perfect cleavage
Like all micas, muscovite has perfect cleavage, splitting into thin sheets. The sheets can be peeled off with a fingernail.

Biotite

Lepidolite

Yellowish green prehnite crystals on matrix

Talc

tectosilicates

Tectosilicates are the most complex silicates. Their silica tetrahedrons are joined together at all four points to form frameworks that are sometimes called "cages." Tectosilicates include minerals in the quartz, zeolite, and feldspar groups. Labradorite, microcline, and amazonite are all feldspars. The rigid framework structure of tectosilicates makes them hard, rating 6–7 on Mohs scale, and also gives them a low density.

The quartz group includes crystalline quartz, such as the colorless rock crystal, and varieties such as chalcedony that are cryptocrystalline (formed of microscopic crystals). Agate is a banded form of chalcedony. The many colors of the quartz family are due to trace elements—small amounts—of various positive cations that bond with the silica, such as iron in purple amethyst and yellow citrine, and manganese in rose quartz. The crystals are usually six-sided prisms ending in a point made of six faces. They have a conchoidal (shell-like) fracture and are often twinned, with no cleavage. They may also contain inclusions of other minerals, such as the brass-colored rutile in rutilated quartz and black tourmaline in tourmalinated quartz.

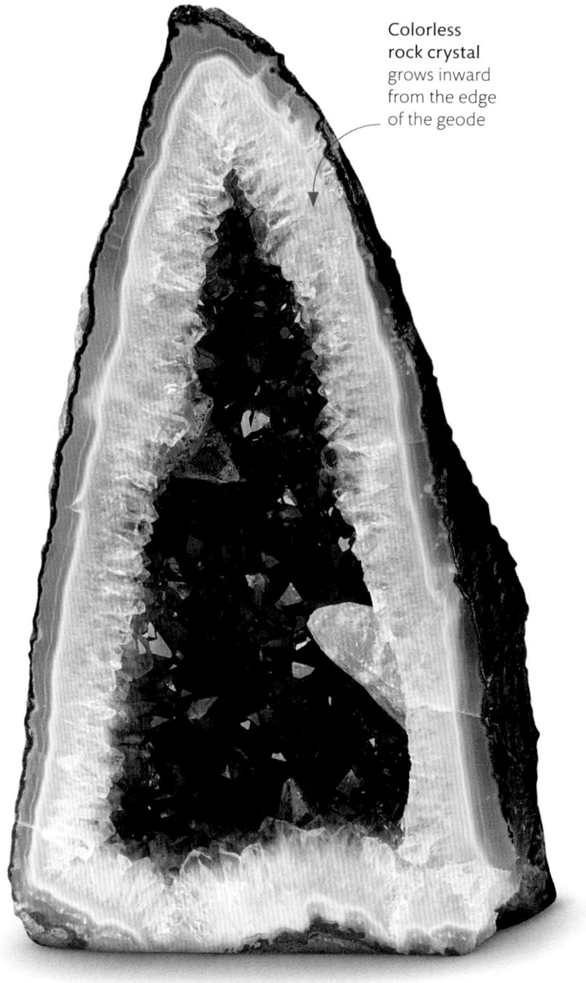

Distinct banding
A geode is a rock cavity lined with crystals that form when mineral-rich waters fill the space inside. Purple amethyst, colorless rock crystal, and gray-blue banded agate line this amethyst geode.

Labradorite

Amazonite

Rock crystal

Rose quartz

oxides

Minerals that have oxygen as the only negatively charged ion (anion) are classified as oxides. They are called hydroxides when the anion is oxygen combined with hydrogen. When the positive cations are smaller than the anion, they form compact, hard crystals. Ruby and sapphire, in the corundum (aluminum oxide) family, are dense and very hard, rating 9 on Mohs scale. They can be scratched only by diamond, which has a hardness of 10. Chrysoberyl has a hardness of 8.5, and its transparent variety is used as a gemstone. Alexandrite, a variety of chrysoberyl, is also used as a gemstone. It exhibits a color change, appearing emerald-green in daylight and red in artificial light.

Hematite, another oxide, is the most important metallic ore of iron. Powdered hematite has been popular as a red pigment since prehistoric times and was widely used in cave art. Cuprite is an important ore of copper, and cassiterite is the main ore of tin. Other oxides include rutile, chromite, spinel, magnetite, and radioactive uraninite, also known as pitchblende. Diaspore is an aluminum hydroxide and an important constituent of bauxite, an aluminum ore.

Gem-quality crystal
The water-worn blue sapphire crystal shown here is of gemstone quality. It has a pyramidal form and exhibits color zoning (uneven distribution of color) that is common in sapphires.

Chrysoberyl

Hematite

Rutile

Spinel (dark crystals) in quartz

anhydrous sulfates

Sulfates get their name from the sulfur that they contain as a complex anion (negatively charged ion). They are classified as anhydrous when they do not contain water in their formula. Barium sulfate, or barite, is characterized by its high density and has a specific gravity of 4.5. It gets its name from the Greek word *barus*, which means heavy. Other minerals within the barite group include celestine (strontium sulfate), anglesite (lead sulfate), and anhydrite (calcium sulfate). Celestine, also known as celestite, is even denser than barite. Its name is derived from the Latin word *caelestis*, meaning celestial—a reference to the pale blue, skylike color of some specimens. It is heavy but soft, with a hardness of only 3–3.5 on Mohs scale, and fragile due to its perfect basal cleavage, which means that it breaks easily and cleanly along planes of weakness parallel to the bases of the crystals. Anglesite is usually colorless, white, yellow, grayish, green, or blue. Its crystals are heavy but fragile and soft, with perfect cleavage.

Anhydrite is an evaporite mineral that forms by repeated evaporation of shallow water at temperatures above 108 °F (42 °C). It is used in the production of sulfuric acid and in the manufacturing of paper. Glauberite is also an evaporite, which forms in both marine and salt-lake environments.

Perfect cleavage
Anhydrite, such as the reddish specimen pictured above, breaks into tabular crystals and blocky rhombohedral crystals as a result of three well-developed cleavages.

Barite
(white crystals)

Anglesite
(colorless crystal)

Celestine

Glauberite

hydrous sulfates

Hydrous sulfates contain both sulfur (S) and water (H_2O) in their formulas. Hydroxide sulfates include oxygen and hydrogen (OH). The sulfur bonds with four atoms of oxygen to give a sulfate (SO_4). The best-known hydrous sulfates are gypsum and epsomite. They form by repeated evaporation of shallow water containing salts in deserts, in lakes, and at sea margins. As the saline water evaporates, gypsum is the first mineral to be precipitated, followed by anhydrite and then halite. Anhydrite can be rehydrated and altered to form gypsum.

Gypsum, used to make plasterboard, is hydrated calcium sulfate. It rates only 2 on Mohs scale and is so soft that it can be scratched with a fingernail. Transparent crystals of gypsum are called selenite, fibrous aggregates of long, satin-like crystals of the mineral are called satin spar, and compact masses of gypsum with a waxy appearance are known as alabaster. Epsomite, commonly known as Epsom salts, is hydrated magnesium sulfate. It is soft, light, and fragile. It forms small, acicular (needlelike) crystals and crusts. The bluish green minerals brochantite, chalcanthite, and linarite are other examples of hydrous sulfates—they are all hydrous copper sulfates.

Uniquely shaped mineral
This is a specimen of daisy gypsum, named for its radiating crystals that spread out to form shapes resembling stars or flowers.

Epsomite

Linarite (dark blue crystals)

Brochantite

Chalcanthite

anhydrous carbonates

Minerals that contain the negatively charged ion $(CO_3)^{2-}$ are called carbonates. Anhydrous carbonates do not include water molecules. The main anhydrous carbonate is calcium carbonate ($CaCO_3$), the mineral calcite. A soft mineral, with a hardness of 3 on Mohs scale, calcite can be scratched with a knife. It has three perfect cleavage planes, which form a rhomb that breaks into smaller rhombic fragments. The transparent and colorless form of calcite is called Iceland spar. Calcite occurs in rocks such as marble, limestone, and chalk. It can also have an organic origin: it forms the hard parts of some mollusks and corals; and bones in the human body are partly composed of calcium carbonate. Aragonite, which has the same chemical composition as calcite, also forms the skeletons of some marine organisms.

Other anhydrous carbonates include dolomite, a calcium magnesium carbonate; magnesite, a magnesium carbonate; rhodochrosite, a manganese carbonate; smithsonite, a zinc carbonate; and cerussite, a lead carbonate. Rhodochrosite is fragile, soft, and has perfect cleavage. It makes an attractive pink gemstone and is used to decorate small carvings and inlays.

Hanging tubes
Stalactites and stalagmites in limestone caves are made of calcium carbonate. Thin stalactites, called soda straws, can be seen here forming on the ceiling of the Inner Space Cavern in Texas.

Aragonite

Rhodochrosite

Smithsonite

Cerussite

hydrous carbonates

Hydrous carbonates contain the carbonate anion $(CO_3)^{2-}$, and hydrogen and oxygen as water (H_2O) or as hydroxide (OH). Azurite, a blue hydrous copper carbonate, was particularly popular as an artists' pigment in the 16th century, during the Renaissance. It is normally associated with the green mineral malachite, another hydrous copper carbonate. The name malachite comes from the Greek word *molochitis*, meaning mallow plant. The green malachite pigment is called "green mountain" and the blue azurite "blue mountain." Polished slabs of malachite are used as decorative stones for carvings, inlays, boxes, and tables, as well as to cover walls and facades. Azurite is less stable in air than malachite and can be replaced by malachite. Malachite interbanded with azurite is called azurite-malachite and, when cut and polished, makes an attractive decorative stone or gemstone.

The blue or green aurichalcite forms in oxidized zones of copper and zinc deposits with minerals such as malachite and azurite. The translucent to opaque hydrozincite, or zinc carbonate hydroxide, forms in the altered parts of zinc-bearing veins. It has a pearly to silky, or sometimes dull, luster.

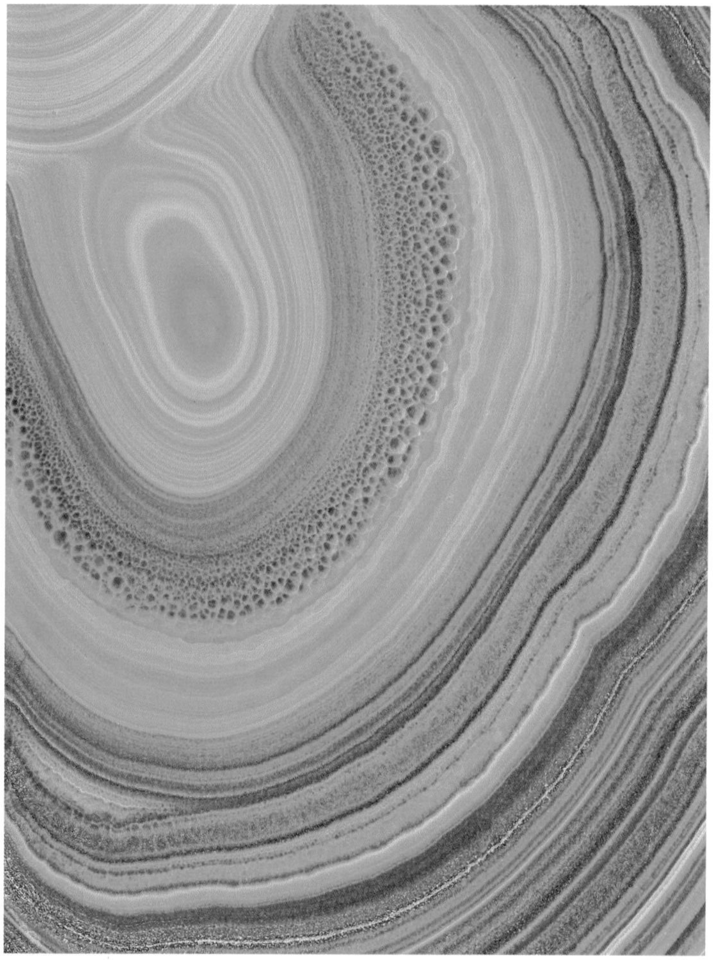

Concentric banding
This cross-sectional view of cut and polished malachite shows the mineral's typical bands of light green, dark green, and black.

Azurite (dark blue crystals) on goethite

Azurite-malachite (dark blue and green crystals) in rock matrix

Aurichalcite (light blue crystals) on limonite

Hydrozincite

sulfides

Minerals in which sulfur, a nonmetal, combines either with a metal or semi-metal are called sulfides. They form in places where there is insufficient oxygen for the negatively charged sulfur anion to combine with other elements to form a sulfate.

Sulfides form the major ore minerals of many metals, such as lead, zinc, iron, and copper. Galena, one of the main sources of lead, is often found alongside pyrite, chalcopyrite, and sphalerite. Chalcopyrite is an important copper ore, sphalerite (or zinc blende) is the most common ore of zinc, cinnabar is the principal ore of mercury, and realgar and orpiment are both arsenic sulfides. Crushed powders of realgar, orpiment, and cinnabar have been used in artists' pigments for the colors yellow, red, and a brilliant vermilion. The metal ores have high density but are usually soft—mostly 1.5–4 on Mohs scale.

Pyrite and marcasite are iron sulfides, with a hardness of 6–6.5 on Mohs scale. Pyrite, also known as fool's gold, is distinguished from marcasite by its higher density and cubic crystal symmetry—marcasite has orthorhombic crystal symmetry.

Metallic luster
A copper and iron sulfide, chalcopyrite is opaque and brassy yellow when mined, but it often develops an iridescent tarnish when exposed to air.

Galena

Sphalerite

Orpiment

Pyrite

halides

Halides are named after the group of elements called halogens, which includes fluorine, chlorine, bromine, and iodine. Fluorite is a halide and so are the salts halite and sylvite. Most halides form as cubic crystals with perfect octahedral cleavage, producing triangular shapes at the corners of the cubic crystals.

Fluorite ranges in color from yellow to pale green, pink, blue, and the purple variety from England known as Blue John. The optical phenomenon fluorescence is named after fluorite, which gives off a cold, incandescent light—usually blue or violet—under ultraviolet light. Some fluorite also phosphoresces—the color remains for a short while after the light source has been removed.

Halite (sodium chloride) is common edible salt. Halite, sylvite (potassium chloride), and carnallite (potassium magnesium chloride) are evaporites—they form by the repeated evaporation of saline waters, in the same way as the minerals gypsum and anhydrite (see pp.332–333). Apart from being used as table salt, halite is mainly used in the chemical industry and for gritting roads. Sylvite is used as a fertilizer, is more soluble in water than halite, and is less common as an evaporite.

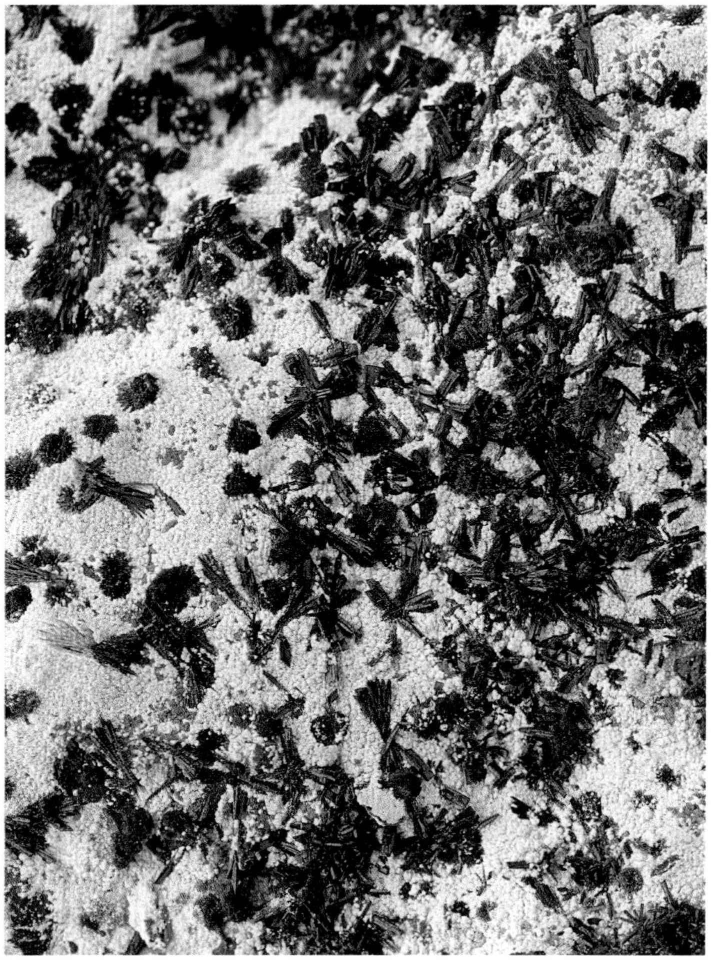

Pointed crystals
Atacamite, shown here, is a copper halide mineral that has crystallized in an arid environment to form small, greenish blue and dark acicular (needlelike) crystals.

Fluorite

Halite (colourless crystals)

Sylvite

Carnallite

phosphates

Phosphates have the negative anion $(PO_4)^{3-}$. They form as a tetrahedron that can occur singly or, like silica tetrahedrons, combine to form pairs, rings, chains, and frameworks.

The most common phosphates are the minerals in the apatite group. They are hard (5 on Mohs scale), dense, and have poor cleavage. The members of the apatite group are classified by the main negatively charged anion. Fluorapatite, with fluorine, is the most common apatite; hydroxylapatite, with hydroxide, is less common; and chlorapatite, with chlorine, is rare. Tooth enamel is made of 96 percent hydroxylapatite and is the hardest substance in the human body.

Turquoise, a complex hydrated copper aluminum phosphate, was discovered in 5000 BCE in Mesopotamia (present-day Iraq). It gets its name from the French word *turquois*, meaning Turkish, as turquoise was transported to Europe via Turkey. It varies in color from sky-blue to bluish green. Amblygonite and brazilianite are other common apatite gemstones. Pyromorphite is a minor ore of lead, and can be green, orange, gray, brown, or yellow in color.

Hexagonal crystals are typically transparent or translucent

Fragile crystals
Pyromorphite has fragile, prismatic crystals that break easily and leave irregular fractured ends, as can be seen in this green specimen from Guangxi, China.

Apatite (dark brown crystals)

Turquoise

Amblygonite

Brazilianite

mineraloids

Mineraloids are naturally occurring, mineral-like geological material. They are noncrystalline and not definite enough in chemical composition or in physical properties to be considered minerals. Hydrocarbons, tektites, volcanic glass, palagonite, chrysocolla, amber, and opal are all mineraloids.

Chrysocolla, a secondary copper mineraloid, is typically found as glassy masses and unknown as crystals. What appear to be crystals are actually pseudomorphs (a crystal consisting of one mineral but having the shape of another). In this case, the chrysocolla takes the shape of a previously existing mineral. Precious opal is a hardened, noncrystalline amorphous silica gel that is also never found as crystals. Jet, a hard, black, gem-quality hydrocarbon, is the fossilized remains of fir trees that lived 180 million years ago.

Glassy mineraloids form where cooling has been too fast for crystals to form. These are known as volcanic glasses and include obsidian, pitchstone, and pumice. Mahogany and snowflake are varieties of obsidians. Pele's hair and Pele's tears are volcanic glasses named after the Hawaiian goddess Pele, for their resemblance to strands of hair and tear-shaped glassy droplets respectively.

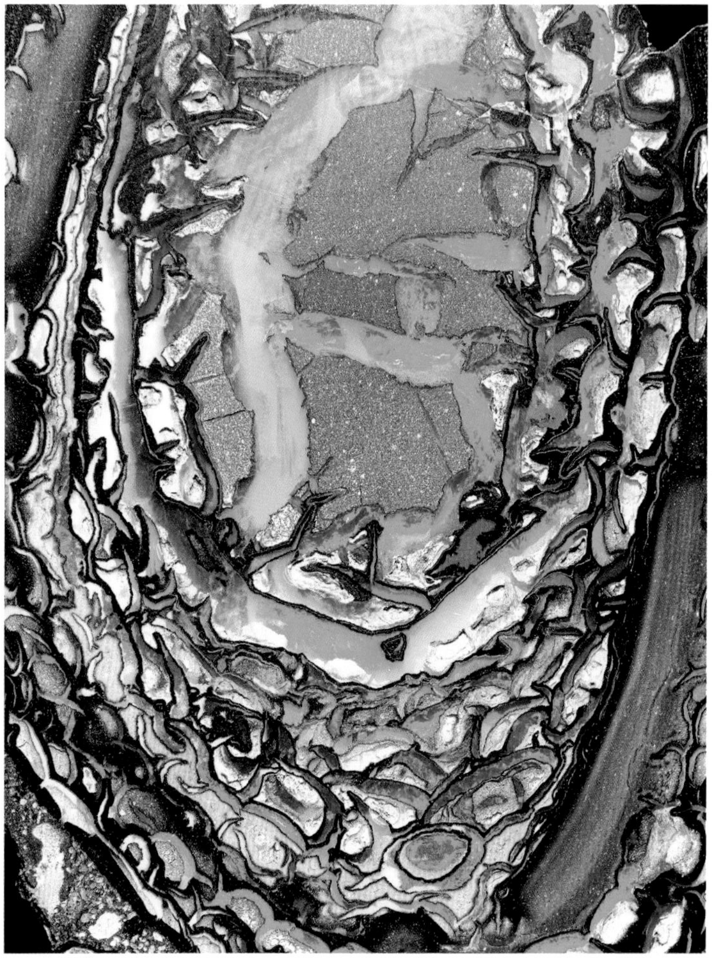

Colorful variety
In this cross section of an Australian nodular opal, the cracks in the sandstone nodule are filled with opal. This variety is called common opal as it does not show the flashes of color (known as play of color) of precious opal.

Chrysocolla

Jet

Mahogany obsidian

Snowflake obsidian

rocks

Rocks are generally defined as natural aggregates, or combinations, of one or more minerals. Most rocks contain more than one mineral—pure marble, made of calcite, is one of the few rocks composed of a single mineral. Geologists who study the classification and description of rocks, as well as their composition, texture, and origin, are called petrologists.

Rocks are classified by their method of formation into three types: igneous, sedimentary, and metamorphic rocks. Each type is further subdivided: igneous rocks are split into intrusive and extrusive igneous rocks; sedimentary rocks are classified by the size, shape, and arrangement of their clasts (grains) or by their composition; and metamorphic rocks are defined by the different temperatures and pressures at which they formed, with gradual boundary zones between them.

By studying rocks and their arrangement, petrologists can reconstruct the history of how the processes of plate tectonics, weathering, erosion, transportation, and deposition occurred. They can also understand past climate change and developing landscapes. The study and interpretation of present-day geological processes allows petrologists to interpret geological events that took place millions of years ago.

History of rocks
This 1817 engraving from French mining engineer and mineralogist François-Michel de Rozière's *Description de l'Egypte* (*Description of Egypt*) shows rock samples collected close to ancient copper and lead mines at Mount Baram, Egypt.

Bedrock of Earth
Crystals of white quartz, pink feldspar, and black tourmaline can be seen in this specimen of granite, an intrusive igneous rock. Granite makes up a large proportion of Earth's continental crust.

1

2

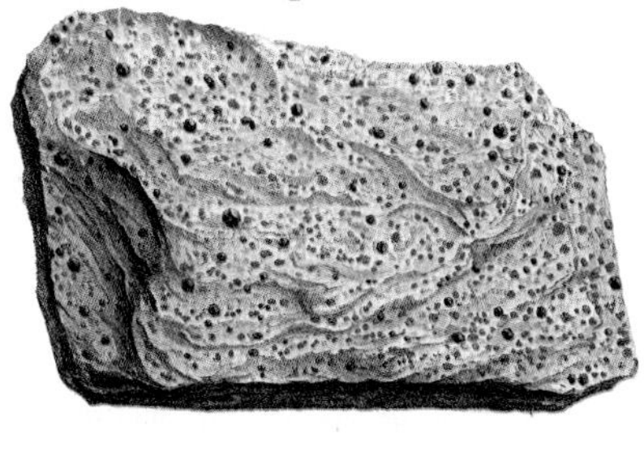

3

4

5

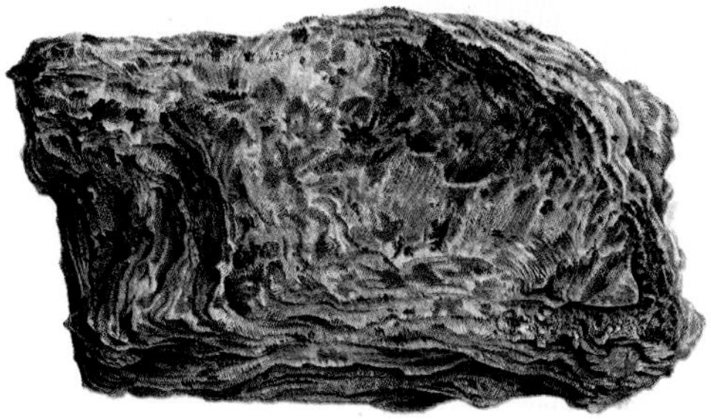

6

7

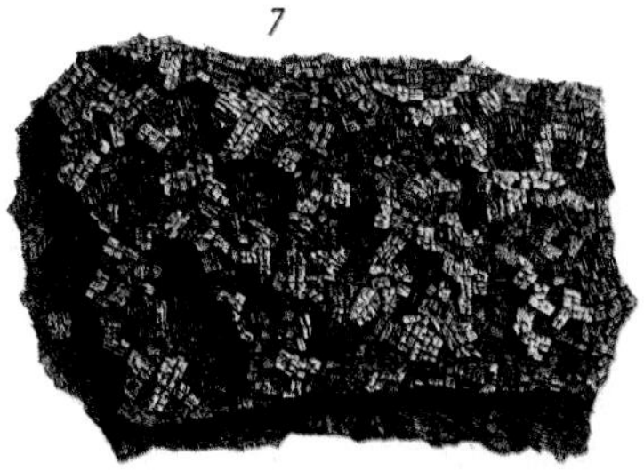

8

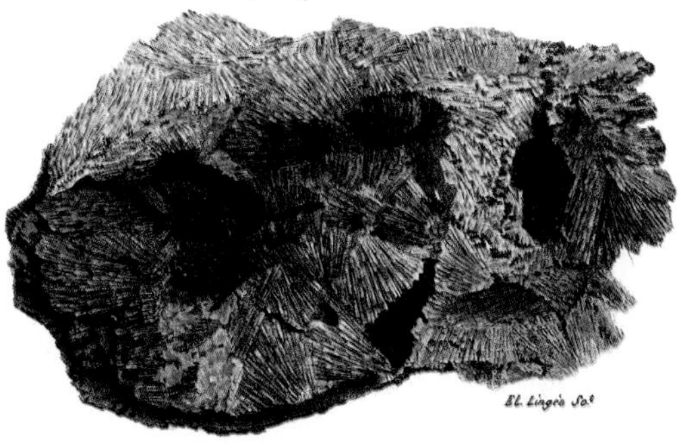

El. Lingé Sc.t

intrusive igneous rocks

Geologists commonly classify igneous rocks according to how they were formed. Magma forms intrusive igneous rocks when it cools and solidifies beneath Earth's surface. Igneous rocks form at high temperatures of more than about 1,100 °F (600 °C). Plutonic igneous rocks solidify deep beneath the surface, and form large features called batholiths. Hypabyssal igneous rocks solidify at shallower depths as smaller igneous intrusions such as sills and dykes. The crystals in igneous rocks are interlocked, like a mosaic, and usually do not show any preferred alignment.

Intrusive rocks are also classified by the sizes of their crystals or mineral grains. Slow cooling of magma at great depth allows large crystals to form in rocks such as granite, gabbro, and peridotite. Individual minerals are large enough to be identified with the naked eye. Sometimes the last remnant of magma solidifies so slowly that very large crystals, up to several yards long, can form. These are called pegmatites. At shallower depths, faster cooling allows only medium-size crystals to form in rocks such as dolerite and microgranite. Crystals are recognizable with the naked eye, but the individual minerals are hard to identify without a x10 magnifying lens.

Towering columns
Columns of intrusive dolerite at Cape Pillar, on the Tasman Peninsula in Tasmania, Australia, rise to a height of 1,000 ft (300 m) above sea level. They are the highest cliffs in Australia.

Diorite

Granite

Gabbro

Peridotite

extrusive igneous rocks

When magma reaches Earth's surface in volcanic eruptions, it is called lava, which cools to form extrusive igneous rocks. These are classified by mineral grain size and mineral content. Basalt is the most common extrusive igneous rock. It is mainly formed at mid-oceanic ridges, where tectonic plates move apart and erupted lava cools to form new oceanic crust.

Volcanic rocks are also classified as types of eruption product, such as lava, ash, volcanic bombs, or ignimbrites from pyroclastic flows. Lava may be runny with low viscosity or thick with high viscosity. Basaltic lava has low viscosity and generally forms less explosive, low-lying volcanoes such as those in Hawaii. Basalts have low silica content and are dark-colored and dense. Lavas with a higher silica content are more viscous and more explosive, forming steeper-sided volcanoes such as Mount Fuji, Japan. These lavas cool to form lighter colored, less dense rocks such as andesite. Rhyolitic lava is extremely viscous and results in very explosive eruptions. It cools to form rhyolite.

When lava cools too fast for crystals to form, it results in a glassy rock, such as obsidian. Dissolved gases escaping from the lava may leave bubbles that are preserved in the rock as vesicles. Pumice has so many vesicles that it can float.

Geometric rocks
The Giant's Causeway in Northern Ireland is made of more than 40,000 polygonal basalt columns that formed as the lava cooled and contracted.

Andesite

Rhyolite

Obsidian

Pumice

sedimentary rocks formed on beaches and in seas

Sedimentary rocks are formed by processes on Earth's surface and classified by their composition and texture (the size, shape, and arrangement of their clasts, or grains), which is determined by the environment in which they are deposited. Once deposited, sediments become compacted and cemented to form sedimentary rocks. By convention, grain sizes are defined as boulder (100–250 mm), cobble (65–100 mm), pebble (4–65 mm), granule (2–4 mm), sand (2 mm–62.5 µm—which is 1/16 mm), silt (62.5 µm–4 µm), and clay (less than 4 µm). Gravel includes cobbles, pebbles, and sand. Mud includes silt and clay.

Several kinds of sedimentary rock are formed from sediments deposited in or near sea water. In a high-energy environment such as a storm beach, large boulders, cobbles, and pebbles are deposited and sorted by tides. They may form conglomerates with rounded clasts. In lower energy environments, such as a sand beach, sand is deposited to form sandstone. In very low-energy environments, such as an estuary or deep sea, muds and clays are deposited, forming mudstone and shale. In warm, shallow seas, corals and microorganisms precipitate calcium carbonate, which accumulates to form chalk and fossiliferous limestone. Oolitic limestones are chemically precipitated in shallow water affected by wave action.

Fossiliferous limestone
Limestones often contain fossils, mainly of animals, such as this sample found in Estonia, from the Ordovician period. It shows specimens of trilobites, brachiopods, and bryozoans.

Black shale

Graywacke

Chalk

Oolitic limestone

sedimentary rocks formed in fresh water

Rock fragments, mineral grains, and dissolved minerals are transported by water, including fresh water flowing in rivers. When the energy of the water is reduced, rock fragments and mineral grains will be deposited. Geologists classify sedimentary rocks as having a chemical origin like halite, a biological (organic) origin like chalk and coal, or a mechanical (clastic) origin like conglomerates and breccias. Clasts are rock fragments held within a matrix of finer grained sediment such as sand, silt, or clay and cemented, typically by calcite or silica.

The well-rounded boulders and pebbles formed by erosion during transit form conglomerates in river channel deposits. Sandstones are also common river deposits, while flood-plain deposits are typically laminated mudstones, which may show cracks when their surfaces dry out. The low-oxygen conditions found in swamps allow large volumes of plant material to accumulate. Over time, and with increasing temperature and pressure due to burial, peat becomes lignite, a brown coal. Further coalification changes the brown coal to a harder, black coal called bituminous coal, and then to the hardest and best-quality coal called anthracite. Lakes are very low energy environments and are typified by very fine-grained sediments such as muds and clays, which form mudstones and shales.

Conglomerate cliff
Distinctive in their appearance, conglomerate rocks consist of small, smooth stones cemented together. The weathered rock cliff in Greece pictured here has rounded clasts within the conglomerate rock.

Sandstone

Bituminous coal

Anthracite coal

Shale

sedimentary rocks formed in deserts

Sandstone is formed by the deposition and accumulation of sand. As younger sediments are added, older, deeper sediments become compacted and lithified (turned to stone) in a process called diagenesis. Sedimentary rocks can be classified as argillaceous rocks, such as shale, that are made of very fine-grained clay minerals; arenaceous rocks, such as sandstone and siltstone, that mainly comprise quartz grains that are 0.002– 0.07 in (0.06–2 mm) in size; and rudaceous rocks, such as breccia, that have clasts more than 0.07 in (2 mm) in diameter with a sand or clay matrix.

Desert sandstones have well-rounded grains, as they rub against each other while being transported by desert winds. They are reddish brown due to a coating of hematite, an iron oxide. Desert breccias are a mix of large rock fragments and sand, and are formed by occasional flash floods in deserts. They are not transported far and are deposited quickly when the flood subsides, so they are not rounded or sorted.

Playas are shallow lakes formed in deserts as a result of flash floods or where sea water becomes trapped in coastal lakes. When salt-laden waters evaporate in the high heat of deserts and the water becomes saturated, the salts precipitate to form rocks such as gypsum, halite, and evaporite.

Layered wonder
Weathering and erosion reveal the different colored layers and cross-bedding of the desert sandstone in the Vermilion Cliffs of Arizona.

Desert sandstone

Red siltstone

Breccia

Evaporite

regional metamorphic rocks

Metamorphic rocks form when an existing rock undergoes changes in its structure and mineral composition due to high heat or pressure, or both. The changes occur due to recrystallization without melting. The resulting type of metamorphic rock depends on the original rock type and the metamorphic processes that it has undergone. During regional metamorphism, the rock transforms due to both heat and pressure. When continents collide at tectonic plate boundaries, rocks are both buried and compressed, leading to regional metamorphism on a large geographic scale.

Regional metamorphism of quartz sandstone produces metaquartzites, while limestones transform into marble. Mudstones and shales change to slate, which can further metamorphose into phyllite, schist, gneiss, and migmatite, as temperature and pressure increases. Slate is a fine-grained metamorphic rock. It has slaty cleavage and splits into thin sheets. Schist is medium-grained and typically composed of mica and garnet. The mica sheets are aligned perpendicular to the direction of pressure by the metamorphism. Gneiss is coarse-grained with mineral grains more than 0.2 in (5 mm) in diameter. It has dark bands of minerals such as biotite, and light bands of quartz and feldspar.

Altered rock
Metaquartzite forms when the silica in quartz sandstones dissolves and is redeposited as quartz veins due to the high pressure and temperature of regional metamorphism.

Slate

Garnet mica schist (dark crystals)

Gneiss

Migmatite

contact metamorphic rocks

Contact metamorphism—also known as thermal metamorphism—occurs due to an increase in temperature, but not in pressure, so the minerals in the altered rock are not aligned. When heat from the magma from an igneous body—such as the large granitic Cornubian batholith in the UK, or smaller dykes and sills elsewhere—alters the surrounding rocks, contact metamorphic rocks are produced. The degree of metamorphism is greater in rocks such as eclogite, which form nearest to the intrusion, since they are subjected to higher temperatures than rocks farther away.

The rocks surrounding the original rock, called country rock, may contain mineral ores, which occur in a series of zones determined by the temperature at which they crystallized. For example, cassiterite (tin ore) may form nearest to the intrusion, followed by chalcopyrite (copper ore) and galena (tin ore). Contact metamorphism of shale forms a hard, flinty rock called hornfels, with andalusite hornfels and spotted rock forming farther away from the intrusion. It also changes sandstones to metaquartzites and limestone to marble and skarn. Fossils and sedimentary structures in limestones are destroyed by metamorphism.

Hidden gem
Garnet and pyroxene are often found in skarn rocks. Here, a dark garnet crystal is within a brown skarn layer, which is bounded by two gray layers of gray rock.

Eclogite

Hornfels

Metaquartzite

Marble

dynamic and burial metamorphic rocks

Dynamic metamorphism typically occurs below Earth's surface, where directed pressure crushes and distorts rocks. The rocks produced range from fault breccias—with large, angular shattered rock fragments—to fine-grained rocks with obvious banding, such as mylonite. Dynamic metamorphism often occurs along thrust faults associated with intense compression, which can occur when continents collide. Burial metamorphism occurs when temperature and pressure increase due to burial rather than compression. For example, jadeite jade, found in Myanmar, was formed by the collision of the Indian and Asian plates about 50 million years ago.

Blueschists form at low temperatures. Greenschists form at higher temperatures and lower pressures than blueschists. Both occur at subduction zones, where one tectonic plate moves beneath another. For example, when an oceanic plate collides with a continental plate and the denser oceanic crust is subducted beneath the less dense continental crust, the sediments and basalts of the oceanic plate are metamorphosed to form schists. Serpentinite, found in Cornwall, UK, formed from the metamorphism of peridotite rocks found in the oceanic crust.

Mantle rock
Serpentinite is formed by the metamorphism of silica-poor (ultramafic) rocks found in Earth's mantle. It has a slippery feel due to the presence of the mineral talc.

Mylonite

Jadeite

Blueschist

Greenschist

meteorites and impactites

Meteors are fragments of rock and space debris that enter Earth's atmosphere. They usually burn up or vaporize as shooting stars, but some survive and land on Earth as meteorites. Passage through Earth's atmosphere melts their surfaces to form a thin layer called a fusion crust, which protects the interior of the fragments. Meteorites are categorized as iron, stony, or stony-iron meteorites based on their mineral composition and structure.

Although many thousands of meteorites land on Earth every year, most go unnoticed. They are called falls when they are seen to fall, and finds when they are not observed falling but discovered later. There have been about 1,500 witnessed meteorite falls and almost 66,000 well-documented meteorite finds worldwide. All the fragments of a fall share the same name, usually the name of the place where they are found. It is more complicated to match finds that may have been discovered many years apart and thousands of years after they landed. Each meteorite is identified by its name and weight and, following the introduction of the *Guidelines for Meteorite Nomenclature* by the Meteorological Society's Nomenclature Committee in 1980, each is also given a unique number.

If a meteorite is large enough and traveling fast enough, it may form a crater—such as Meteor Crater in Arizona—throwing terrestrial rocks in the air and scattering them. The intense heat, pressure, and shock of the impact can also melt the surrounding rocks to form impactites, including glassy tektites such as moldavite and Libyan desert glass (see p.353).

Prophetic role
Meteors have been considered omens of both good and bad events since as far back as Biblical times. The Apocalypse foretold in the words "And the stars of heaven fell to Earth" in the Bible (Revelation 6:12–17) is shown in this print from a 1498 woodcut by German artist Albrecht Durer.

Pock-marked, glassy surface

Glassy tektite
Indochinites are dark-colored tektites named after the region where they are found. Tektites occur in various colors, shapes, and sizes. They may be as small as a few micrometers or as large as 4 in (10 cm) in diameter, and range in color from green or dark brown to black.

meteorites

Many thousands of meteorites land on Earth every year. Most originate from the asteroid belt, which orbits the sun between the planets Mars and Jupiter. The belt is made of rocky bodies that remained when the planets in the solar system, including Earth, formed almost 4,600 million years ago. Like Earth, some large asteroids have a core, mantle, and crust. Asteroids are knocked out of their orbits due to collisions with other asteroids or space debris. The type of meteorite that lands on Earth depends on which part of an asteroid it originated from. Iron meteorites come from the core of an asteroid and are made of iron and nickel, like Earth's core. Stony-iron meteorites are fragments from the core and mantle, and stony meteorites from the crust. Stony meteorites are mainly composed of silicate minerals.

About 80 percent of the meteorites that reach Earth are stony meteorites. They are classified as chondrites, or carbonaceous chondrites, if they contain tiny grains called chondrules, and achondrites, such as shergottite, if they do not. Stony-iron meteorites are divided into mesosiderites and pallasites, and constitute less than 2 percent of all known meteorites. Pallasites contain crystals of the gemstone peridot (olivine).

Iron meteorite
The Canyon Diablo is an iron meteorite that landed in present-day Arizona some 50,000 years ago. It weighs more than 30 tons (27 metric tons).

Chondrite

Shergottite

Mesosiderite

Pallasite

impactites

The impact of a very large meteorite may form a crater, and the heat, pressure, and shock waves produced can alter the rocks, sand, and soil to form new rock types called impactites. Tektites, natural glass objects, are a type of impactite. They are named after the area in which they are found. The most abundant tektites are the dark-colored Indochinites found scattered over a vast area that includes parts of China, Thailand, and Cambodia. Their impact crater has not been identified, nor has the crater for Asian Chinites or Australian Australites. Libyan desert glass, which is translucent yellow, formed about 29 million years ago on what is now the Egyptian–Libyan border when desert sands were melted by the heat of a meteorite impact.

Moldavite, an olive-green translucent natural glass tektite, is found across a large area spanning Moravia and Bohemia in the Czech Republic. It is used as a gemstone and may have formed as a result of the meteorite impact that formed the Nördlinger Ries crater in present-day Germany about 15 million years ago. This impact also resulted in the formation of impact breccias—angular fragments of shattered rocks formed on land, thrown into the air by the impact, that fell back to Earth in and around the crater and over time became cemented.

Moldavite
One of several types of tektite, moldavite forms when silica-rich rocks melt due to the heat produced by meteorite impacts. The rocks cool and solidify, often as irregular or droplet shapes.

Indochinite

Australite

Libyan desert glass

Impact breccia

Initially interpreted as jellyfish imprints, these structures are now believed to be holdfasts where organisms were anchored to the seafloor

Life in the Proterozoic
Most organisms in the Proterozoic eon were microscopic, so their fossils usually consist of chemical traces. From the Cryogenian period (c. 720–635 MYA), evidence of animals is known, foreshadowing the earliest large fossils such as *Eoporpita* (above) in the Ediacaran.

the geological timescale

Scientists represent Earth's c. 4.6-billion-year history using a standardized geological timescale, created by measuring the ages of rocks directly or using the relative order in which rock layers were laid down. The ages of rocks are usually determined from the steady decay of certain radioactive elements—a process called radiometric dating (see p.28). Where these elements are not available, the fossils contained within the rocks can help link them with layers of known age. This approach shows that Earth began as a hostile, volcanic environment. As it cooled and oceans formed, single-celled life was able to originate in the oceans. About 2.4 billion years ago, photosynthetic bacteria caused oxygen to build up slowly in the atmosphere. More complex organisms, including multi-celled algae, arose afterward and survived reorganizing continents and climate swings. By the Ediacaran period, which began c. 635 MYA, we have evidence of a diversity of animals.

MILLIONS OF YEARS AGO (MYA)	EON	ERA	PERIOD	EPOCH
0.0117	PHANEROZOIC	CENOZOIC	QUATERNARY	HOLOCENE
2.58				PLEISTOCENE
5.3			NEOGENE	PLIOCENE
23				MIOCENE
34			PALEOGENE	OLIGOCENE
56				EOCENE
66				PALEOCENE
101		MESOZOIC	CRETACEOUS	LATE
145				EARLY
164			JURASSIC	LATE
174				MIDDLE
201				EARLY
237			TRIASSIC	LATE
247				MIDDLE
252				EARLY
299		PALEOZOIC	PERMIAN	
359			CARBONIFEROUS	
419			DEVONIAN	
444			SILURIAN	
485			ORDOVICIAN	
539			CAMBRIAN	
635	PROTEROZOIC	NEOPROTEROZOIC	EDIACARAN	
720			CRYOGENIAN	
1,000			TONIAN	
1,600		MESOPROTEROZOIC		
2,500		PALEOPROTEROZOIC		
4,000	ARCHEAN			
4,600	HADEAN			

Geological intervals
Scientists subdivide geological time into smaller intervals that reflect major events in tectonics, climate, or the evolution of life. Many of the Phanerozoic intervals are defined by changes in index fossils, which often reflect extinction or diversification events.

Index fossils
This 1837 dictionary of natural history demonstrates our long fascination with ammonites. Ammonites came to be recognized as important index fossils because they are abundant, widespread, and diverse. Certain ammonite species can link rock layers. For example, the beginning of the Jurassic period is marked at the first appearance of the ammonite species *Psiloceras*.

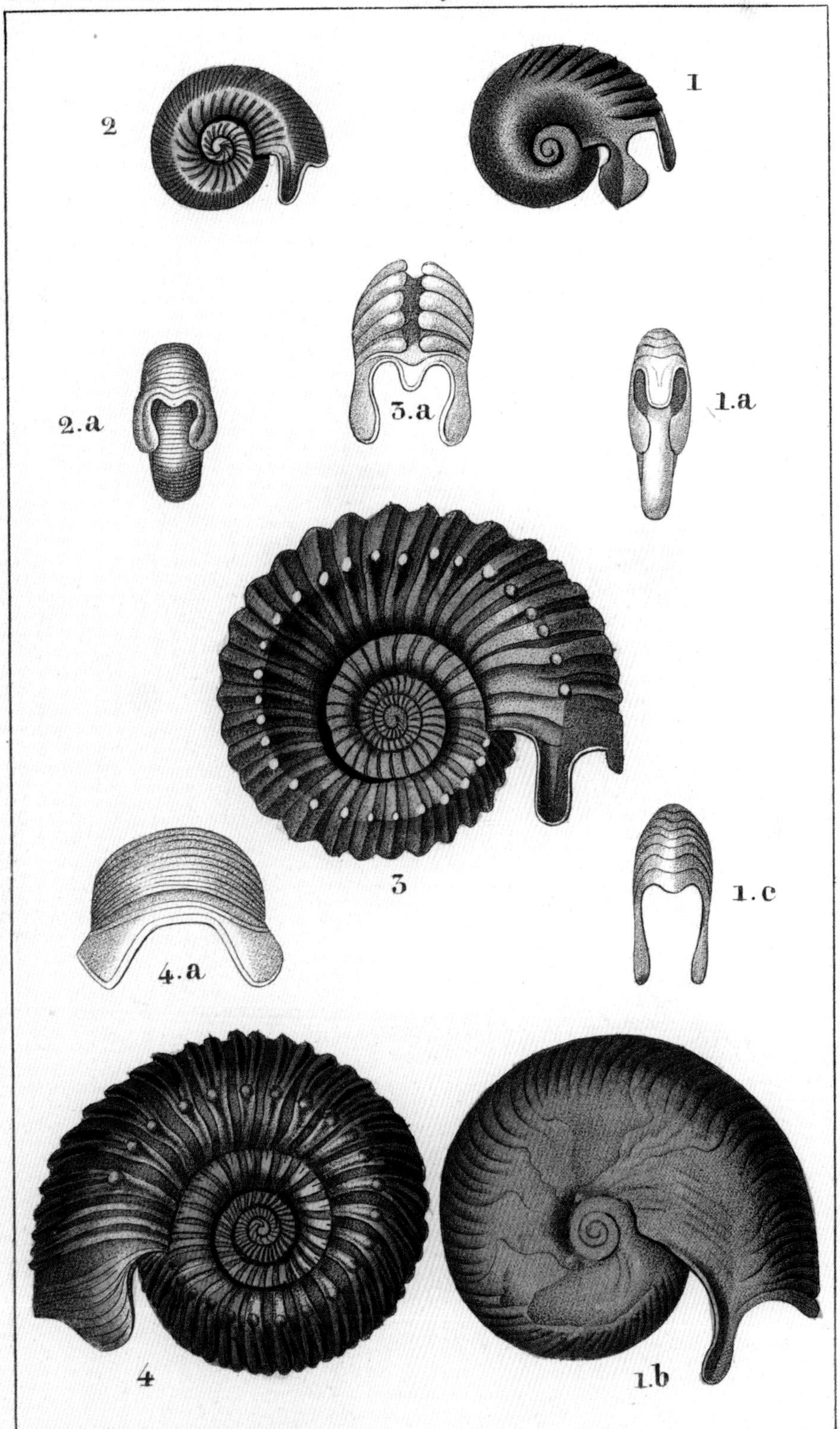

Pretre dip. Corsi inc.

1. AMMONITE di Caen (Def.) *Giovane individuo?*
1.a. *Id. ved. di faccia* 1.b *Id. più avanzata* 1.c. *Id. ved. di faccia.*
2. AMMONITE di Deslonchamps (*Def.*) *giov. individuo?* 2.a *Id. ved. di fac.*
3.____ id.____ di Bayeux (*Def.*) 3.a *Id. ved. di faccia.*
4.____ id.____ di Braikenridge (*Sow.*) 4.a *Id. ved. di faccia.*

Ediacaran period

DATES: 635–539 MYA **ERA:** Neoproterozoic **EON:** Proterozoic

The Ediacaran was the last period of the Proterozoic eon, part of the so-called "Precambrian," before the evolution of complex life. By the Ediacaran, oceans and continents had formed, the atmosphere had abundant oxygen, and single-celled life forms had already proliferated in the oceans.

The Ediacaran began as Earth thawed after a global ice age in which nearly the entire surface of the planet was covered in ice. Shortly after, the first multicellular organisms appeared, represented by body fossils and enigmatic trace fossils (such as trackways) in a variety of shapes. Roughly 600 million years old, these were the earliest fossils of organisms visible to the naked eye. Scientists still debate whether some of these, such as *Charnia*, represent true animals, and even whether some distinctive traces are of living organisms at all. However, the geochemistry of some of these fossils, for example *Dickinsonia*, is consistent with animals. Exceptionally preserved internal features in others draw links to later Cambrian groups, particularly the broad set that includes mollusks, brachiopods ("lamp shells"), and worms. Regardless, these organisms would vanish with the radiation of more complex animals in the Cambrian.

Under discussion
The symmetry of many Ediacaran fossils, such as *Tribrachidium*, suggests that they are the remains of multicellular organisms. Like many other fossils, the relationship of *Tribrachidium* to living organisms is still debated.

Spriggina
Uncertain affinity

Dickinsonia
Uncertain affinity

Mawsonites
Uncertain affinity

Charnia
Uncertain affinity

Cambrian period

DATES: 539–485 MYA **ERA:** Paleozoic **EON:** Phanerozoic

The beginning of the Cambrian period and the Phanerozoic eon was marked by the first fossil burrows (remains of holes or tunnels in the sea floor)—clear evidence of complex life forms and true animals. Within just a few million years, myriad unique creatures had evolved, in an event known as the Cambrian Explosion. The evolution of hard body parts, such as shells and exoskeletons, played a major part in the diversification. Beyond providing the animals with protection against the earliest-known predators, these hard parts enhanced their chances of being fossilized, likely accounting for the plethora of fossils from this period.

Among these early pioneers were the hugely successful arthropods, the group that includes trilobites and modern-day insects, arachnids, and crustaceans. Some of these appear to have ventured onto land, but they would have found it barren: land plants evolved 45 million years later. The first mollusks arose, along with brachiopods, which look superficially similar to clams. Also appearing were many worms; urchin-like echinoderms; early fishlike vertebrates; and less familiar animals with no living descendants, such as the five-eyed *Opabinia*.

Origin of animals
Fossils, such as the branched sponge *Vauxia*, found in the Burgess Shale fossil site in Canada, provide some of the earliest records of evolutionary radiation in animals. Most modern animal groups originated during the Cambrian.

Acadoparadoxides briareus
Trilobite

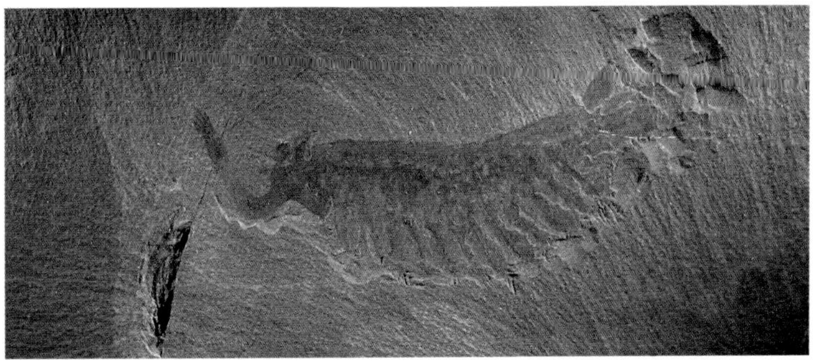

Opabinia
Arthropod relative

Anomalocaris (mouthpart)
Arthropod relative

Ordovician period

DATES: 485–444 MYA **ERA:** Paleozoic **EON:** Phanerozoic

The Great Ordovician Biodiversity Event, the second most important evolutionary event after the Cambrian Explosion, occurred in the Ordovician. Several unusual Cambrian life forms were replaced by groups that would rule the oceans all through the Paleozoic. These included many brachiopods, echinoderms such as sea lilies and starfish, corals, mollusks, and crustaceans, and each group doubled or tripled in diversity. These animals formed rich reef ecosystems, while their planktonic larvae, floating near the surface, filled the oceans with food. Multi-level food webs were established for the first time. Fishlike vertebrates began to capitalize on this abundance and developed mineralized skeletons and teeth that protected them and helped them catch prey.

The first land plants began to colonize the continents. These were simple mossy plants, lacking roots, leaves, or seeds, and they could only survive in moist environments. Nevertheless, they had a profound impact on life on Earth: they changed the patterns of nutrient cycling between land and the oceans, and caused global temperatures to drop, plunging Earth into an ice age. As a result, life experienced its first major mass extinction at the end of the Ordovician.

Evolving adaptations
Alongside other arthropods, the trilobites, including the stalk-eyed *Asaphus kowalewskii*, diversified in the Ordovician. Trilobite eyes evolved into an array of sizes and shapes as they moved into different habitats.

Stenaster obtusus
Starfish

Dinorthis anticostiensis
Brachiopod

Tetragraptus
Graptolite (extinct filter-feeding animal)

Pleurocystes
Cystoid echinoderm

Silurian period

DATES: 444–419 MYA **ERA:** Paleozoic **EON:** Phanerozoic

After the extinction at the end of the Ordovician, life quickly recovered. Unlike later major mass extinctions, there were few changes in the structure of the recovering Silurian ecosystems. Like in the Ordovician, brachiopods, echinoderms, corals, arthropods, and mollusks diversified and ruled the sea floors. However, an important change started happening in the waters above: vertebrates began to turn parts of their gill arches (skeletal struts between the gill openings) into the first jaws. These jawed fish began to diversify, splitting into two main lineages: the cartilaginous fish, including sharks and rays, and early bony fish, which became hugely successful. From the Silurian onward, vertebrates became much more abundant and ecologically important, evolving myriad diets and lifestyles.

On land, plants developed roots that allowed them to draw nutrients from rocks and break the rocks down into soil. However, these plants were still relatively simple, lacking leaves and reproducing with spores instead of seeds. Although simple, these early land ecosystems supported the first fully terrestrial animals: arthropods, including the ancestors of living centipedes, spiders, and insects.

Long life
Sea lilies, such as *Sagenocrinites expansus*, are closely related to starfish and urchins. They were a major part of marine ecosystems throughout the Paleozoic, and still survive today.

Ateleaspis
Jawless fish

Pterygotus bilobus
Sea scorpion

Syringopora
Tabulate coral

Baragwanathia
Land plant

Devonian period

DATES: 419–359 MYA **ERA:** Paleozoic **EON:** Phanerozoic

In the Devonian, shelled animals including brachiopods and mollusks diversified, and among the mollusks, the first ammonoids evolved. However, brachiopods were more abundant and ecologically dominant throughout most of the Paleozoic, and reached peak diversity in the Devonian. Jawed fish became more common, and groups such as sharks, bony fish, and the armored placoderms—for example *Bothriolepis*—were highly successful throughout the period. At the same time, a peculiar group of jawed fish, relatives of *Eusthenopteron*, moved into shallow waters with their fleshy fins, which began to develop fingers and toes. While they likely remained primarily aquatic throughout the Devonian, these four-legged fish, or "tetrapods," would eventually make the profound leap to life on land. Multiple extinction events toward the end of the Devonian affected trilobites, brachiopods, and armored fish.

On land, more complex plants with roots and leaves arose and dispersed. However, some plant communities were dominated by the enormous fungus *Prototaxites*, which formed treelike structures up to 30 ft (9 m) tall.

Fishy ancestry
Bony fish diverged into ray- and lobe-finned fish in the Devonian. Lobe-finned fish, such as *Eusthenopteron*, would later give rise to four-legged animals, including human ancestors.

Paralejurus
Trilobite

Prototaxites logani
Fungus

Pteraspis
Jawless fish

Bothriolepis
Placoderm

Carboniferous period

DATES: 359–299 MYA **ERA:** Paleozoic **EON:** Phanerozoic

Throughout the Carboniferous, marine groups continued to diversify. Reefs proliferated, now built by three distinct types of corals. Sea lilies, brachiopods, bivalves, and new kinds of plankton diversified, but trilobites declined.

Life on land flourished in the Carboniferous. Complex plants, loosely called seed ferns since they reproduced with seeds and pollen, diversified. Horsetails, ferns, and relatives of club moss became widespread and grew into enormous treelike forms. These decayed into the rich coal beds that give the period its name—Carboniferous means "coal-bearing." Insects and arachnids thrived in these lush habitats and grew to incredible sizes in the oxygen-rich atmosphere. Early tetrapods gave rise to two main branches of terrestrial vertebrates. One of these, the amphibians, became successful on land, but had to return to the water to lay their jellylike eggs. In contrast, this link was severed in the amniotes, which developed shelled eggs that would not dry out on land. In the Late Carboniferous, the amniotes spawned three major groups: the synapsids, whose skulls had an opening behind the eye for the jaw muscles; the diapsids, with two such openings; and the parareptiles, with none.

Mixed lifestyle
Amphibians, such as *Branchiosaurus amblystomus*, were among the first vertebrates on land, and became highly adapted for terrestrial life. Nonetheless, their life cycle was closely linked to water, as their earliest stages were still aquatic.

Lithostrotion
Coral

Odontopteris stradonicensis
Seed fern

Archimylacris eggintoni
Cockroach

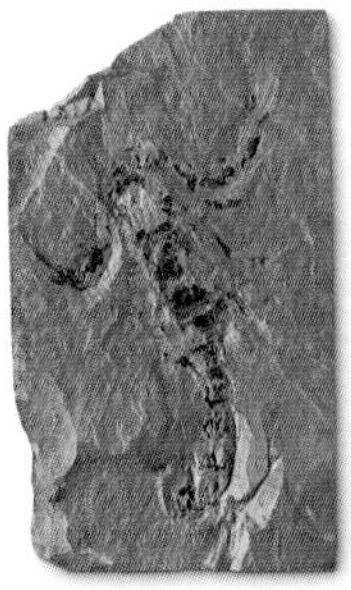

Paraisobuthus
Scorpion

Permian period

DATES: 299–252 MYA **ERA:** Paleozoic **EON:** Phanerozoic

A major diversification event in the oceans led to a rich assemblage of marine life during the Permian period. Brachiopods reached a second peak in diversity, and new types of shelled microorganisms became abundant as plankton. Several groups of ammonoids flourished, and crustaceans as well as bony fish also began to diversify.

Synapsids (which later gave rise to mammals) rapidly rose to dominance on land, specializing into enormous herbivores feeding on the forests of tree ferns and early conifers, and sail-backed carnivores such as *Dimetrodon*. They lived in the first complex terrestrial ecosystems, alongside amphibians such as *Diplocaulus*, diapsids, and parareptiles (both traditionally regarded as reptiles, although paleontologists avoid the term).

At the end of the Carboniferous, all land had assembled into a single supercontinent called Pangaea, which became arid and warm in the Permian. Many small extinction events disrupted Permian life, but the end-Permian extinction was the most severe. Huge volcanic eruptions caused global warming and acidification of the ocean, leading to the loss of more than four-fifths of life on Earth.

Returning to water
Shortly after emerging from the seas, some vertebrates, such as the parareptile *Mesosaurus*, returned to life in the water. *Mesosaurus*'s slender body would have made it an effective swimmer and the needlelike teeth would have helped it grab prey.

Procolophon skull
Parareptile

Xenodiscus
Ammonoid

Unknown species
Tree fern

Diplocaulus magnicornis
Amphibian

Early Triassic epoch

DATES: 252–247 MYA **PERIOD:** Triassic **ERA:** Mesozoic **EON:** Phanerozoic

Life recovered slowly after the end-Permian mass extinction. A harsh climate and further small extinction events in the Early Triassic prevented stable ecosystems from developing. Marine communities were not very diverse, but hot spots of diversity are known, where surviving species coexisted with new groups. These sites show the beginnings of communities including abundant ammonites and other mollusks, starfish and other echinoderms, sharks, and bony fish—a pattern that survives today.

On land, simple plants dominated for a brief period before the recovery of leafy, woody plants. Once established, the more advanced plants, including seed ferns, cycads, ferns, horsetails, and conifers, would remain common throughout the Mesozoic. Synapsids, such as the tusked dicynodont *Lystrosaurus* and mammal-like cynodonts, lived alongside amphibians and diapsids. Diapsids included the archosauromorphs, an active, agile group of animals with more erect limbs. They would later give rise to crocodiles, pterosaurs, and dinosaurs that would rule the Jurassic and Cretaceous periods. Other diapsids returned to the sea and evolved into marine reptiles that ruled the oceans throughout the Mesozoic.

Important flora
Throughout the Triassic, seed ferns such as *Dicroidium* were abundant, and set the stage for the recovery of rich terrestrial ecosystems after the end-Permian mass extinction.

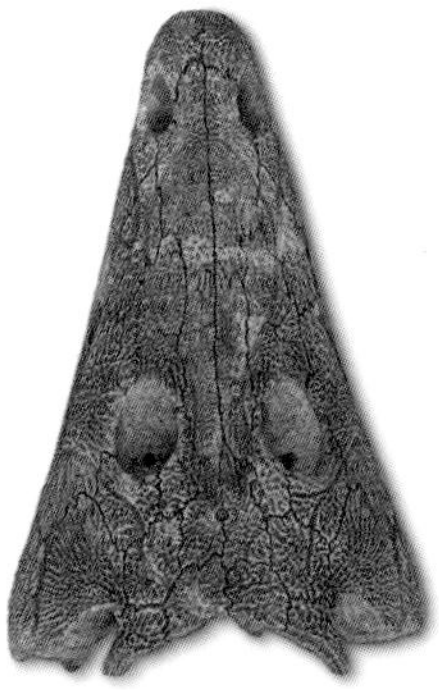

Benthosuchus skull
Amphibian

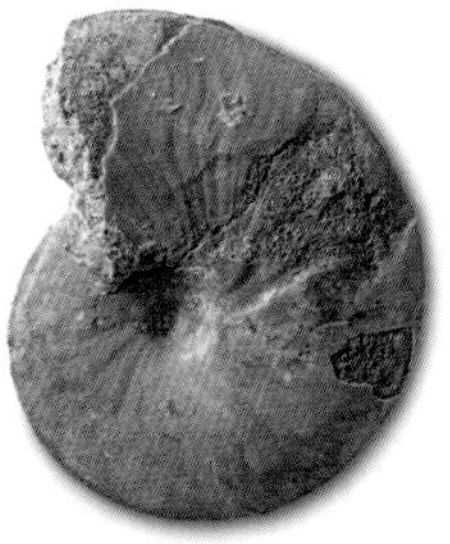

Meekoceras gracilitatus
Ammonite

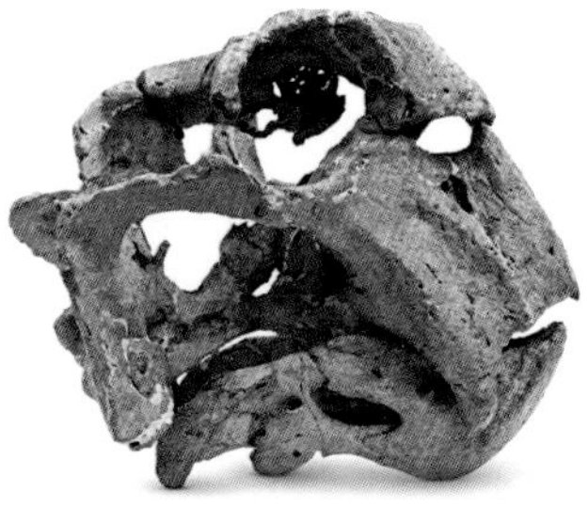

Lystrosaurus murrayi skull
Dicynodont synapsid

Voltzia coburgensis
Conifer

Middle Triassic epoch

DATES: 247–237 MYA **PERIOD:** Triassic **ERA:** Mesozoic **EON:** Phanerozoic

Recovery from the end-Permian extinction (see p.362) continued into the Middle Triassic, and ecosystems began to rival the abundance and diversity that had flourished in the Permian period. In the oceans, complex food webs were reestablished, but with new groups filling each role. Marine reptiles such as ichthyosaurs and the lizard-like nothosaurs had already evolved into top predators despite only returning to the seas a few million years earlier, and remained so in the Middle Triassic.

On land, ecosystem recovery took longer, although archosauromorphs began to diversify into a variety of new shapes and groups, such as rhynchosaurs, enabled by stable plant communities. The first precursors of dinosaurs arose, but they remained small and uncommon. Cynodont synapsids became progressively smaller and developed complex teeth, fur, and a sophisticated inner ear—all signs of an endothermic (warm-blooded) metabolism.

Major changes in global geography were beginning, too. The east coast of what would become North America began to detach from northern Africa, foreshadowing the breakup of Pangaea and the opening of the Atlantic Ocean.

Aquatic predator
Like all nothosaurs, *Lariosaurus* had an elongated neck and webbed feet. It lunged after fish in the water and caught them with its needle-sharp teeth.

Cynognathus skull
Cynodont

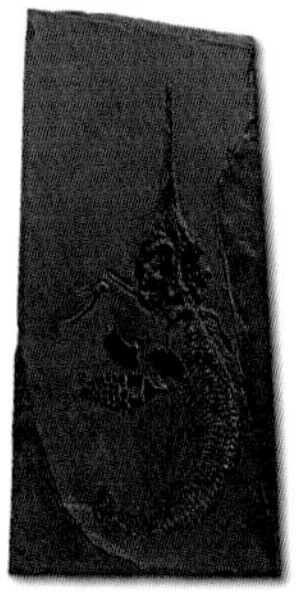

Mixosaurus
Ichthyosaur

Dicellopyge
Bony fish

Rhynchosaurus skull
Rhynchosaur archosauromorph

Late Triassic epoch

DATES: 237–201 MYA **PERIOD:** Triassic **ERA:** Mesozoic **EON:** Phanerozoic

Distinct plant zones were established in the Late Triassic separated by an arid belt around the equator. While the southern hemisphere had a relatively uniform habitat dominated by the seed fern *Dicroidium*, there was more regional variation in plants in the northern hemisphere, where cycads, ferns, and ginkgoes were common.

The first true dinosaurs and pterosaurs arose at the beginning of the Late Triassic, but at this time, other archosauromorphs remained the dominant land animals. However, dinosaurs soon overtook them and became much larger, and more diverse, and expanded into herbivorous diets. The first mammals—tiny, shrewlike animals descended from cynodonts—also appeared.

The Triassic–Jurassic boundary is marked by one of the five major mass extinctions. This event was more profound in marine ecosystems than on land, although many terrestrial synapsid and diapsid groups went extinct, paving the way for dinosaurs and pterosaurs to become the dominant land animals. The cause of the extinction is usually attributed to intense volcanism caused by the rifting of Pangaea and the opening of the Atlantic Ocean.

End of the line
Despite success throughout the Triassic epoch, many synapsids, such as the cynodont *Massetognathus*, went extinct during the Triassic–Jurassic extinction event.

Morganucodon jaw and teeth
Mammal-like synapsid

Baiera munsteriana
Ginkgo

Proganochelys skull
Turtle

Plateosaurus foot
Dinosaur

Early Jurassic epoch

DATES: 201–174 MYA **PERIOD:** Jurassic **ERA:** Mesozoic **EON:** Phanerozoic

Marine organisms were significantly affected by the two pulses of extinction at the Triassic–Jurassic boundary, but bivalves (clams), brachiopods, belemnites, and especially ammonites diversified from the survivors. Conodonts—jawless eellike vertebrates—went extinct early in the Jurassic, and plesiosaurs (four-flippered marine reptiles) became much more common. They may have competed with ichthyosaurs for resources, but both groups lived together until the Late Cretaceous. Toward the end of the Early Jurassic, another minor mass extinction occurred, greatly affecting shell-forming marine organisms such as mollusks, but not fish or the marine reptiles.

Plants were affected very little by the Triassic–Jurassic extinctions, and conifers diversified in the Early Jurassic. Lush conifer forests supported the evolution of many herbivorous dinosaurs, including the enormous, long-necked sauropods and the plodding, armored dinosaurs. Large amphibians and archosauromorphs that ruled the Triassic were replaced as top carnivores by theropod dinosaurs. Crocodilians ventured into the oceans and became highly specialized marine predators.

Early pterosaur
Dimorphodon had the typical body structure of early pterosaurs. It had a toothy skull and a long tail—both features that would be lost in later pterosaurs.

Asteroceras obtusum
Ammonite

Pachyteuthis abbreviata
Belemnite

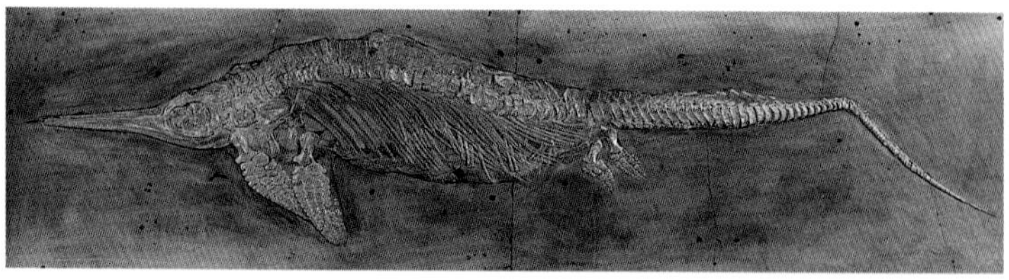

Stenopterygius megacephalus
Ichthyosaur

Middle Jurassic epoch

DATES: 174–164 MYA **PERIOD:** Jurassic **ERA:** Mesozoic **EON:** Phanerozoic

During the Middle Jurassic, the northern and southern halves of the supercontinent Pangaea—called Laurasia and Gondwana respectively—began to separate. From this point onward, life on each landmass began to diverge, resulting in the different communities that we see today.

Rare terrestrial fossil sites from the Middle Jurassic reveal novel plants, such as *Ginkgo*, *Araucaria* (monkey puzzle), and Bennettitaleans, whose reproductive structures were similar to flowers. While some theropod dinosaurs grew to be enormous, others became small and would eventually give rise to birds. Meanwhile, plant-eating dinosaurs such as the armored stegosaurs and the bipedal ornithopods proliferated. Mammals developed specializations for burrowing, tree-climbing, swimming, and gliding. Pterosaurs grew large for the first time, and continued to become more efficient fliers.

In the seas, reptiles, enormous fish, and ammonites flourished, and a subtle change—the rise of the shell-forming plankton—had major consequences. The tiny shells of these photosynthetic organisms act as a buffer against changes in ocean acidity, and have helped to stabilize marine ecosystems since the Middle Jurassic.

Long-necked predator
Two groups of plesiosaurs—one long-necked and one large-headed—terrorized the Middle Jurassic seas. Long-necked plesiosaurs such as *Cryptoclidus* were likely ambush predators.

Ginkgo biloba
Ginkgo

Zamites gigas
Bennettitalean

Pentasteria cotteswoldiae
Sea star

Araucaria (monkey puzzle)
Conifer

Late Jurassic epoch

DATES: 164–145 MYA **PERIOD:** Jurassic **ERA:** Mesozoic **EON:** Phanerozoic

The continued breakup of Pangaea throughout the Jurassic resulted in major changes in the climate and environment. Smaller continents meant fewer arid interior deserts, and humid tropical conditions spread throughout both Laurasia and Gondwana. Shallow inland seas filled the growing rifts between the continents, and rich coral reef ecosystems multiplied in the warm, sunlit waters. Most groups of sharks living today had arisen by the Late Jurassic, alongside many modern crustaceans and a wide array of bony fish.

Titanic, long-necked sauropod dinosaurs lumbered across the land, hunted by enormous theropods, such as *Allosaurus*. These giants were the largest herbivores and carnivores to have ever lived on land at that time. Although dinosaurs were the preeminent land vertebrates, mammals and amphibians multiplied, too. By the Late Jurassic, three groups of mammals living today had diverged: egg-laying monotremes; pouched marsupials; and the placentals, which today include humans. However, members of each of these groups remained small, nocturnal insect-eaters. Overhead, the earliest birds took to the skies, alongside a new group of pterosaurs with short tails and longer wings.

Avian ancestor
Birds evolved from small theropod dinosaurs, such as *Compsognathus*, whose skeleton shows remarkable similarity to the earliest birds.

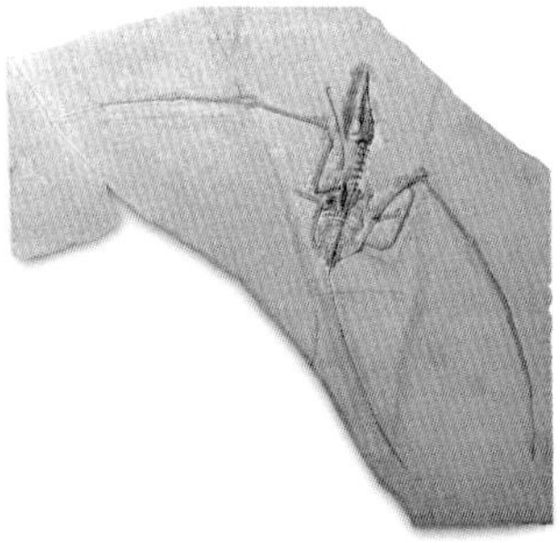

Rhamphorhynchus gemmingi
Pterosaur

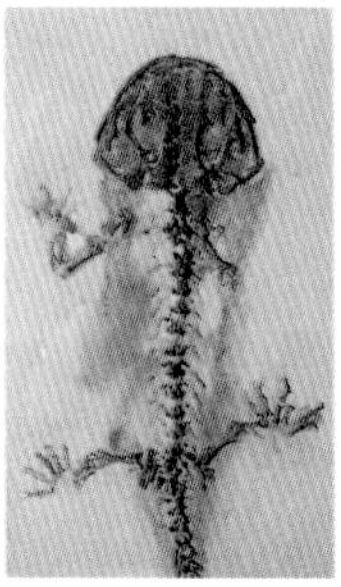

Karaurus sharovi
Salamander

Allosaurus claw
Theropod dinosaur

Camptosaurus skull
Ornithopod dinosaur

Early Cretaceous epoch

DATES: 145–101 MYA **PERIOD:** Cretaceous **ERA:** Mesozoic **EON:** Phanerozoic

A major change occurred on land during the Early Cretaceous: angiosperms (flowering plants) rapidly overtook conifers and ferns to become the dominant plants. The rise of flowers led to the diversification of their pollinators—insects such as bees, flies, and beetles—that provided an abundance of food for small, insect-eating mammals, lizards, snakes, and birds. Likewise, the new vegetation supported many new herbivorous dinosaur groups, such as the duck-billed hadrosaurs, the horned ceratopsians, armored ankylosaurs, and even toothless examples of the more usually carnivorous theropods. This diversity marked a turning point in the history of life on Earth: the number of species on land surpassed that in the ocean, which remains the case today.

Changes were also happening in the oceans, where rising sea temperatures led to expansive reefs formed by hardier bivalves instead of corals. The continued spreading of the continents widened the Atlantic Ocean. The warm climate and absence of polar ice caps resulted in high sea levels and the development of many shallow inland seas. Here, plesiosaurs started replacing ichthyosaurs as top predators in a rich ecosystem of fish, turtles, and ammonites.

Early ceratopsian
Although its horns were less elaborate, *Psittacosaurus mongoliensis* was a ceratopsian and a relative of the more famous *Triceratops*. Its name means "parrot lizard," a reference to its parrotlike beak.

Aralia saportana
Flowering plant

Hoplites dentatus
Ammonite

Iguanodon tooth
Ornithopod dinosaur

Polacanthus skin impression
Ankylosaur dinosaur

Late Cretaceous epoch

DATES: 101–66 MYA **PERIOD:** Cretaceous **ERA:** Mesozoic **EON:** Phanerozoic

The bountiful flowering plants of the Late Cretaceous supported rich communities of dinosaurs, mammals, birds, and pterosaurs. Plant-eating ceratopsians and hadrosaurs radiated across the northern supercontinent, Laurasia, preyed upon by fearsome tyrannosaurs. In Gondwana in the south, sauropods became the largest land animals of all time, but had to contend with huge predatory dinosaurs called carcharodontosaurids, cousins of *Allosaurus*. Flying overhead were toothless birds and gigantic pterosaurs. Mammals developed grinding and cutting surfaces on their teeth, and so evolved a wide array of diets, despite remaining small-bodied. The lizard-like mosasaurs emerged as top predators in the shallow inland seas, and they pushed the ichthyosaurs to extinction by the end of the Cretaceous.

The Mesozoic ended with a catastrophic asteroid impact c. 66 million years ago. Debris ejected by the impact spread around the world, blocking the sun and collapsing plant communities. All dinosaurs, except birds, went extinct, as did most marine reptiles. Even the surviving groups suffered heavy losses, and many species from all major groups of life, such as sharks, went extinct.

Fearsome look
The imposing horns and frill of *Triceratops* might appear to be a perfect defense shield against *Tyrannosaurus rex*, but are more likely to have been display structures, like the antlers of a moose today.

Platecarpus coryphaeus
Mosasaur

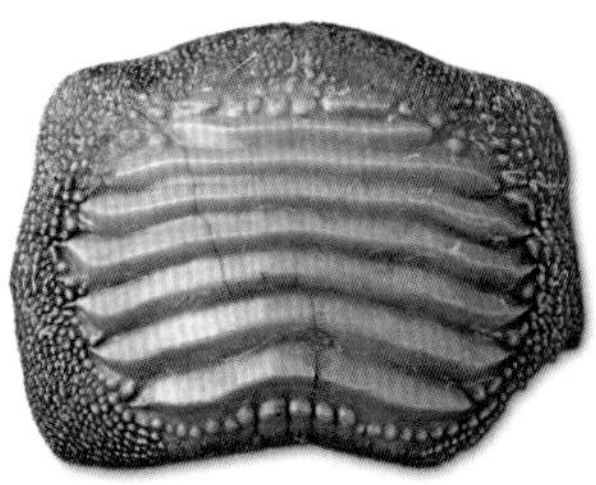

Ptychodus latissimus tooth
Shark

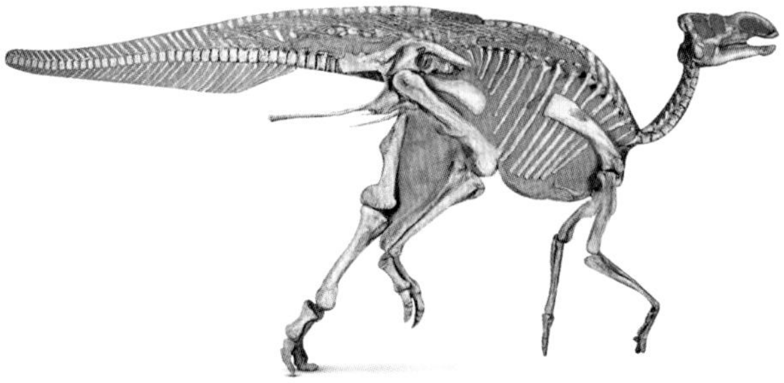

Gryposaurus
Hadrosaur

Paleocene epoch

DATES: 66–56 MYA **PERIOD:** Paleogene **ERA:** Cenozoic **EON:** Phanerozoic

In the aftermath of the end-Cretaceous extinction, life recovered relatively quickly, and rich marine ecosystems emerged in the impact crater within 30,000 years. However, these ecosystems were starkly different from those of the Cretaceous: ammonites, marine reptiles, mosasaurs, and many other species had disappeared. In their place, ray-finned fish exploded in diversity, evolving a wide variety of body shapes, diets, and habitat preferences.

On land, mammals and birds quickly filled the space left by the dinosaurs and pterosaurs. Within a few hundred thousand years, mammals had already grown into larger body sizes than they had throughout the entire Mesozoic, and developed a variety of lifestyles. By the end of the Paleocene, mammals dominated ecosystems around the globe and included stocky herbivores, large carnivores, and modern groups such as primates. The rodentlike multituberculates were eventually replaced by true rodents. It is likely that most mammal groups alive today originated during this period. Birds also radiated, and nearly all of the modern groups evolved during this period, including the earliest penguins and owls.

Dominant plant species
In the absence of herbivorous dinosaurs, thick forests grew throughout the Paleocene. They were dominated by just a few species, including ginkgos, redwoods, and sycamores such as *Platanus*.

Phenacodus jaw bone and teeth
Ungulate (hoofed mammal)

Striatolamia macrota tooth
Shark

Sand dollar
Echinoderm

Taeniolabis taoensis jaw bone and teeth
Multituberculate (extinct mammal group)

Eocene epoch

DATES: 56–34 MYA **PERIOD:** Paleogene **ERA:** Cenozoic **EON:** Phanerozoic

At the beginning of the Eocene, global temperatures spiked, leading to drastic changes in land vegetation and minor extinctions in the oceans. After this spike, the climate remained warm but stable, and for the most part there were no icecaps at the poles. Instead, lush subtropical forests spread as far north as Alaska, while Antarctica was home to boreal forests and tundra.

Mammals thrived in the warm climate, and many of the earliest members of modern groups were found in the Eocene. These included several ungulate groups—hoofed mammals such as sheep, deer, cows, rhinoceroses, and horses. These animals developed complex ridges and grooves on their teeth, which made effective grinding surfaces when eating the abundant foliage. A slate of less familiar carnivores, such as bear-dogs, hyaenodonts, and false saber-toothed cats, flourished alongside these herbivorous groups. More specialized groups of mammals also appeared, including bats and early whales, such as *Dorudon*. Birds began to resemble their modern diversity more closely, and carnivorous hawklike birds and the earliest songbirds evolved during this time.

Surviving species
Ectothermic (cold-blooded) animals, such as the soft-shelled turtle *Trionyx*, thrived in the warm climate of the Eocene. They even expanded their habitat range to high latitudes in the Arctic.

Nypa burtinii fruit
Monocot flowering plant

Protorohippus teeth
Early horse

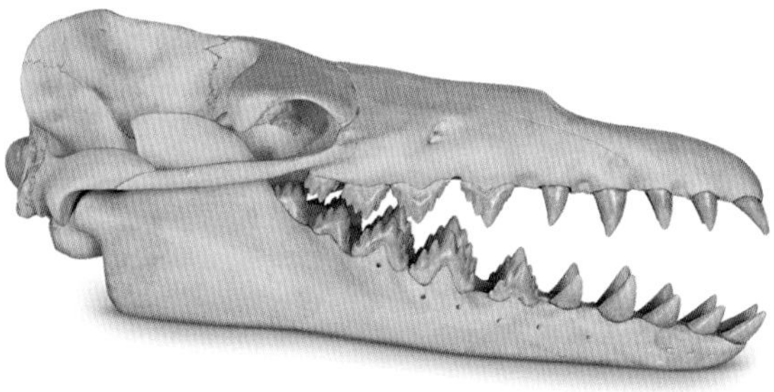

Dorudon skull
Early whale

Oligocene epoch

DATES: 34–23 MYA **PERIOD:** Paleogene **ERA:** Cenozoic **EON:** Phanerozoic

Cooler temperatures prevailed in the Oligocene as sea levels fell and an icecap formed at the South Pole. All this had a major effect on plant life as the colder, drier climate promoted the decline of tropical forests and gave rise to grasslands.

These environmental changes had major consequences for life on land. Songbirds dispersed around the world and exploded in diversity. In South America, carnivorous terror birds radiated, and became top predators alongside crocodiles and marsupials. Around the world, ungulates (hoofed mammals) evolved adaptations that allowed them to run across the open, grassy plains that had developed. This led to changes in the habits of predatory animals, as modern carnivores such as dogs, cats, and bears became more successful. As sea levels fell, land bridges were established, bringing once-separate regions together, influencing the elimination and replacement of species in the mammal world.

In the oceans, the cooling climate does not appear to have had any profound immediate impact, although it may have helped establish more abundant plankton-based ecosystems. Marine mammals continued to diversify, and modern groups of whales appeared, as did seals.

Shaped for running
The slender legs and two-hoofed toes of *Leptomeryx* were adaptations suited to running. Many hoofed mammals that inhabited the expanding grasslands of the Oligocene evolved similar adaptations.

Hyaenodon horridus skull
Carnivorous mammal

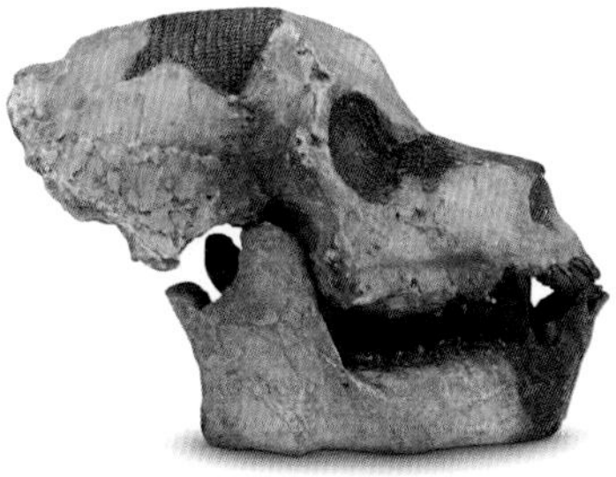

Aegyptopithecus skull
Primate

Paraceratherium bugtiense jaw and teeth
Hoofed mammal

Carcharocles tooth
Shark

Miocene epoch

DATES: 23–5 MYA **PERIOD:** Neogene **ERA:** Cenozoic **EON:** Phanerozoic

The Miocene marked the beginning of relatively modern ecosystems. In the Arctic, the sea would have been covered with ice throughout the winter, but ice-free in the summer. Grasslands had spread across the world, and by the end of the Miocene, formed expansive prairies and savannas extending across North America, Africa, Europe, and much of Asia.

The cooling climate promoted an increase in size in many types of mammals, including elephants, rhinoceroses, and the unusual chalicotheres—large, ground sloth–like relatives of horses. Other giants arose, too, including the largest crocodiles and turtles ever. Apes evolved from other primates, giving rise to the group that now includes gorillas, chimpanzees, and humans. The human line diverged from its common ancestor with chimpanzees sometime in the late Miocene, but the relationships of these ancient apes are still debated.

Oceanic kelp forests bloomed in the Miocene and hosted a rich diversity of life, including grazing urchins such as *Lovenia*, fish, sharks, crabs, and many marine mammals such as otters, seals, and whales.

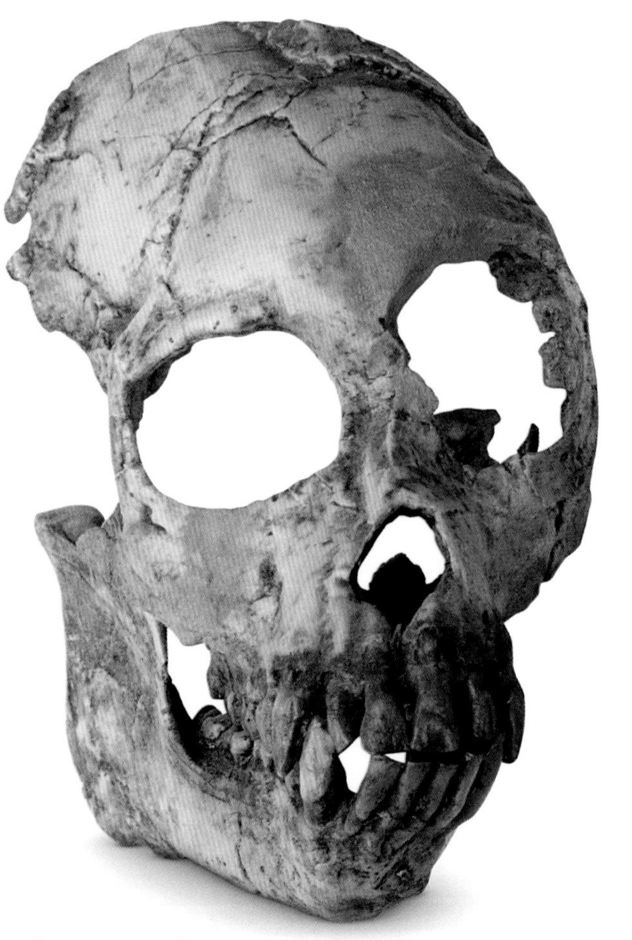

Earliest representative
Considered to be one of the oldest apes, *Ekembo* was the size of a modern baboon or gibbon. It had a slightly larger brain size relative to its body, which suggests evolving intelligence.

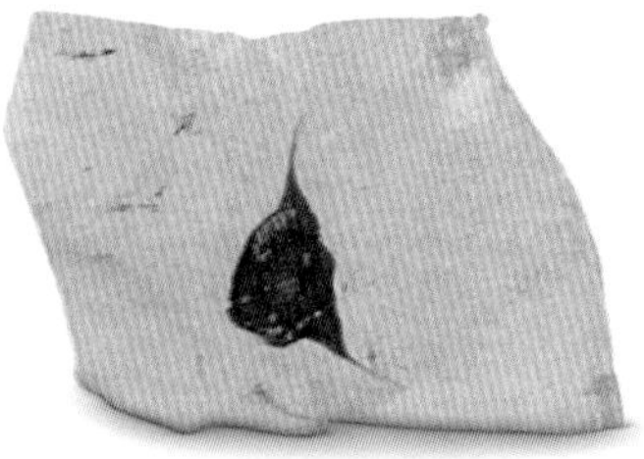

Water chestnut seed
Flowering plant

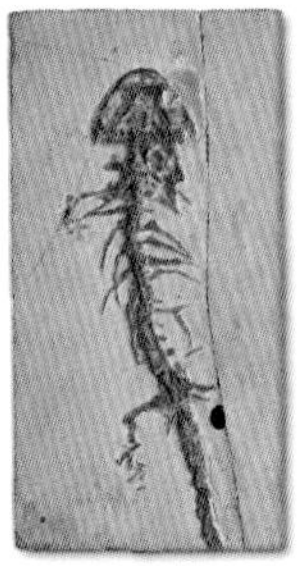

Andrias scheuchzeri
Giant salamander

Lovenia forbesi
Heart urchin, an echinoderm

Archaeogeryon peruvianus
Crab

Pliocene epoch

DATES: 5–2.6 MYA **PERIOD:** Neogene **ERA:** Cenozoic **EON:** Phanerozoic

In the Pliocene, a permanent northern icecap developed, leading to further global cooling that would lead to later glaciations. The continents were essentially in the same positions as today, but North and South America were still separate. They were joined at the end of the Pliocene, resulting in a major interchange of land animals. Giant ground sloths, terror birds, and enormous armored glyptodonts—relatives of the armadillos—as well as many smaller species headed north. Today, only a few opossums, armadillos, monkeys, and porcupines survive from the southern invaders. Species from North America were far more successful immigrants, and many kinds of hoofed mammals, elephants, dogs, cats, bears, and rodents displaced native South American groups.

The human evolutionary line saw several stages of progress in the Pliocene. The first bipedal apes, the australopiths, arose. Early australopiths were adapted for life both in the trees and on the ground, but the link to the trees was quickly lost, alongside changes to the face and an increase in brain size. The emergence of our own genus, *Homo*, followed toward the end of the epoch.

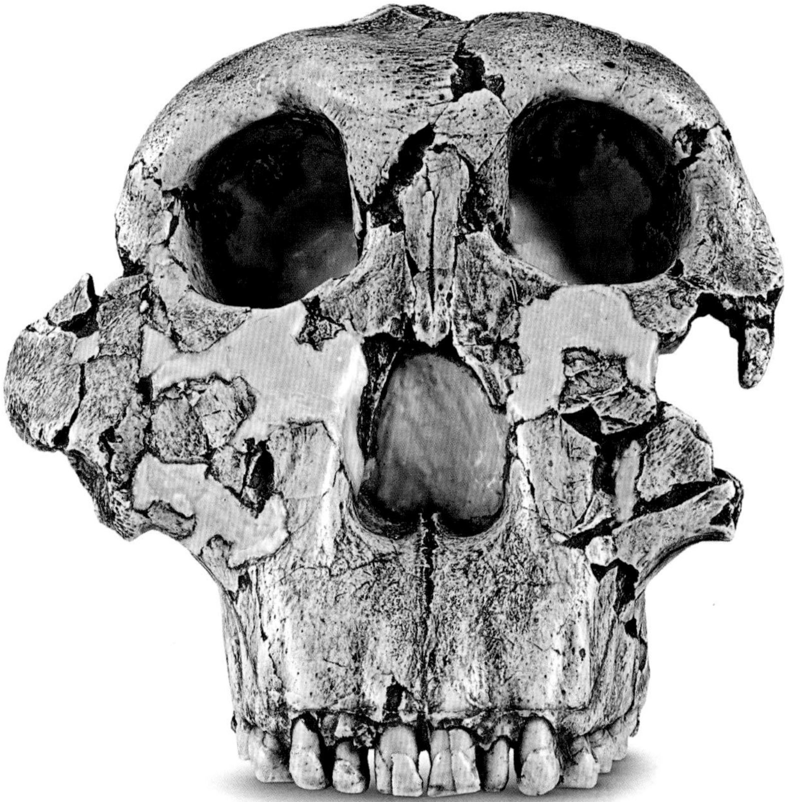

Upright ape
The broad skull and strongly built jaw of *Paranthropus boisei* were suited to powerful chewing, giving this australopith the nickname "nutcracker man."

Megatherium claw
Giant sloth

Sivatherium giganteum skull
Giant giraffe

Cancellothyris platys
Brachiopod

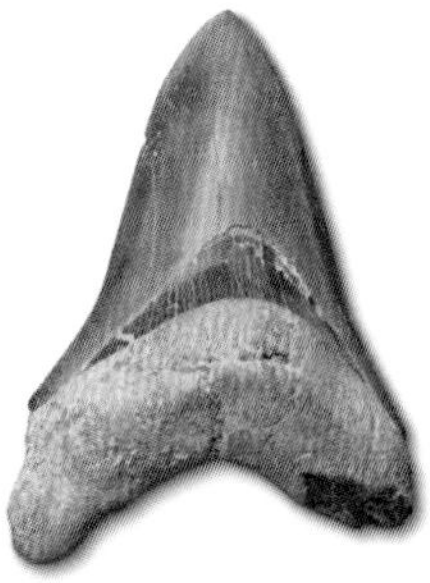

Carcharocles megalodon tooth
Shark

Pleistocene epoch

DATES: 2.6 MYA–11,700 YA **PERIOD:** Neogene **ERA:** Cenozoic **EON:** Phanerozoic

The climate cooling trend that had started in the Eocene–Oligocene (see pp.372–373) reached its culmination in the Pleistocene. Permanent icecaps at the North and South poles repeatedly expanded and contracted in a series of glacial and interglacial periods. During glacial periods, enormous Arctic ice sheets—some more than 2 miles (3 km) thick—covered the land as far south as New York, making vast areas uninhabitable. These areas emerged clear of the ice during the regular interglacial periods between the glaciations. Far from the ice, in Africa, the first members or our own genus, *Homo*, emerged, with some species dispersing as far as China and Indonesia.

Amid this climatic instability, plants and animals assembled into ecosystems that were essentially identical to those today. A major difference, however, was the presence of "megafauna"—large mammals and birds, including the woolly mammoth, the cave bear, and the moa. These species, alongside many other large animals, went extinct in the Late Pleistocene. In many areas, these extinctions coincided with the dispersal of modern humans, *Homo sapiens*. However, the role of these humans in the extinctions remains debated, and it is likely that multiple factors contributed to the phenomenon.

Ancient handyman
The use of stone tools allowed *Homo habilis* access to new foods that were richer in energy and protein. This may have been a factor in allowing *H. habilis* and its descendants to develop a larger brain.

Ursus spelaeus tooth
Cave bear

Bison priscus skull roof
Bison

Deinotherium giganteum cheek tooth
Proboscidean, or elephant relative

Coral

Holocene epoch

DATES: 11,700 YA–Present **PERIOD:** Neogene **ERA:** Cenozoic **EON:** Phanerozoic

The Holocene encompasses the interval from the last glacial period to the present, although some scientists recognize an extra epoch, the Anthropocene (see p.317), based on the profound impact of industrial civilization on Earth. Throughout the Holocene, humans have invented increasingly complex tools and cultures, and have developed agriculture around the world.

The predominant change in Holocene ecosystems has been the destruction of vast expanses of natural habitat by human activities such as farming, forestry, and fishing. This has caused regional as well as global extinction of many species. In fact, some scientists recognize this as the beginning of the sixth major mass extinction. The use of fossil fuels since the 19th century has rapidly increased the levels of atmospheric carbon dioxide. This has been accompanied by a rise in average global temperatures, in turn melting the polar icecaps and leading to changes in ocean currents.

Nevertheless, advances in energy, food science, and conservation efforts provide the means to rectify some of the damage done, and prevent further decline in life on Earth.

Prime target
The elephant bird of Madagascar, *Aepyornis*, was a victim of human dispersal. It was hunted to extinction just a few hundred years ago, possibly because of its large eggs.

Homo sapiens upper jaw
Modern human

Mammuthus primigenius hair
Mammoth

Raphus skull and lower jaw
Dodo

Aurochs skull
Wild cattle

Kilauea

Kilauea is the most active of five volcanoes that make up the easternmost, and largest, of the Hawaiian Islands. It is located above a deep-rooted mantle plume that is currently punching through the crust to reach the surface, as the Pacific tectonic plate moves slowly over it. Over about 70 million years, this hot spot has formed not just the Hawaiian Islands but also the 3,700-mile- (6,000-km-) long Hawaiian-Emperor Seamount chain of undersea volcanoes. Kilauea's volcanic cone has built up some 16,400 ft (5,000 m) from the ocean floor for about 210,000–280,000 years, emerging above sea level 100,000 years ago. Although there have been episodes of explosive eruption, Kilauea's outbursts in the last 200 years have involved the steady, effusive eruptions of hot, runny basaltic lava that is typical of broad, gently sloping shield volcanoes.

Cinder cone
The Pu'u 'O'o vent, on the southeastern flank of Kilauea, has been erupting since 1983. The buildup of pyroclastic fragments has led to the formation of a conical hill with a crater at the top—called a cinder cone—over the vent.

Rocky Mountains

The Rocky Mountains, or Rockies, extend from northern Canada to New Mexico and include more than 100 summits over 10,000 ft (3,000 m) high. Parts of the range are covered in coniferous forests, while glaciers occupy some of the higher peaks. The Rockies form North America's Continental Divide: to the east, rivers flow into the Atlantic and Arctic oceans; to the west, those that reach the coast flow into the Pacific.

The Rockies formed about 80–55 million years ago as the oceanic Farallon Plate subducted beneath the North American Plate, propelled by the rapidly spreading Pacific Ocean. The continental crust was deformed and thickened, giving rise to a long, broad plateau. Erosion stripped away the higher and weaker rocks, forming a mountain landscape that has been further sculpted by glaciation over the last 2.5 million years.

Rising high
This panoramic view of the Canadian Rockies shows the spectacular Sunburst Peak overlooking the Cerulean Lake and Mount Assiniboine (left) partially hidden by clouds.

Pacific Northwest Rainforest

The coastal mountains of Washington and British Columbia rise more than 7,000 ft (2,100 m) above sea level. Frequent storms coming in from the Pacific Ocean give up much of their moisture as precipitation when the warm, wet air rises and cools, sustaining a temperate rainforest on the slopes and valleys of the mountains. High rainfall, 12–14 ft (3.5–4.2 m) per year in places, and moderate temperatures throughout the year, provide ideal conditions for the growth of large, long-lived trees, such as Douglas fir and coastal redwoods. Some trees reach 250 ft (75 m) in height and 60 ft (18 m) in circumference, and are many hundreds of years old. Mosses, ferns, lichen, and fungi grow on and between the trees. The abundance of living and dead plant material results in a rich biomass comparable to that of the tropics.

Magnificent remnant
Located in a part of the Pacific Northwest Rainforest that once stretched along the Pacific Coast from Alaska to California, the Hoh Rainforest is one of the finest remaining old-growth temperate rainforests in the US.

Great Basin Desert

The Great Basin is a vast internal drainage area of about 200,000 square miles (520,000 square km) between the Rocky Mountains to the east and the Sierra Nevada to the west. Since there is no outlet to the sea, the water flowing off the surrounding high ground either drains into lakes, sinks underground, or evaporates. Lying in the rain shadow of the Sierra Nevada and Cascade Ranges, the area receives little rainfall. Most of the basin is a desert and the majority of its lakes are saline.

The Great Basin is a series of basins and mountain ranges that formed where Earth's crust has been stretched and fractured. Alluvial deposits and salt flats drape the valley floors, where sagebrush is the dominant vegetation. Islands of forest and alpine plants survive on the mountains between.

Hot and dry
Death Valley, straddling the border between the Great Basin and Mojave deserts, is one of the hottest and driest places in the world.

Great Plains

The Great Plains are an immense area of temperate grassland stretching 2,000 miles (3,200 km) from Texas in the south to the Canadian provinces of Alberta and Saskatchewan in the north. Situated between the Rocky Mountains and the central lowlands of the Mississippi River basin, the plains lie 2,000–5,500 ft (600–1,700 m) above sea level. The plains are semi-arid and receive less than 24 in (60 cm) of rain annually. The low rainfall restricts trees to river valleys and hills, while high levels of sunshine and deep, fertile soils make parts of the plains suitable for livestock grazing and grain production. Although largely flat, there is variety in the landscape—the Black Hills, Badlands, Sand Hills, and the volcanic plug known as Devil's Tower in the north contrast with the canyons cut by the Rio Grande and its tributaries in the south.

Grassy hills
Rolling hills of shortgrass prairie dominate the landscape of Badlands National Park in South Dakota. The Great Plains are the largest area of mixed-grass prairie in the country and originally supported vast herds of bison.

Great Lakes

The Great Lakes comprise five interconnected bodies of water—the lakes Superior, Michigan, Huron, Erie, and Ontario. Together, they contain about one-fifth of Earth's surface fresh water. The lakes formed at the end of the last ice age when the great Laurentide Ice Sheet, covering the northern part of North America, began to melt. Thick glaciers at the edges of the ice sheet carved out a series of broad depressions along existing river valleys and fault systems. As the ice withdrew about 10,000 years ago, glacial meltwater filled these basins. The St. Lawrence River connects the five lakes to the Atlantic Ocean and allows ships to travel 2,200 miles (3,540 km) inland from the ocean. The Great Lakes modify the regional climate and make the winters slightly warmer and summers slightly cooler than they would be otherwise.

Strong waves
Although unaffected by tides, the Great Lakes can exhibit sealike features. Sustained winds produce rolling waves and strong currents, such as these in Lake Superior.

Mississippi–Missouri

The Mississippi–Missouri is the largest river system in North America. The Mississippi River flows 2,350 miles (3,780 km) across the eastern US from its source in Lake Itasca, Minnesota, to its mouth in the Gulf of Mexico. It is exceeded in length by its tributary, the Missouri River, which is about 100 miles (160 km) longer. The other main tributaries of the Mississippi are the Ohio River to the east and the Alabama River to the west. The Mississippi–Missouri drainage basin extends from the Appalachian Mountains in the east across the Great Plains to the Rocky Mountains in the west, covering about 40 percent of the continental US. The Mississippi River experiences seasonal floods each year that lead to the formation of a broad, fertile floodplain, and has an evolving "crow's foot" delta extending into the Gulf of Mexico.

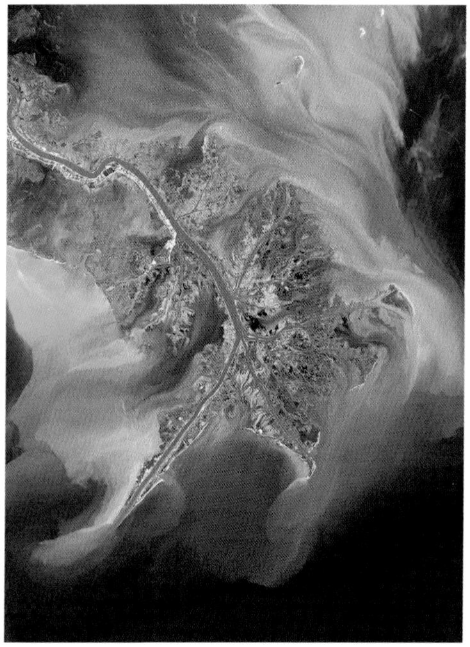

Fertile floodplain
The crow's foot shape of the Mississippi Delta can be seen in this satellite image. Each year, the river deposits more than 550 million tons of nutrient-rich sediment into the Gulf of Mexico, leading to the formation of algal blooms.

Everglades

The Everglades is a large area of subtropical wetland occupying most of the southern Florida peninsula. The water drains out from the south side of Lake Okeechobee in a broad flow—about 60 miles (100 km) wide and 100 miles (160 km) long—across a dense, relatively impermeable limestone shelf.

With a shallow gradient, the sheet of water moves slowly, at about a 3 ft per hour. The Everglades has been described as "a river of grass." The area is a unique ecosystem but is at risk from human development. Its immense diversity supports a mixture of dense forest, open grassland, swamps, and croplands as well as more than 400 species of birds and some 50 species of reptiles. In some places, peat deposits have built up to a depth of 18 ft (5.5 m). Slightly higher ground to the east and west channels the wetland's water flow to the southwest, into the Gulf of Mexico.

Winding through
Fresh water flows slowly through winding side channels across saw grass prairies in the Everglades before emptying into the brackish waters of an estuary.

Greenland Ice Sheet

Although Greenland's coastline is ice-free, the vast majority of the island's surface is buried under a thick ice sheet. Containing about 8 percent of Earth's fresh water, the Greenland Ice Sheet is more than 9,800 ft (3,000 m) thick in places. The immense mass of the ice depresses Earth's crust, so a significant part of Greenland's interior bedrock lies below sea level. The ice sheet is constrained by coastal mountains, which channel the ice-flow into numerous glaciers, most of which reach the sea. The Jakobshavn Glacier in western Greenland is the world's fastest-flowing glacier. The Greenland Ice Sheet is currently losing 183 billion tons (166 billion metric tons) of ice a year, causing sea levels to rise about ½ in (1 cm) since 1990. If all of Greenland's ice were to melt, the world's sea levels would rise by 24 ft (7.2 m), flooding many major cities.

Magnificent icebergs
Towering icebergs in unique shapes and sizes abound in Greenland's Southern Fjords. Icebergs form when pieces of ice calve (break off) from glaciers or ice shelves.

Andes

The Andes are Earth's longest land-based mountain range, running 4,350 miles (7,000 km) down the western side of South America. The highest peak in the range is Mount Aconcagua in Argentina, which rises to a height of 22,838 ft (6,961 m) above sea level. The Andes have formed over about the last 30 million years due to the collision between the South American and Nazca tectonic plates. The convergence of the plates caused the crust to deform and thicken, and the oceanic Nazca Plate to subduct beneath the South American Plate, creating an arc of volcanoes in the mountain range above. The Andes display a wide range of landscapes, including the high plains of the Altiplano, the Atacama Desert—one of the most arid places on Earth—the world's highest volcanoes, and the extensive ice-fields and glaciated peaks of southern Chile and Argentina.

Towering above
Mount Fitz Roy, the highest point of Los Glaciares National Park in Argentina, is 11,171 ft (3,405 m) high. It is surrounded by the Southern Patagonian Ice-Field between Chile and Argentina.

Altiplano

The Andean Altiplano—Spanish for "high plain"—is the second-largest mountainous plateau in the world after the Tibetan Plateau. It lies between the Cordillera Occidental and Cordillera Oriental sections of the Andes mountains, where the range extends into a broad plateau, mostly in Bolivia but partly in southern Peru and northern Chile. Here, the thickness of the crust and sediments, totaling some 46 miles (75 km), has elevated the plateau to 12,300 ft (3,750 m) above sea level. The climate is cool and semiarid.

The Altiplano is an inland drainage basin containing several large lakes such as Lake Titicaca (see p.384) in the north, which is more humid, and salars (salt flats) in the drier south. Salar de Uyuni is the world's largest salt flat, with a crust of salt several yards deep. The plateau is also dotted with volcanoes, many of them active.

Tree-shaped rock
The Altiplano's strong winds have produced some unusually shaped rock formations. Among the most famous of these is the Árbol de Piedra, meaning Stone Tree, pictured here.

Atacama Desert

The Atacama Desert is a stony coastal desert stretching 1,000 miles (1,600 km) along the western shore of South America and some way up into the Andes mountains. The Pacific Ocean's Humboldt Current brings cold water from the Antarctic. There is little evaporation from the sea's surface, so winds blowing over the Atacama are usually dry. Parts of the desert receive only about 2/5 in (5 mm) of rain in a year. The Atacama is the driest place on Earth, and possibly its oldest desert, having been hyperarid for at least 3 million years. However, coastal fogs and rare heavy rains allow some plants and animals to survive. The driest part of the desert is a plateau between two high mountain ranges—the Andes and the Chilean Coast Range—that block incoming moisture from both the Pacific and Atlantic oceans, making the Atacama a double-rain-shadow desert.

Mars-like landscape
Although known as the Valley of the Moon, part of the Atacama in northwest Argentina so closely resembles the surface of Mars that it has been used to test rovers designed to explore the red planet.

Lake Titicaca

The largest lake in South America, Lake Titicaca covers about 3,230 square miles (8,300 square km) and is one of the oldest and highest navigable lakes in the world. Situated 12,507 ft (3,821 m) above sea level, the lake originated several million years ago. It is fed by more than 25 tributaries, but it has just one outflow, the Desaguadero River. Lying on the Peru–Bolivia border in the Altiplano basin, the lake is divided into two sub-basins connected by the Strait of Tiquina.

Most of Lake Titicaca's water is lost through evaporation driven by the region's fierce sun and strong winds. About 5 percent drains out through the Desaguadero, which feeds the large salt-water Lake Poopó to the south. The flow of the Desaguadero is no longer enough to compensate for evaporative losses from the surface of Lake Poopó, which dried out in 2015 and shows no signs of recovery.

Home to many
More than 40 islands, some of them densely populated, dot the waters of Lake Titicaca. The largest of them is the Isla del Sol (Island of the Sun) in Bolivia.

Amazon Rainforest

The drainage basin of the Amazon River and its surrounding areas contain the world's largest tropical rainforest. Found mostly in northern Brazil, but also covering Guyana, Suriname, and parts of Venezuela, Colombia, Peru, Ecuador, and Bolivia, the Amazon Rainforest spans 2.6 million square miles (6.7 million square km) and constitutes about half the world's remaining tropical rainforest. The area has a rich biodiversity and is home to 10 percent of all known animal species. The plant life is so dense and vigorous that it creates its own climate: trees in the Amazon pump more water up from the ground into the atmosphere by the process of transpiration than flows through any river. The area and its indigenous peoples are threatened by deforestation—17 percent of the rainforest has been lost in the last 50 years.

Snaking through
The world's largest river by flow, the Amazon meanders through the most extensive tropical rainforest on Earth toward the northeastern coast of Brazil, where it discharges about 53 billion gallons (200 billion liters) of water every second into the Atlantic Ocean.

Pantanal

The world's largest tropical wetland, the Pantanal—meaning "big swamp" in Portuguese—occupies a sedimentary basin on the borders of Brazil, Bolivia, and Paraguay. Here, a broad depression in Earth's crust predating the formation of the Andes Mountains causes tributaries of the Paraguay River to come together and form an inland river delta. The rivers often overflow, depositing their sediments and creating a patchwork of shallow lakes, marshes, grasslands, and forests spanning 58,000 square miles (150,000 square km). About 80 percent of the area is underwater in the rainy season. The region supports a diverse range of plant species, including those found in the surrounding rainforest, semiarid woodland, and savanna biomes. The Pantanal is also rich in animal life and is home to species such as the jaguar, caiman, giant river otter, capybara, and giant anteater.

Floating gardens
Aquatic plants such as water hyacinth and water ferns carpet the swamps, islands, and pools that form in the Pantanal as water levels rise by up to 16 ft (5 m) during the wet season (December–March).

Pampas

The Pampas of Uruguay, eastern Argentina, and southern Brazil is a vast, fertile, temperate lowland, similar to the North American prairies or the European steppes. Its name is derived from the Quechua peoples' word for "plain." Covering about 295,000 square miles (760,000 square km), the area is draped with thick deposits of loess—fine, wind-borne sediment—laid down in the last 2 million years. Frequent wildfires have ensured that smaller vegetation types dominate. Grasses thrive in the mild climate, with rainfall distributed throughout the year. Much of the area is now given over to grain production or grazing livestock, and the original coarse grasses have been replaced with grass species more suitable to livestock. The plains are interrupted by only a few low hills, the highest being the Ventana Hills, which rise to 4,265 ft (1,300 m).

Resilient species
The ombu tree, with its umbrellalike canopy that spreads up to 50 ft (15 m) across, is one of the few tree species to grow in the arid Pampas. Its trunk contains water storage tissue, making it resistant to fires.

Atlas Mountains

The highest mountain range in North Africa, the Atlas rises to 13,671 ft (4,167 m) above sea level at the summit of Mount Toubkal. The range stretches 1,600 miles (2,500 km) across Morocco and Algeria into Tunisia, separating plains and hills along the Mediterranean Coast from the Sahara Desert. The mountains are often snow-capped in winter but show no signs of glaciation—their steep valleys are V-shaped, having been eroded by rivers.

The Atlas range is split into several subranges that formed at different times. The Anti-Atlas in the south formed when Precambrian rocks were folded by Africa's collision with North America about 300 million years ago. The central High Atlas, containing hard volcanic rocks, was uplifted 66–1.8 million years ago, along with the eastern Saharan Atlas, created by Africa's collision with the Iberian Plate.

Ancient range
The Anti-Atlas range in Morocco is a complex assemblage of multiple rock layers, including bands of sandstone and limestone. It contains some of the Atlas range's oldest rocks.

Sahara Desert

Stretching 3,000 miles (4,800 km) from the Atlantic Ocean in the west to the Red Sea in the east, the Sahara is the world's largest hot desert, covering as much as 31 percent of Africa. It lies under the subtropical high-pressure belt of descending dry air. It separates Africa's Mediterranean shore from the savanna of the Sahel region, as well as the rest of the continent, to the south. The Sahara is mostly a stony plateau, but there are some large expanses of sand piled up in dune fields. The largest of these fields are called ergs.

The Sahara receives little rainfall—half the region gets less than 1 in (25 mm) in a year, the rest up to 4 in (100 mm). However, it was not always so dry. Variations in the tilt of Earth's rotational axis and atmospheric changes at the end of the last glacial period, about 12,000 years ago, brought higher rainfall and savanna-like conditions in the past.

Desert specialist
Animal species endemic to the Sahara, such as the sand viper (*Cerastes vipera*), show various adaptations to the extreme heat of the desert. The sand viper is mainly nocturnal, and sinks itself in the sand to escape the scorching sun.

Nile

From its source in the heart of Africa to the streams of its delta on the shores of the Mediterranean Sea, the Nile River flows for 4,130 miles (6,650 km), making it Earth's longest river. Historically, it has been an important transportation route as well as a provider of the irrigation water and sediment that has turned desert into fertile agricultural land. The Nile was critical to the development of the ancient Egyptian civilization. The river valley was regularly flooded, enriching the soil and extending its delta. Now, however, since the construction of several large dams in the 20th century, less silt flows north. Upstream of Khartoum, Sudan, the Nile splits into two branches: the Blue Nile rises in the highlands of Ethiopia and the White Nile rises somewhere near Lake Victoria (see p.388), although the exact location remains uncertain.

Desert lifeline
The Nile winds its way through the arid landscape of north Sudan, near the city of Dongola. The river is a vital source of drinking water for the region.

Great Rift Valley

A series of fissures, or rifts, in Earth's crust runs over 2,200 miles (3,500 km) through East Africa from the Afar Depression (see p.390) in the north to Mozambique in the south. The Great Rift Valley has been opening up for about the last 25 million years as the Nubian and Somalian parts of the African tectonic plate pull apart.

Spectacular escarpments (steep slopes that form due to faulting or erosion) drop to the floor of the valley, which is partly filled with river sediments and, particularly in its western branch, a series of large lakes. The valley and its flanks are dotted with volcanoes, including Africa's highest mountain, Kilimanjaro, which rises to a height of 19,341 ft (5,895 m). The valley's topography has given rise to a variety of ecosystems that have played host to human evolution and enabled wildlife diversity to flourish.

Rift lakes
The eastern arm of the Great Rift Valley contains relatively small, shallow lakes, such as Lake Abijatta and Lake Shalla in Ethiopia. The two lakes are surrounded by striking escarpments.

Lake Victoria

The largest lake in Africa and the second largest in the world, Lake Victoria is the biggest inland source of fish in Africa and enables electricity generation for some 30 million people living on its shores. It lies in a shallow depression between the two arms of the Great Rift Valley (see p.387) in East Africa, straddling the borders of Uganda, Kenya, and Tanzania. With a surface area of 23,146 square miles (59,947 square km), Lake Victoria occupies most of its drainage basin, making its water level highly variable. To the north, it drains into the White Nile (see p.387) through a dam, which helps stabilize its water level. Lake Victoria is shallow, with an average depth of 130 ft (40 m). It is a young lake, dating back only 400,000 years. At the time of its formation, the rising walls of the Great Rift Valley blocked drainage to the west and created the basin where Lake Victoria formed.

Access to water
This satellite image shows Lake Victoria's lengthy and irregular shores. The lake is a source of fresh water for millions of people, including those inhabiting several large African cities such as Kampala, Uganda's capital.

Serengeti

The Serengeti is a large area of rolling plains stretching across northern Tanzania and southern Kenya, between the Great Rift Valley and Lake Victoria. About 12,000 square miles (30,000 square km) of the plains are protected as a national park and game reserve. The Serengeti is a savanna ecosystem of grassland, thornbush, and woodland, home to a large diversity of animals. Rainfall patterns spur vast herds to undertake the Great Migration—an annual journey around the Serengeti led by 1.7 million wildebeest seeking the best grazing grounds and water supplies in the semiarid region. The ecosystem owes its vitality partly to recent geological activity. The Serengeti stands on a plateau above the Great Rift Valley, where volcanic eruptions of ash and lava have preserved ancient hominid fossils, including in Olduvai Gorge.

Towering trees
Tall acacias dominate parts of the Serengeti's landscape. They are among the few tree species found in the region and provide food for many animal species, including giraffes.

Okavango Delta

Rising in the highlands of Angola, the Okavango River flows across 600 miles (965 km) of bush, but does not reach the sea. Instead, it runs into a depression in Botswana's Kalahari Desert, where it splits into numerous streams and swamps, forming the Okavango Delta, an inland river delta about 90 miles (150 km) across. More than 97 percent of the water that enters the delta evaporates or eventually seeps into the sands of the Kalahari Desert, and the remaining drains into Lake Ngami to the west. Annual rains in the Angolan highlands cause the river to surge, and this wetland oasis more than doubles in size. The flooding triggers a spectacular transformation of the landscape that attracts one of Africa's greatest concentrations of wildlife, including lions, leopards, buffaloes, elephants, and rhinoceroses. Antelope, fish, and birds also abound.

Ever-changing landscape
The pattern of the waterways in the Okavango Delta changes every year as channels become blocked by aquatic plants and the slow-moving water finds new routes.

Victoria Falls

Forming part of the border between Zambia to the north and Zimbabwe to the south, the Victoria Falls are one of the world's most spectacular waterfalls. Crossing a basalt plateau, the Zambezi River drops 354 ft (108 m) across a broad front of 5,600 ft (1,700 m) into a zigzag canyon that it has cut along faults in the hard basalt. The Kololo name for the falls is Mosi-oa-Tunya, meaning "the smoke which thunders," due to the immense clouds of spray that are thrown up as the water cascades down the rock face. The spray may sometimes rise as much as 1,300 ft (400 m) above the top of the falls and can be seen from up to 25 miles (40 km) away. The flow varies throughout the year, reaching up to 105,944 cubic ft (3 million liters) per second during the rainy season.

Curtain of water
The falls form the centerpiece of the Victoria Falls National Park on the Zimbabwean side of the Zambezi River. This waterfall is often described as having the largest sheet of falling water in the world.

Afar Depression

There are few places on Earth where it is possible to walk on newly formed crustal rock. The Afar Depression (which includes parts of Ethiopia, Eritrea, and Djibouti), created by tectonic activity in East Africa, is one of them. About $\frac{1}{4}$–$\frac{3}{4}$ in (1–2 cm) of new crust forms every year in this low-lying region. Here, heat rising from the mantle below lifts Earth's crust, which splits, erupting basalt lavas and forming a triple-junction between the Arabian Plate to the north and two sections of the African Plate—the Nubian part to the west and the Somalian part to the southeast. As the crust rifts apart the intervening land sinks, dropping 426 ft (130 m) below sea level in several places. The rifts to the north and east already lie beneath the Red Sea and the Gulf of Aden respectively. The third branch of the triple-junction is the Great Rift Valley (see p.387) to the south.

Hostile zone
Parts of the Afar Depression lie below sea level, making it among the hottest places on Earth. Dallol (shown here) in the north of the region, is 150 ft (45 m) below sea level and the site of hot springs that deposit salt and sulfur.

Pyrenees

The Pyrenees Mountains stretch over 250 miles (400 km) from the Atlantic Ocean to the Mediterranean Sea. Their rugged peaks form the natural border between France and Spain, and separate the Iberian Peninsula (Spain and Portugal) from the rest of Europe. The Pyrenees were uplifted at about the same time as the Alps, 55–20 million years ago, when the Iberian microcontinent collided with the Eurasian tectonic plate, folding and uplifting coastal sediments dating back 500 million years. Glaciation during the last ice age left sharp peaks, steep U-shaped valleys, hanging valleys, and waterfalls. The landscape of the Pyrenees includes its tallest peak, Pico d'Aneto, rising to 11,168 ft (3,404 m) above sea level, the Ordesa Canyon, which extends for 7 miles (11 km), and more than a thousand small glacial lakes.

Glaciated valley
The glaciated, amphitheater-like hollow known as the Cirque de Gavarnie overlooks a verdant valley in this panoramic view of the Pyrenees. Some of Europe's most spectacular waterfalls are visible in the distance.

Alps

The highest mountain range to lie wholly within Europe, the Alps contain more than 100 peaks higher than 13,000ft (4,000m). The range stretches in an arc for about 750 miles (1,200km), covering about 80,000 square miles (207,000 square km) across eight countries between the French town of Nice and the Austrian capital, Vienna. The Alps have a complex geological structure, which includes intricate folding, faulting, and large-scale overthrusting of deep-water marine sediments that were squeezed between the African and Eurasian tectonic plates as the ancient Tethys Sea closed. Their uplift began about 50 million years ago and continues even today. The Alps have been sculpted into a spectacular landscape by many episodes of glaciation over the last 2 million years, and glaciers still drape the higher slopes.

Border peak
Standing on the French–Italian border, Mont Blanc is the highest peak in the Alps. It rises to a height of 15,778ft (4,809m) above sea level.

Mount Etna

One of the world's most active volcanoes, Mount Etna has been almost constantly erupting in modern times. Rising to 10,892ft (3,320m) on the eastern shore of the Italian island of Sicily, it is a broad stratovolcano—one with steep sides. Its conical slopes, built from eruptive deposits, produce fertile soils that are intensively farmed. Etna's thick, sticky lava—combined with one of the highest gas contents of any volcano—makes its eruptions unpredictable and often spectacular, with lava, volcanic ash, and molten rocks thrown out in fountains of fire. There are reports of volcanic events going as far back as 1500 BCE, and evidence of almost all types of volcanic eruptions through the ages, including gentle lava flows, pyroclastic flows, catastrophic collapse of flanks, and explosive events that sent ash as far as Rome, 500 miles (800km) to the north.

Slow-moving fire
Etna's lava is rich in silica, making it thick and slow-moving. It flows only a few miles before solidifying, resulting in a volcano with steep sides.

Danube

Central Europe's longest river, the Danube runs 1,770 miles (2,850 km) from its source in Germany's Black Forest to its outflow through a broad delta on the Romanian shore of the Black Sea. It weaves its way between the Alps and the Carpathian Mountains, draining an area of 309,447 square miles (801,463 square km) that is home to 83 million people. It forms the borders between Slovakia and Hungary, Serbia and Croatia, Serbia and Romania, Romania and Bulgaria, as well as Romania and Ukraine. Four capital cities lie on its banks—Vienna, Bratislava, Budapest, and Belgrade. The Danube flows through more countries than any other river in the world and is a major international trade route, linked by a canal to the Rhine River and the North Sea.

High cliffs
The banks of the Danube narrow into a series of towering limestone cliffs near the Serbia-Romania border, forming a gorge known as the Iron Gate.

Białowieża Forest

Occupying the border between Poland and Belarus, the Białowieża Forest is one of the few large areas of primeval, or old-growth, forest left in Europe. This mix of lowland forest, with conifer and broad-leaved trees, meadows, and wetlands, covers about 835 square miles (2,162 square km) of land on the watershed between the drainage basins of the Baltic and Black seas. Extensive undisturbed areas in the forest have left a lot of standing and fallen deadwood, making it an area of high biodiversity rich in fungi, insects, birds, and wildlife. As well as otters, lynx, and wolves, the Białowieża Forest supports some 900 European bison bred from captive animals after their near extinction in 1923. The forest area is protected by national parks on both sides of the border and has been a UNESCO World Heritage Site since 1979.

Dense ground cover
Decaying trees, brushwood, and fallen leaves cover large parts of the forest floor, accounting for about a quarter of the total tree biomass, an unusually high proportion.

Arabian Desert

The Arabian Peninsula is almost entirely occupied by a vast subtropical hot desert—the largest in Asia. The Arabian Desert stretches from Yemen in the south, across Saudi Arabia, to Jordan in the north. Mountains rise to the desert's west and south, and the Persian Gulf lies to the east. At its heart is Ar Rub 'al-Khali, or the Empty Quarter, one of the largest and driest expanses of sand on Earth, containing dunes that reach up to 1,000 ft (300 m) in height. Typically, the desert records an annual rainfall of about 4 in (100 mm), but it may also receive as little as 1/2 in (30 mm). The temperature usually exceeds 104 °F (40 °C) in summer. Despite the extreme climate, sparse low-lying vegetation is found across the region, along with grazing animals such as gazelles and oryx, and predators such as foxes and caracals.

Landscape of shifting sand
In Ar Rub 'al-Khali strong winds known as *shamals* move huge amounts of sand twice a year, reshaping the dunes.

Siberian Taiga

The Siberian taiga is the world's largest boreal forest, stretching about 3,600 miles (5,800 km) from the Ural Mountains in the west; across the floodplains of the Ob, Yenisey, and Lena rivers; the Central Siberian Plateau; and the Verkhoyansk and Stanovoy mountains to the Pacific Ocean in the east. The taiga biome lies south of the Arctic Circle, between temperate forests and the tundra. The dense forests are mostly evergreen conifers such as pine and spruce along with hardy deciduous larch, which can survive the winter freeze. However, deforestation is a threat to the environment. Permafrost or bedrock are usually not far below the surface, restricting soil drainage and creating shallow boggy areas. Blanketed in snow in the winter, the growth of taiga trees in the summer absorbs significant amounts of carbon dioxide from the atmosphere.

Insulating layer
Winter in the Siberian taiga lasts 6–7 months. However, the snow-covered trees act as an insulator, keeping the ground warmer than the air above the forest.

Lake Baikal

The world's largest lake by volume, Lake Baikal contains about 20 percent of the world's surface fresh water. It is also the deepest and oldest lake. Situated to the south of the Central Siberian Plateau, near Mongolia, it occupies a rift valley created 30–25 million years ago, which plunges to a depth of 5,387 ft (1,642 m). The rift is still widening, at a rate of about ¾ in (2 cm) a year. In some parts, the sediments on the lake bed are up to 4.3 miles (7 km) thick, placing the rift floor 5.3 miles (8.6 km) below the surface—the deepest continental rift. Lake Baikal supports a range of endemic animals from the Baikal seal to some giant crustaceans. While hundreds of rivers feed the lake, it drains from a single outlet, the Angara River, into the Arctic Ocean. Lake Baikal's water is so clear that it is possible to see objects 130 ft (40 m) below the surface.

Winter freeze
Lake Baikal is completely covered with ice for 4–5 months a year. Due to the huge volume of water in its basin, it freezes later in the Siberian winter than the region's other lakes.

South China Karst

A humid tropical to subtropical climate has enabled the development of a spectacular karst landscape in southern China over about the last 270 million years. Karst areas are characterized by sinkholes, caves, and underground drainage systems formed as limestone or dolomite is dissolved by naturally acidic rainwater. As the sinkholes expand and merge, they create sunken regions, and any high ground left between them becomes linked to form cone-shaped hills called *fengcong*. Erosion by surface water can transform these hills into isolated towers called *fenglin*. Among the best known features of the South China karst are the Wulong Plateau with its giant sinkholes, natural bridges, and caves; the stone forests of Shilin; and the cones and towers of Libo. The aesthetic appeal of these areas has long been celebrated in Chinese art.

Karst landscape
In southern China, rocky outcrops of limestone and dolomite adorn an area of about 190,000 square miles (500,000 square km) between the high Yunnan-Guizhou Plateau and the lowland plain of the Guangxi Basin.

Deccan Traps

The Deccan Traps are a series of stepped escarpments stretching 200,000 square miles (500,000 square km) across western India. They are composed of flood basalt lava flows that accumulated to a thickness of more than 6,500 ft (2,000 m) during one of the largest volcanic events in Earth's history. About 65 million years ago, India passed over a deep mantle hot spot now sitting under the Indian Ocean. Fissure eruptions poured out about 120,000 cubic miles (500,000 cubic km) of runny basaltic lava over perhaps half a million years. Huge amounts of volcanic gas, including sulfur dioxide and carbon dioxide, would also have been released—enough to have a profound impact on the climate. Global average temperatures may have dropped 9 °F (5 °C), which in turn may have played a role in the extinction of the dinosaurs.

Volcanic stairway
The hills of Matheran on the Sahyadri range in Maharashtra, India, form a part of the Deccan Traps. The word "traps" comes from the Swedish word for stairway, as the many layers of basalt form steps up the hillsides.

Yangtze

Known in China as the Chang Jiang—"Long River"—the Yangtze is the longest river in Asia and the third longest in the world. From its source high on the Tibetan Plateau, it runs 3,900 miles (6,300 km) across China, through steep valleys, towering gorges, lake-strewn plains, and farmland, to its outlet into the East China Sea near the port city of Shanghai. The Yangtze has cut several spectacular gorges over the course of a 13,120-ft (4,000-m) drop from its source, through the Sichuan Basin, to its floodplain. Its drainage basin covers one-fifth of China's land area and is home to one-third of the country's population. Flooding of the lower reaches of the river has historically killed tens of thousands of people, but this risk has been reduced by the building of a series of dams upstream to control the flow.

One river, many names
Pictured here is the first bend of the Yangtze River at Shigu, Yunnan Province. Here, the Yangtze is known as the Jinsha Jiang—the "Gold Dust River"—due to the alluvial gold panned from its waters.

Mount Fuji

At 12,388 ft (3,776 m), Mount Fuji is Japan's tallest mountain. It is located 55 miles (90 km) southwest of Tokyo. Fuji lies above a subduction zone and is a steep-sided, conical stratovolcano. Its last eruption was in 1707. Fuji's slopes show distinct vegetation zones, with forests on the lower slopes giving way to thinner brush and grass higher up, ending with bare lava and ash surrounding the summit. The altitude of the snow line varies with the season and the year, but the mountain is usually snow-capped five months of the year. The crater at Fuji's summit is 2,650 ft (800 m) across and 650 ft (200 m) deep. Volcanic activity has created the landscape around Mount Fuji, including the scenic Fuji Five Lakes recreation area and the older volcanic caldera of Hakone, which last erupted in 950 BCE.

Snow-capped cone
Mount Fuji's symmetrical cone looks its picturesque best, framed by bright red maple leaves in the fall. On a clear day, the mountain can be seen from as far away as 90 miles (150 km).

Mount Pinatubo

Situated on the island of Luzon, Philippines, the stratovolcano of Mount Pinatubo lies above a subduction zone and was dormant for nearly 500 years before its catastrophic eruption in June 1991. When gas-charged magma reached the surface, a series of huge explosions threw 2 cubic miles (10 cubic km) of rock and ash 25 miles (40 km) up into the air. So much material was ejected from the volcano that its peak collapsed, leaving a wide caldera. Huge pyroclastic flows burned land up to 10 miles (17 km) away, and lahars (ash flows mixed with typhoon rain) destroyed villages 37 miles (60 km) away. Thousands of people were evacuated ahead of the eruption, but it still killed 300. Volcanic gases and fine dust spread around the world within a few weeks, producing spectacular sunsets and a dip of 0.9 °F (0.5 °C) in global average temperatures over the next three years.

High and deep
The summit caldera created by Pinatubo's 1991 eruption is 1½ miles (2.5 km) wide and is now filled by a lake. With a depth of 2,000 ft (600 m), it is the deepest lake in the Philippines.

Simpson Desert

The Simpson Desert is a red sandy plain extending 68,100 square miles (176,500 square km) across South Australia, Queensland, and the Northern Territory. It is an erg (sand sea), containing the world's longest parallel sand dunes. Summer temperatures here can soar to 122 °F (50 °C), and the area usually receives only about 6 in (15 cm) of rain a year. The dunes are held in place by drought-resistant shrubs and grasses, and occasional heavy rains can bring the desert into bloom.

The desert is part of the Lake Eyre inland drainage basin. Several of its rivers dry up in the Simpson Desert, including the Finke, which left its mark on Australia's Precambrian rocks about 400–300 million years ago, making it one of Earth's oldest rivers—a testament to how geologically ancient the red heart of Australia is.

Parallel dunes
This aerial view of the Simpson Desert shows the size and extent of its parallel sand dunes. The dunes can reach as high as 100 ft (30 m) and can extend for 120 miles (200 km) or more.

Great Dividing Range

The Great Dividing Range stretches 2,175 miles (3,500 km) down Australia's east coast from the Cape York Peninsula to Victoria. It is a series of hills, plateaus, and mountains of varying heights divided into several segments, including the Australian Alps in the south and the Blue Mountains in New South Wales. The Snowy Mountains within the Australian Alps contain the country's highest peak, Mount Kosciuszko (7,310 ft/2,228 m), and are the source of its longest river, the Murray (1,558 miles/2,508 km).

As its name suggests, the Great Dividing Range divides eastward-flowing, Pacific-draining rivers from those draining to the west, north, and south. The range has a long, complex geological history, from continental collisions about 300 million years ago to more recent hot spot volcanism. The last volcanoes in the northern range erupted just 6,000 years ago.

Sandstone formation
The rocky outcrop known as the Three Sisters overlooks the Jamison Valley in the Blue Mountains. The formation, made of sandstone, was created around 200 million years ago.

Great Barrier Reef

The Great Barrier Reef extends more than 1,200 miles (2,000 km) along the northeastern coast of Australia and up to 155 miles (250 km) out into the Coral Sea, to the edge of Australia's continental shelf. It is the world's largest coral reef system and is often described as the largest structure made by living organisms. Here, about 3,000 individual reefs and 900 small islands make up some of the most spectacular scenery on Earth, as well as a habitat of great biodiversity. The Great Barrier Reef is home to 400 species of coral, 1,500 fish species, 4,000 mollusk species, and 240 species of birds. During the last glacial period, sea levels were lower, exposing flat-topped limestone hills where the reefs now grow, but there is evidence of coral growth during at least four preceding interglacial periods. Coral bleaching from rising water temperatures threatens the reef's survival.

Ocean barrier
The Great Barrier Reef is so-called because it acts as a barrier between the Australian Coast and the large waves of the Pacific Ocean.

Rotorua

Rotorua is a volcanic caldera on New Zealand's North Island. It is occupied by a lake about 13½ miles (22 km) in diameter. The Rotorua volcano last erupted 220,000 years ago, when its magma chamber collapsed, creating the current caldera. Volcanic steam and gases seep from the shores of the lake even today. Just north of the caldera lies the active geothermal area of Wai-o-tapu, containing numerous colorful hot springs, geysers, and boiling mud pools. Rotorua and Wai-o-tapu are part of the wider Taupo Volcanic Zone, which extends from White Island, just offshore to the north, to Mount Ruapehu, 110 miles (180 km) to the south. The zone has produced some of the most violent volcanic eruptions in history, such as the Mount Tarawera eruption in 1886 and the Taupō Volcano eruption that occurred around 186 CE.

Bright deposits
The orange coloration at the edges of this hot spring, one of the many in Wai-o-tapu, are caused by deposits of toxic minerals containing arsenic and antimony.

Southern Alps

The Southern Alps are the backbone of New Zealand's South Island, running from north to south along almost its full length. Although not as high as their European namesake, the Southern Alps rise to 12,217 ft (3,724 m) above sea level at the summit of Aoraki, or Mount Cook. Formed in about the last 60 million years, the relatively young Southern Alps are still tectonically active and continue to rise by up to several millimeters annually. The range lies along a transform boundary between two tectonic plates. A prominent fault, called the Alpine Fault, marks their western edge, with a narrow coastal strip lying on the Australian Plate, while the rest of the island is part of the Pacific Plate. There is a pronounced rain shadow to the east, where the land extends into a broad plain before reaching the Pacific shore.

Wind and snow
Prevailing westerly winds from the Tasman Sea bring moist air that sustains snowfields and glaciers (including the Godley Glacier, seen here) and a temperate rainforest.

Antarctic Ice Sheet

About 60 percent of Earth's fresh water is frozen in the Antarctic Ice Sheet. It covers nearly the whole continent of Antarctica, occupying almost 5.4 million square miles (14 million square km)—about the size of the United States and Mexico combined—and reaching up to 15,669 ft (4,776 m) thick. Over the last 800,000 years and more, ice has accumulated from high in the interior and flowed to the coast and hundreds of miles over the sea as floating ice shelves. Mountains appear above the ice in places, while underneath the ice in parts of East Antarctica and across most of West Antarctica, bedrock and meltwater lakes appear below sea level. With the air so cold and dry, the ice sheet is technically a desert. A few areas, such as the Dry Valleys in Victoria Land, remain free of snow and ice.

Mammoth size
The world's largest ice shelf, with an ice cliff along its edge, the Ross Ice Shelf ranges in thickness from about 1,150 ft (350 m) at its front end to around 2,460 ft (750 m) farther back.

glossary

Terms that appear in **bold text** refer to other entries in the glossary

ABYSSAL A term applied to the depths of the ocean. The abyssal plain is the almost flat plain of the deep ocean floor beyond the continental margin. The abyssal zone is the zone of water 13,000–20,000 ft (4,000–6,000 m) deep.

ACCRETION The process by which material is added to a continental plate at a **subduction** zone.

AFTERSHOCK A smaller **earthquake** or tremor following a larger earthquake.

ALGA (pl. ALGAE) Any organism that **photosynthesizes** but is not a true plant, including seaweeds and many microscopic forms. **Cyanobacteria** are usually now excluded from the definition.

ALLUVIAL A term applied to sedimentary material deposited by rivers. An alluvial fan is a deposit laid down where a stream leaves a highland area and spreads out over a plain. See also **Sediment**.

AMMONITE The fossilized coiled shell of an extinct cephalopod **mollusk** (a relative of squid) that died out about 66 million years ago. Members of the broader group of animals that includes ammonites are called ammonoids.

AMNIOTE A **vertebrate** whose embryo is protected by membranes and sometimes a shell. Reptiles, birds, and **mammals** are all amniotes.

ANGIOSPERM A flowering plant (as distinct from a **conifer**, fern, moss, or other plant). Angiosperms include many tree species.

ANION A negatively charged ion. See also **Cation**, **Ion**.

ANTICLINE An archlike, upward **fold** of originally flat strata formed by horizontal compression. See also **Syncline**.

ANTICYCLONE A weather system in which winds circle around an area of high pressure. See also **Cyclone**, **Hurricane**, **Tropical cyclone**.

ARACHNID A member of the group of arthropods that includes mites, spiders, scorpions, and relatives.

ARCHOSAUROMORPH A member of the **clade** of **diapsid** reptiles that includes crocodiles, **dinosaurs**, and birds.

ASTEROID One of thousands of bodies of rocky material orbiting the sun. Their diameters range from a few yards to about 600 miles (1,000 km).

ASTHENOSPHERE The layer of the **mantle** immediately below the rigid **lithosphere**. It is sufficiently nonrigid to flow slowly in a solid state and plays a key part in the movement of **tectonic plates**.

ATOLL A ring-shaped group of coral islands surrounding a **lagoon**, usually on top of a **seamount**.

AURORA A display of lights in the night sky in Arctic and Antarctic regions, caused by the interaction between electrically charged particles of the sun and Earth's magnetic field.

AUSTRALOPITH One of several species of upright apes, including *Australopithecus*, living 1.2–4.4 million years ago.

BASALT Earth's most common volcanic rock, which usually originates as solidified **lava**. Basalt is glassy to fine-grained (composed of very small **crystals**).

BATHOLITH A huge, irregularly shaped mass of **igneous rock**, 60 miles (100 km) or more across, formed from the intrusion of **magma** from deep underground. See also **Pluton**.

BEDDING The manner in which a **sedimentary rock** is laid down in layers. A bedding plane is a surface in the rock that separates these layers.

BIOGENIC Produced by the activity of living creatures or biological processes.

BIOMINERALIZATION The process by which living creatures produce minerals.

BIOTIC Relating to, or produced by life and living organisms.

BIVALVE An aquatic **mollusk**, such as an oyster or clam, with two hinged shells that cover its whole body.

BLACK SMOKER A type of **hydrothermal vent** in which the hot water is colored dark by sulfide minerals. A white smoker emits other minerals, such as silica and barite.

BRACHIOPOD A major group of marine **invertebrates**. Their two-part shell makes them look superficially similar to **bivalve mollusks**, but they are unrelated. They were abundant and diverse in the **Paleozoic** and Mesozoic.

BRACKISH A term to describe water that is slightly salty.

BRECCIA A **sedimentary rock** made up of angular fragments, called **clasts**, cemented by **minerals**. See also **conglomerate**.

BRINE A highly concentrated solution of salt in water, often found as sea water.

CALCAREOUS Containing or made of the **mineral** calcium carbonate.

CALDERA A bowl-shaped volcanic depression larger than a **crater**, typically greater than 0.6 miles (1 km) in diameter, caused by the collapse of a volcano into its emptied **magma** chamber.

CALVING The process by which a **glacier** creates icebergs by shedding ice blocks into a sea or lake.

CAPILLARY WAVE A wave at the boundary of a fluid that is dominated by the **surface tension** of the fluid, rather than gravity.

CARAT A term used to describe the proportion of gold in a gold alloy, with 24-carat gold being pure gold. It is also a unit of weight, equivalent to 0.007 oz (0.2 g), used for diamonds and precious stones.

CARTILAGINOUS FISH A class of fish whose skeleton is made of cartilage, not bone. It includes sharks and rays.

CATION A positively charged ion. See also **Anion**, **Ion**.

CHEMOSYNTHESIS The process by which an organism can grow and multiply using the energy stored in simple chemicals, such as hydrogen sulfide or methane. Chemosynthesis functions in contrast to **photosynthesis**, which depends on energy coming from the sun. Many bacteria can carry out chemosynthesis, notably those living around **hydrothermal vents**.

CINDER CONE A relatively low-lying volcano formed from cinders and ash that have fallen from an eruption cloud.

CLADE A group of species consisting of all the evolutionary descendants of a given ancestor.

CLAST A fragment of **mineral** or rock, especially when incorporated into newer, **sedimentary rock**.

CLEAVAGE The way some **minerals** break along preferred planes (known as cleavage planes) dictated by their atomic structure.

CLIMATE CHANGE Long-term changes to weather patterns and average temperatures on a global or regional scale.

COMET A body composed of rock and ice that orbits the sun, usually with a highly eccentric orbit. As it approaches the sun, vaporization of the ice produces an extensive cloud, called the coma, and one or more tails.

COMPOUND A substance containing atoms of two or more **elements**.

CONDENSATION NUCLEUS A tiny particle in the air around which a raindrop or snow crystal may form. See also **Precipitation**.

CONGLOMERATE A **sedimentary rock** that is made up of rounded, gravel-size **clasts** cemented together by **minerals**. See also **Breccia**.

CONIFER A cone-bearing tree. Nearly all conifers, such as pines and firs, have needlelike leaves and are evergreen. Some, such as larch, lose their needles in the winter months.

CONTACT METAMORPHISM The processes by which rocks are transformed by the effects of contact with hot **magma**. See also **Dynamic metamorphism**, **Regional metamorphism**.

CONTINENTAL CRUST The part of Earth's **crust** that forms the continents. It is less dense and thicker than **oceanic crust**.

CONTINENTAL ISLAND A body of land surrounded by water that lies on the **continental shelf** of a continent.

CONTINENTAL RISE The region between deep ocean seabed and the **continental slope**.

CONTINENTAL SHELF The relatively flat and shallow seafloor surrounding a continent, which is considered geologically part of that continent. See also **Continental slope.**

CONTINENTAL SLOPE The sloping seabed that leads from the edge of a **continental shelf** down to the **continental rise**.

CONTOUR CURRENT A slow-moving current that flows parallel to the **continental rise**.

CONVECTION Circulating currents (for example, in Earth's atmosphere or **mantle**) caused by differences in temperature. This results in the warmer, less dense air or fluid rising upward, which drives the circulation.

CONVERGENT BOUNDARY The line along which two **tectonic plates** that are moving toward each other meet.

CORE Earth's innermost layer. It consists of a liquid outer core and a solid inner core, both made of nickel-iron. See also **Mantle**, **Crust**.

CORIOLIS EFFECT The tendency for winds and currents moving in a northerly or southerly direction to be deflected and move at an angle because of the effect of Earth's rotation. The deflection is to the right in the northern hemisphere and to the left in the southern hemisphere.

CRATER A bowl-shaped depression through which an erupting volcano discharges gases, rock fragments, and **lava**. The crater walls form by accumulation of ejected material. Also refers to a circular depression in the landscape caused by a large meteorite impact. See also **Caldera**.

CRATON A stable area of Earth's **continental crust** that is made of old rocks largely unaffected by mountain-building activity since the Precambrian. Also called a shield.

CROSS-BEDDING Layering in **sedimentary rock** that is at an angle to the main bedding plane. See also **Bedding**.

CRUST Earth's outermost rocky layer. The continents and their margins are made of thicker but less dense **continental crust**, while thinner, denser **oceanic crust** underlies the deep ocean floors. See also **Mantle**, **Tectonic plate**.

CRYSTAL Any solid in which the individual atoms or molecules are arranged in a regular geometrical pattern compared to unordered solids, such as glass. There are seven basic patterns of crystal growth, called crystal systems. See also **Geode**, **Habit**.

CYANOBACTERIA An abundant group of microscopic **photosynthesizing** bacteria, formerly called blue-green **algae**. Their tiny cells lack nuclei, as do those of other bacteria.

CYCAD A member of the group of non-angiosperm seed plants that are superficially similar to palms, but produce seeds and pollen in cones. See also **Angiosperm**.

CYCLONE A pressure system in which air circulates around an area of low pressure. See also **Anticyclone**, **Hurricane**, **Tropical cyclone**.

CYNODONT A member of a group of advanced **synapsids** that arose in the Late Permian. **Mammals** are their living descendants.

DELTA The area of gently sloping **sediment** built up by a river when it enters the sea, a lake, or another river. Its shape depends on the sediment carried by the river, and the currents and tides it meets. See also **Estuary**, **Submarine fan**.

DENDRITIC A term applied to **minerals** occurring naturally in a branching, treelike form within rocks.

DEPOSITION The laying down of material such as sand and gravel in new locations, usually by wind, water, or ice.

DESERT PAVEMENT A rocky or stony surface layer found in many deserts.

DIAPSID A major division of **amniotes** whose name refers to the two openings on each side of the skull. Traditionally regarded as reptiles, diapsids include the **archosauromorphs** (including **dinosaurs**, birds, and crocodilians) and lepidosaurs (including lizards and snakes). See also **Synapsid**.

DICYNODONTS A group of herbivorous **synapsids** (the ancestors of **mammals**) with two tusks and a blunt beak.

DIFFERENTIATION A process in which heavy **elements** in a body of **magma**, such as the early Earth or a magma chamber, sink while lighter elements float.

DIKE A sheetlike intrusion of **igneous rock** that cuts across existing rock structures.

DINOSAUR A major group of **archosauromorph amniotes** that had an erect gait with walking limbs directly beneath the body. They became extinct at the end of the Cretaceous, with the exception of their descendants, birds.

DIVERGENT BOUNDARY The line along which two **tectonic plates** are moving away from each other.

DIVERSIFICATION Multiplication of species over evolutionary time, often in the process of filling ecological niches.

DOWNWELLING The sinking down of water from the surface of the ocean. Large-scale downwelling in certain regions gives rise to a process known as **thermohaline circulation**. See also **Upwelling**.

DYNAMIC METAMORPHISM A type of metamorphism in which a rock is transformed by pressure being applied in a particular direction during large-scale movement of Earth's crust. See also **Contact metamorphism**, **Regional metamorphism**.

EARTHQUAKE A sudden shaking of Earth's surface caused by **seismic waves** traveling through the **lithosphere**.

ECHINODERM A member of the marine **invertebrate** phylum that includes starfish, sea urchins, sea cucumbers, and relatives. These animals feature tube feet used for feeding and movement.

ECOSYSTEM The entirety of living and nonliving features of a region and the interactions between them.

ECTOTHERMIC A term to describe a "cold-blooded" animal, one that regulates its body temperature through external heat sources, such as sunlight.

ELEMENT A substance that cannot be broken down into simpler substances.

ENDOTHERMIC A term to describe a "warm-blooded" animal, one that regulates its body temperature through internal metabolic processes.

EPICENTER The point on Earth's surface directly above the point of origin of an **earthquake**. See also **Seismic**.

EROSION The processes by which rocks or soil are loosened and worn or scraped away from a land surface. The main agents of erosion are wind, moving water, and ice, and the rock particles that they carry.

ESTUARY A broad, tidal, funnel-shaped stretch of **brackish** water found where many large rivers meet the sea.

EXTRUSIVE ROCK A rock formed from **lava** that flowed onto Earth's surface or was ejected as **pyroclastic** material.

FAST ICE A continuous sheet of **sea-ice** that is frozen onto the coast, or land-formed ice at its landward edge.

FAULT A fracture where the rocks on either side have moved relative to one another. If the fault is at an angle to the vertical, and the hanging wall block (above the fault plane) has slid downward, it is called a normal fault. If the hanging wall block has slid upward (in relative terms) it is called a reverse fault. A strike-slip fault is one where movement is horizontal.

FELDSPAR A common type of **silicate** mineral found in **igneous rock**.

FETCH The extent of open water across which a wind or water wave has traveled.

FJORD A former glacial valley on the coast that has become an inlet of the sea.

FLOODPLAIN A flat plain next to a river that is liable to be covered with water when the river floods.

FOLD A geological structure in which originally flat-lying layers of rock have been bent under compression. They may bend upward in the middle to form a ridge (**anticline**) or downward to form a trough (**syncline**). In an overturned fold, one side (or limb) of the fold moves more than the other and extends over the other limb.

FOLIATION The arrangement of **minerals** in parallel bands in some deformed **metamorphic rocks**.

FOSSILIFEROUS A term applied to **sedimentary rocks** containing fossils.

FROST SHATTERING **Weathering** caused by the alternate freezing and thawing of water in cracks in a rock.

FULGURITE Naturally formed tubes of glassy **minerals** formed when lightning strikes the ground.

FUMAROLE In volcanic regions, a small opening in the ground through which steam and hot gases can escape.

GASTROPODS The largest class of **mollusks**, which includes all snails and slugs, as well as limpets and sea slugs.

GEMSTONE A jewel or stone used in jewelery that is prized for its durability, beauty, and rarity.

GEODE A hollow cavity within a rock that is lined with **crystals**.

GEYSER A jet of boiling water and steam that rises at intervals from the ground. It is powered by hot rocks heating the surrounding **groundwater**.

GINKGO A seed-bearing tree with fan-shaped leaves that is native to China.

GLACIER A mass of ice that flows slowly downhill from an **icecap** or mountainous region. A valley glacier flows between the walls of a valley. A tidewater glacier is a valley glacier that flows into the sea.

GONDWANA An ancient supercontinent that included South America, Africa, Antarctica, Australia, and India.

GORGE A narrow, deep valley, usually with vertical cliffs on either side.

GRAPTOLITE Member of an extinct group of colonial and mostly planktonic **invertebrates**, typically growing in the form of long, narrow colonies of small individuals, supported by a skeleton.

GREASE ICE The first stage of formation of **sea-ice**. Ice **crystals** form in the sea water, giving it a thick, oily texture.

GROUNDWATER Water occurring below the surface of the land and held in spaces (interstices) between the grains of rocks. The upper limit of such a groundwater zone is called the **water table**.

HABIT The external shape of a **crystal** (or group of crystals), including the forms and faces of individual crystals.

HADROSAUR A member of the group of duck-billed herbivorous **dinosaurs** that lived during the Late Cretaceous.

HANGING VALLEY A valley, usually carved by a **glacier**, that enters high up the side of a larger valley.

HARDNESS Of a **mineral**, the extent to which it can resist scratching or abrasion. See also **Mohs scale**.

HEADWATER The upper portion of any river or stream, close to its source.

HORSETAIL A group of spore-bearing plants whose above-ground stems have a jointed appearance, with rings of small branches bearing tiny, scalelike leaves. Some extinct forms were as big as trees.

HOT SPOT A long-lived zone of volcanic activity thought to originate deep in Earth's **mantle**. **Tectonic plates** passing over hot spots are marked by linear chains of volcanoes that become progressively older with increased distance from the hot spot. See also **Mantle plume**.

HUMUS A dark-colored substance found in soils, derived from dead plants, microorganisms, and animals.

HURRICANE A huge, circular tropical storm in which there are wind speeds of 74 mph (119 kph) or more. It is also called a **tropical cyclone** and (especially in east Asia) a **typhoon**. Its energy comes from the latent heat of water that has evaporated from warm oceans and later condenses.

HYDROTHERMAL VEIN A thin, sheetlike body of minerals, laid down by the circulation of hot, mineral-bearing waters in Earth's **crust**.

HYDROTHERMAL VENT A fissure in a volcanically active area of the ocean floor, where water rich in chemicals issues from the rock at high temperatures. See also **Black smoker**, **Chemosynthesis**, **Hydrothermal vein**.

ICECAP A dome-shaped mass of ice that submerges a landscape from a single mountain to an entire polar region.

ICE FLOE A large area of floating **sea-ice**.

ICE SHEET A very large, flowing mass of ice that permanently covers a land area, as in Antarctica or Greenland.

ICE SHELF A floating extension of an **ice sheet** or **glacier** over the ocean.

IGNEOUS INTRUSION A body of **igneous rock** that formed when **magma** cooled and solidified underground. See also **Batholith**.

IGNEOUS ROCK Rock that originates from the solidification of molten **magma**. See also **Metamorphic rock**.

INTRUSIVE ROCK **Igneous rock** that has solidified below the surface, cooling slowly enough to allow larger **crystals** to form. A body of intrusive igneous rock is called an **igneous intrusion**.

INVERTEBRATE An animal without a backbone, such as an insect, snail, or worm. See also **Vertebrate**.

ION An atom or group of atoms that has lost or gained one or more electrons to become electrically charged. The molecules of most minerals include a combination of **cations** (positively charged ions) and **anions** (negatively charged ions).

ISOSTATIC REBOUND The rise of a land mass after the huge weight of an **ice sheet** is removed.

ISOTOPE One of two or more forms of the same chemical **element** that have equal numbers of protons but different numbers of neutrons in the nuclei of their atoms.

KARST A characteristic landscape formed of water-soluble rocks, especially limestone, that have been weathered by water flowing underground. A karst coast occurs where limestone has been weathered and later inundated by the sea.

LAGOON An area of sheltered sea water, almost cut off from the open ocean.

LARVA (pl. LARVAE) The young stage of an animal when it is very different in structure from an adult.

LAURASIA An ancient supercontinent that included North America, Europe, and most of Asia (apart from India).

LAVA Molten rock (**magma**) that flows onto Earth's surface.

LIGHT MICROGRAPH A photograph taken through a microscope that uses visible light.

LITHOSPHERE The layer of Earth that includes the **crust** and the upper **mantle**.

LOBE-FINNED FISH A mainly extinct group of fish having fleshy, muscular bases to their paired front and back fins. They are first known from the Silurian period and include the present-day lungfish and coelacanth.

LOESS An unconsolidated sedimentary deposit made up of wind-blown silt. One of the largest examples is the Loess Plateau in northern China. See also **Sedimentary rock**.

LUSTER The way a **mineral** reflects light and the extent of its sheen.

MACROCRYSTALLINE A term applied to **crystals** that are large enough to be recognized with the naked eye. See also **Microcrystalline**.

MAGMA Liquid, molten rock in Earth's **mantle** and **crust**. It cools to form **igneous rock**. It may crystallize beneath the surface or be erupted as **lava**.

MAGNETOSPHERE The region around Earth (or other planet) dominated by the planet's magnetic field.

MAMMAL A class of warm-blooded **vertebrates**, almost all of which give birth to live young and nourish them using milk produced by the female.

MANTLE The rocky layer of Earth between the **crust** and the **core**. It makes up 84 percent of Earth's volume.

MANTLE PLUME A column of hot rocks rising through the **mantle** and **crust**, giving rise to a **hot spot** at Earth's surface.

MARGINAL SEA A partially enclosed sea bordering a continent.

MARSUPIAL A **mammal**, such as a koala or kangaroo, in which offspring are born in a relatively undeveloped state; they often continue their growth within an external pouch on the mother. See also **Placental**.

MASSIF A well-defined mountainous region whose rocks and landforms tend to be similar across the region.

MASSIVE Of a rock: having a structure that does not show layering or other divisions into smaller segments.

MATRIX The mass of relatively fine-grained material that binds together larger particles in some heterogeneous **sedimentary rocks** and larger **crystals** in volcanic rocks. See also **Porphyritic**.

MESOSPHERE The layer of the Earth's atmosphere between the **stratosphere** and **thermosphere**, at an altitude of about 30–50 miles (50–80 km).

METAMORPHIC AUREOLE A zone of rock around a body of **magma** that has been altered by **contact metamorphism**.

METAMORPHIC ROCK A rock that has been transformed underground by heat or pressure, altering its texture or minerals. For example, marble is metamorphosed limestone. See also **Igneous rock**.

METEOR A small mass of rock from space that vaporizes completely as it falls through Earth's atmosphere, glowing as it does so.

METEORITE A rock from space that has fallen to Earth's surface without completely burning up.

METEOROID A potential **meteor** or **meteorite** before it has entered Earth's atmosphere.

MICROCRYSTALLINE Having **crystals** that are so minuscule that they are detectable only with the aid of a microscope. See also **Macrocrystalline**.

MID-OCEAN RIDGE An undersea mountain range running along the deep ocean floor, and a site where new oceanic **crust** is created.

MINERAL Any solid, naturally occurring inorganic material that has a characteristic **crystal** structure and well-defined chemical composition. Most rocks are mixtures of more than one mineral.

MOHS SCALE A measure of the **hardness** (resistance to scratching) of minerals, ranging from 1 to 10.

MOLLUSK A member of the group of soft-bodied **invertebrates** that includes **gastropods**, **bivalves** and cephalopods (such as octopuses and squid). See also **Ammonite**, **Nautiloid**.

MONOCOT Also called a monocotyledon, a member of the group within the **angiosperms** that includes grasses, orchids, palms, and lilies. They are characterized by having only one cotyledon (seed-leaf) within each seed.

MONOTREME An egg-laying **mammal**, such as the present-day platypus and echidna (spiny anteater). Egg-laying is thought to be the original mode of reproduction for **mammals**. See also **Marsupial**, **Placental**.

MORAINE An accumulation of rock debris resulting from glacial action. Active valley **glaciers** create lateral moraines at their edges, medial moraines where two glaciers merge, and terminal moraines at their ends. Moraines often remain after a glacier has melted. See also **Outwash**, **Striation**.

MOSASAUR A member of an extinct group of Cretaceous marine reptiles that are thought to be relatives of snakes and monitor lizards.

MULTISPECTRAL Relating to two or more wavelengths in the electromagnetic spectrum. Multispectral imaging can be used to make maps of natural and geological features.

MULTITUBERCULATE A member of an extinct group of mainly rodent-like early **mammals** found from the Jurassic to the Paleogene period.

NATIVE ELEMENT A chemical **element** that occurs in a pure, or uncombined, state in nature.

NAUTILOID A member of the group of cephalopod **mollusks** related to the **ammonites** and nautiluses. Their coiled external shell contains gas-filled chambers that lighten the body in the water.

NEAP TIDE A tide in which the difference between high and low tide is at its smallest. See also **Spring tide**, **Tidal bore**.

NODULAR About a rock: containing rounded lumps of minerals or other materials. For example, flint in chalk.

NONCRYSTALLINE A term to describe a solid material that does not have a consistent arrangement of particles.

NUTRIENTS Chemicals, especially salts of **elements** such as nitrogen, phosphorus, and iron, that are essential for living organisms to grow.

OCEANIC CRUST The part of Earth's **crust** that underlies most of the world's oceans. It is thinner and denser than **continental crust**.

ORE A rock from which metal can be profitably mined.

ORNITHISCHIAN A member of one of the two main subgroups of **dinosaurs** (the other being saurischians), the name literally meaning "bird-hipped." Ornithischians include ornithopods, stegosaurs, ankylosaurs, and ceratopsians.

OUTWASH The deposit of sand, gravel, and other materials carried by the meltwater from a **glacier**. See also **moraine**.

OZONE A form of oxygen with three atoms in its molecules. It is present in the upper atmosphere (the ozone layer), where it absorbs ultraviolet radiation.

PALEOZOIC The era of geological time extending from the beginning of the Cambrian to the end of the Permian (539–252 million years ago). It was followed by the Mesozoic.

PANGAEA An ancient supercontinent that included almost all the present-day continents.

PARAREPTILE A member of a diverse group of extinct **amniotes** traditionally called reptiles. They include mesosaurs and several other groups.

PHOTOSYNTHESIS The process by which plants, **algae**, and **cyanobacteria** use the energy of sunlight in the presence of chlorophyll to convert water and carbon dioxide into food. See also **Chemosynthesis**.

PHYTOPLANKTON The tiny, mostly single-celled organisms that live in the surface waters of oceans and lakes, and are the foundation of most aquatic food chains. See also **Algae**, **Plankton**.

PILLOW LAVA Pillow-shaped mounds of rock formed from **lava** that has been ejected underwater.

PLACENTAL Those **mammals** in which the fetus grows to a relatively advanced stage in the mother's womb, nourished by a placenta. Includes all living mammals except **marsupials** and **monotremes**.

PLACODERM A group of extinct jawed fish, widespread in the Devonian period, whose skin was protected by bony armor-plating.

PLANETESIMAL One of millions of rocky objects of variable size believed to have been present in the early history of the solar system, and which later came together to create the planets.

PLANKTON Any living species (plants, animals, or microorganisms) living in open water that drift with the currents. Most plankton are small, but some—such as jellyfish—are larger. See also **Phytoplankton**, **Zooplankton**.

PLANKTON BLOOM A rapid increase in the population of **phytoplankton** in oceans or lakes. It can cause the water to appear blue-green, brown, or even red.

PLUME The column of ash, volcanic particles, and gas released during an explosive volcanic eruption.

PLUNGE POOL The deeper area at the bottom of a waterfall formed by the force of water hitting soft bedrock at its base.

PLUTON A smaller mass of **igneous rock** that has formed beneath the surface of Earth by solidification of **magma**. See also **Batholith**.

PORPHYRITIC A term to describe an **igneous rock** texture in which large **crystals** are set in a finer **matrix**.

POTHOLE A hole formed in the bed of a river by the action of water carrying small pebbles and **sediment**.

PRECIPITATION Water that reaches Earth's surface from the atmosphere, including rain, snow, hail, and dew.

PRIMATE A member of the group of **mammals** that includes monkeys, apes, and humans. Typical characteristics include grasping hands and forward-facing eyes.

PRISMATIC A term to describe **crystals** in which the parallel rectangular faces form prisms.

PROTOZOA Animal-like single-celled organisms, mainly microscopic, that are common in virtually all habitats and may be free-living or parasitic. The thousands of species are in many different subgroups, not all closely related. See also **Zooplankton**.

PTEROSAUR A flying **diapsid** related to the **dinosaurs**, with wings formed by skin stretched over their front limbs. Originating in the Triassic, they went extinct at the end of the Cretaceous.

PYROCLASTIC Consisting of, or containing, volcanic rock fragments. Pyroclastic flows are fast-moving, sometimes deadly clouds of hot gases and debris.

RAIN SHADOW An area of low rainfall downwind of a mountain range, caused by air shedding moisture as it passes over the windward side of the range.

RAISED BEACH A beach that is now above high-tide level because of a change in sea level or uplift of the land.

RAVINE A smaller valley with steep sides. See also **Hanging valley**, **Rift valley**.

RECRYSTALLIZATION The transformation of smaller **crystals** into larger ones that takes place under very high pressure during metamorphism.

REFRACTION The change of direction of waves, including light waves, when they pass from one medium to another. Water waves change direction when they reach shallow water.

REGIONAL METAMORPHISM The transformation of rocks by metamorphism over broad areas, such as in mountain ranges.

RHYNCHOSAUR A member of the group of extinct **archosauromorph** herbivores. One of the most common reptiles of the Triassic.

RIFT VALLEY A large block of land that has dropped vertically downward compared with its surroundings as a result of horizontal extension of the crust and normal faulting. See also **Fault**.

ROGUE WAVE An unexpected and unusually large ocean wave.

SALINITY The concentration of dissolved salts in, for example, water or soil.

SALT A **compound** formed from the reaction of an acid with a base. Also, the common name for sodium chloride.

SALT LAKE A body of landlocked water containing water that is very salty because of evaporation.

SALT PAN A shallow depression, often in a desert, that contains salt deposits.

SAUROPOD A member of the group of huge herbivorous **dinosaurs** (including *Diplodocus* and *Brachiosaurus*) with long necks and long tails that were the largest land animals ever to have lived.

SEAFLOOR SPREADING The process by which new oceanic **crust** is created as **tectonic plates** split apart from one another. See also **Mid-ocean ridge**.

SEA-ICE Ice that arises from sea-water freezing. See also **Fast ice**, **Grease ice**, **Ice floe**, **Ice shelf**.

SEAMOUNT An undersea mountain, usually formed by volcanic activity. See also **Atoll**.

SEA SCORPION An extinct aquatic predatory arthropod, also called a eurypterid. Sea scorpions occurred from the Ordovician to the Permian.

SEA STACK A tall column of rock rising out of the sea near a coastline, a resistant remnant of surrounding cliffs that have otherwise disappeared due to **erosion**.

SEDIMENT Particles carried by moving water, wind, or ice, or deposits formed by such particles, including gravel, sand, silt, or mud. See also **Sedimentary rock**.

SEDIMENTARY ROCK Rock formed when small particles are deposited—by wind, water, volcanic processes, or mass movement—and later compacted and cemented together.

SEED FERN A member of one of the several groups of extinct seed plants that have fernlike leaves despite not being related to ferns.

SEISMIC Relating to **earthquakes**. A seismic wave is a shock wave generated by an **earthquake**.

SEISMOGRAPH An instrument for recording **seismic** waves. A seismogram is a recording made by a seismograph. See also **Earthquake**.

SEISMOMETER The part of a **seismograph** that responds to ground motions, often caused by **earthquakes**.

SHIELD VOLCANO A volcano with shallow slopes that is built from thin, fluid **lava**. See also **Cinder cone**, **Stratovolcano**.

SILICATE Any rock or **mineral** composed of groups of silicon and oxygen atoms in chemical combination with atoms of various metals. Silicate rocks make up most of Earth's **crust** and **mantle**.

SILICEOUS About a rock: containing or consisting of **silicate**.

SILL A roughly horizontal, sheetlike **igneous intrusion** that usually forms when **igneous rock** forces its way between layers of existing **sedimentary rocks**.

SINKHOLE A depression in the surface of a **karst** landscape. It often leads down to an underground drainage system.

SOLAR WIND The stream of charged particles released from the upper atmosphere of the sun.

SPRING TIDE A tide that occurs when the effects of the moon and sun reinforce each other, producing the highest high tide and the lowest low tide. See also **Neap tide**.

STALACTITE A deposit of calcium carbonate hanging down from the roof of a cave or underground passage.

STALAGMITE A deposit of calcium carbonate rising up from the floor of a cave or underground passage.

STRATIGRAPHY The geological study of the order and relative position of strata (layers of **sedimentary rock**) in Earth's **crust**.

STRATOSPHERE A layer of Earth's atmosphere extending from the top of the **troposphere**, at 5–10 miles (8–16 km), up to about 30 miles (50 km). See also **Mesosphere**.

STRATOVOLCANO A conical volcano that is built from alternate layers of **lava** and other rock fragments produced in an explosive eruption. See also **Cinder cone**, **Shield volcano**.

STRIATIONS Grooves and scratches that are left in the bedrock as a **glacier** moves over them.

SUBDUCTION The descent of an oceanic **tectonic plate** under another plate when two plates converge. Subduction zones can be classified as either ocean–ocean or ocean–continent depending upon the nature of the two converging plates. See also **Crust**, **Convergent boundary**.

SUBMARINE FAN **Sediment** at the base of the **continental rise** that has been deposited on the **seafloor**.

SUBSOIL The layer of soil immediately beneath the **topsoil**.

SURF The tops of waves as they break on rocks or the shore. See also **Swell**.

SURFACE TENSION An effect that makes a liquid seem as though it has an elastic "skin." It is caused by cohesive forces that pull molecules at the surface of a body of water inward or sideways.

SWELL A series of regular waves created by weather systems over a long distance.

SYNAPSID A major group of tetrapod **vertebrates** that branched off early in the evolution of **amniotes**, and eventually gave rise to the **mammals**. Synapsids were the largest terrestrial creatures during the Permian. See also **Diaspid**.

SYNCLINE A downward **fold** of strata that were originally flat, resulting from an application of horizontal compression. See also **Anticline**.

TABULAR A term to describe the **habit** of a **crystal** with predominantly large, flat, parallel faces.

TECTONIC PLATE Any of the large rigid sections into which Earth's **lithosphere** is broken. The relative motions of different plates leads to **earthquakes**, volcanic activity, continental drift, and mountain-building. See also **Convergent boundary**, **Divergent boundary**, **Transform boundary**.

TERRESTRIAL Relating to Earth; found on land rather than in the sea.

TERRIGENOUS A term to describe ocean **sediments** that were created by **erosion** on land.

THERMOCLINE Any layer at a particular depth in an ocean where the average temperature changes rapidly with depth. Thermoclines can also occur in the atmosphere.

THERMOHALINE CIRCULATION A worldwide circulation of deep-water currents that is driven by differences in temperature and **salinity** between different water masses. See also **Coriolis effect**.

THERMOSPHERE The layer of Earth's atmosphere above the **mesosphere**. It extends over altitudes of about 50–400 miles (80–640 km).

THEROPOD A member of a major group of bipedal **dinosaurs** that includes top predators such as *Tyrannosaurus rex*, as well as many smaller species, such as *Velociraptor* and the ancestors of modern birds.

TIDAL BORE A single large wave sometimes created when a rising tide enters a narrowing channel such as an **estuary**.

TOPOGRAPHIC Relating to the features and forms of a land surface.

TOPSOIL The outermost layer of soil, containing **minerals** and organic matter in which plants grow. See also **Subsoil**.

TRACTION The process by which coarse **sediment** grains are moved along a surface by water or wind currents.

TRAIN A succession of waves of similar wavelength spaced at regular intervals. See also **Refraction**.

TRANSFORM BOUNDARY A boundary between **tectonic plates** where the plates slide horizontally past each other.

TRILOBITE A member of an enormously diverse group of extinct marine arthropods. Trilobites went extinct at the end of the Permian period.

TROPICAL CYCLONE A large-scale circulating weather system in tropical and subtropical regions that is powered by warm ocean water and produces violent winds and heavy rain. Also known as a **hurricane** or **typhoon**. See also **Cyclone**.

TROPOPAUSE The boundary between the **troposphere** and the **stratosphere**. Air temperature starts to increase with height above the tropopause, which varies from about 10 miles (16 km) at the Equator to 5 miles (8 km) over the poles.

TROPOSPHERE The lowest, densest layer of the atmosphere, where most weather phenomena occur. The height of the upper limit of the troposphere (known as the **tropopause**) varies from the equator to the poles.

TSUNAMI A fast-moving, often destructive sea wave generated by **earthquake** activity. It rises in height rapidly as it reaches shallow water. See also **Tidal bore**.

TURBIDITY CURRENT A dense flow of **sediment**-laden water in lakes, and from the continental margin down onto the ocean floor.

TYPHOON A **hurricane** or **tropical cyclone** occurring in the western Pacific or Indian oceans.

UNCONFORMITY A gap in the geological record indicating that one or more strata have been removed by erosion. See also **Sedimentary rock**.

UNIFORMITARIAN THEORY The theory that the laws and processes that operate on Earth today also operated in a similar way in the geological past.

UPWELLING The rising of deep ocean water to the surface. It can be caused by wind blowing parallel to a coastline or by an underwater obstruction such as a **seamount** interrupting a deep-sea current. Upwelled water often enriches the surface of the ocean with nutrients. See also **Downwelling**.

VERTEBRATE A backboned animal. Vertebrates include fish, amphibians, and **amniotes**, which include reptiles, birds, and **mammals**. See also **Invertebrate**.

VISCOSITY Resistance to flow in fluids. The higher the viscosity of a fluid, the more sluggishly it flows.

VITREOUS A type of **mineral luster** resembling that of glass.

WATER TABLE The upper surface of the **groundwater** zone in places where it is not confined by impermeable rocks. Water soaking into the ground will tend to flow downward until it reaches the water table.

WEATHERING The in-situ breaking down of rocks through contact with ice, water, wind, heat, chemicals, or other agents. See also **Erosion**.

ZOOPLANKTON Animals and animal-like organisms living in the plankton. See also **Plankton**, **Phytoplankton**.

index

Page numbers in **bold** refer to main entries.

M

N

O

S

T

U

V

W

XY

Z

acknowledgments

DK would like to thank:
Lee Skoulding at My Lost Gems, Southwold, UK (mylostgems.com) for help with photoshoots; Ina Stradins for her contribution to design; Gary Ombler for photography; Steve Crozier for image retouching; Peter Bull for the artwork of the water cycle on p.224; ETH Zurich for providing new images of mountain models; Maya Myers for fact-checking; Aarushi Dhawan, Kanika Kalra, Arshti Narang, and Pooja Pipil for design assistance; Vijay Kandwal, Nityanand Kumar, and Mohd Rizwan for hi-res assistance; Mrinmoy Mazumdar for DTP assistance; Ahmad Bilal Khan and Vagisha Pushp for picture research assistance; Rakesh Kumar for DTP design on the jacket; Tom Booth for the glossary; Richard Gilbert for proofreading; and Helen Peters for compiling the index.

DK would like to thank the following people at **Smithsonian Enterprises**:

Product Development Manager
Kealy Gordon
Director, Licensed Publishing
Jill Corcoran
Vice President, Business Development and Licensing
Brigid Ferraro
President
Carol LeBlanc

Smithsonian: National Museum of Natural History
Special thanks to Paige Towler

The publisher would like to thank the following for their kind permission to reproduce their photographs:

(Key: a-above; b-below/bottom; c-center; f-far; l-left; r-right; t-top)

1 Dreamstime.com: Daniel127001. **2 Science Photo Library:** Eye Of Science. **4-5 Getty Images:** Arctic-Images. **8-9 Popp-Hackner Photography OG:** Verena Popp-Hackner & Georg Popp. **10-11 123RF.com:** Roberto Scandola. **12-13 NASA:** ESA / N. Smith (University of California, Berkeley / The Hubble Heritage Team. **13 NASA:** CXC / SAO / JPL-Caltech / STScI (crb). **14 ESO:** ALMA (ESO / NAOJ / NRAO) (tl). **14-15 ©Alan Friedman / avertedimagination.com. 16 Getty Images:** Walter Geiersperger / Corbis Documentary (tl). **Science Photo Library:** Dennis Kunkel Microscopy (cl). **16-17 Getty Images:** Walter Geiersperger / Corbis Documentary (b). **18 Shutterstock.com:** Auscape / UIG (crb). **18-19 Alamy Stock Photo:** Jean-Paul Ferrero / AUSCAPE. **20 NASA. 21 Shutterstock.com:** Michael Wyke / AP (ca). **22 Alamy Stock Photo:** Michael Runkel / robertharding (tr). **22-23 ESA:** DLR / FU Berlin. **25 Alamy Stock Photo:** Composite Image / Design Pics Inc (br). **26-27 Robert Harding Picture Library. 27 Alamy Stock Photo:** B.A.E. Inc. (tr). **28-29 ESA:** Rosetta / NAVCAM. **28 J. W. Valley, University of Wisconsin-Madison:** (cl). **30 Courtesy of Smithsonian. ©2020 Smithsonian:** Chip Clark, NMNH (tl). **31 John Cornforth. 32-33 Jeffrey Sipress. 34-35 Getty Images:** mikroman6 / Moment. **37 Alamy Stock Photo:** Phil Degginger. **38 Alamy Stock Photo:** J M Barres / agefotostock (cra); Natural History Museum, London (cr). **Depositphotos Inc:** Minakryn (fbl). **Dreamstime.com:** Annausova75 (cl). **Science Photo Library:** Phil Degginger (fcr); Charles D. Winters (cla); Natural History Museum, London (bl); Dirk Wiersma (br); Millard H. Sharp (fbr). **Shutterstock.com:** Albert Russ (fcra). **39 Dorling Kindersley:** Holts Gems (cl). **Dreamstime.com:** Bjrn Wylezich (bl). **Shutterstock.com:** Albert Russ (r). **40-41 Shutterstock.com:** Bjoern Wylezich. **40 Alamy Stock Photo:** J M Barres / agefotostock (cb); J M Barres / agefotostock (crb). **Dreamstime.com:** Miriam Doerr (bl); Infinityphotostudio (clb). **Shutterstock.com:** Sebastian Janicki (br). **43 Petra Diamonds. 44 Alamy Stock Photo:** Andreas Koschate / Westend61 GmbH (ca). **Dorling Kindersley:** RGB Research Limited (clb). **Dreamstime.com:** Bjrn Wylezich (cla); Bjrn Wylezich (cb). **44-45 Alamy Stock Photo:** Bjrn Wylezich. **46-47 Crystal Classics LTD / crystalclassics.co.uk. 47 Dreamstime.com:** Bjrn Wylezich (c). **48 Dreamstime.com:** Daniel127001 (fcla); Bjrn Wylezich (cla); Epitavi (fcl). **Science Photo Library:** Millard H. Sharp (fcr). **Shutterstock.com:** Sebastian Janicki (cl); Sebastian Janicki (cr). **49 Dorling Kindersley:** Natural History Museum, London (cra). **Getty Images:** Walter Geiersperger / Corbis Documentary (cla). **Science Photo Library:** Joyce Photographics / Science Source (ca). **Shutterstock.com:** Sebastian Janicki (b). **50-51 Shutterstock.com:** Sebastian Janicki. **51 Alamy Stock Photo:** Susan E. Degginger (cra); Susan E. Degginger (fcra). **Science Photo Library:** Phil Degginger (fcla). **Shutterstock.com:** Henri Koskinen (cla). **52-53 Dreamstime.com:** Bjrn Wylezich. **54 Dreamstime.com:** Bjrn Wylezich (cl). **56 Alamy Stock Photo:** Roland Bouvier (bl). **56-57 Dreamstime.com:** KPixMining. **58 Alamy Stock Photo:** Enlightened Media (clb); Valery Voennyy (cla). **Science Photo Library:** Paul Biddle (crb); Mark A. Schneider (cb); Charles D. Winters (bl). **Shutterstock.com:** Cagla Acikgoz (bc); DmitrySt (ca); luca85 (cra); Minakryn Ruslan (br). **59 Alamy Stock Photo:** Ian Dagnall (cla); Natural History Museum, London (br). **Dreamstime.com:** Tatiana Neelova (cra). **Shutterstock.com:** Jirik V (ca). **60-61 Dreamstime.com:** Cristian M. Vela. **61 Depositphotos Inc:** Dr.PAS (cla). **Dreamstime.com:** Avictorero (cra). **Shutterstock.com:** valzan (ca). **64-65 Getty Images:** Dale Johnson / 500Px Plus. **65 Dreamstime.com:** Michal Baranski (bc). **66-67 Dreamstime.com:** Lukas Bischoff. **67 Dreamstime.com:** Zelenka68 (tl). **68-69 Unsplash:** Robby McCullough. **69 Unsplash:** Yang Song (cb). **70 Shutterstock.com:** MarcelClemens (tl). **70-71 Dreamstime.com:** Zelenka68. **72-73 Bryan Lowry. 72 Alamy Stock Photo:** tom pfeiffer (clb). **74-75 Richard Bernabe Photography. 76-77 Getty Images:** Piriya Photography. **77 Shutterstock.com:** Yanping wang (br). **78-79 Dreamstime.com:** . **79 Alamy Stock Photo:** J M Barres / agefotostock (bc). **80-81 The Light Collective:** Ignacio Palacios. **81 Alamy Stock Photo:** Natural History Museum, London (tc). **82 Dreamstime.com:** Losmandarinas (bl). **82-83 Alamy Stock Photo:** Natural History Museum, London. **84 British Geological Survey:** Permit Number CP22 / 005 British Geological Survey © UKRI. **85 Alamy Stock Photo:** Natural History Museum, London (cl). **Trustees of the Natural History Museum, London:** (t). **87 Alamy Stock Photo:** Siim Sepp (tr). **88-89 Getty Images:** Pete Rowbottom / Moment. **89 Alamy Stock Photo:** J M Barres / agefotostock (tc). **90 Alamy Stock Photo:** Siim Sepp (cra). **90-91 Getty Images:** Dean Fikar / Moment. **92-93 Getty Images:** MelindaChan / Moment. **93 Science Photo Library:** Sinclair Stammers (br). **94-95 Getty Images / iStock:** Meinzahn. **95 Getty Images / iStock:** Pears2295 (crb). **96-97 Alamy Stock Photo:** michal812. **97 Alamy Stock Photo:** kristianbell / RooM the Agency (br). **98 Getty Images:** Sherry H. Bowen Photography / Moment (tl). **98-99 Alamy Stock Photo:** Dennis Hardley. **100-101 Getty Images:** Sergey Pesterev / Moment. **101 Alamy Stock Photo:** Guy Edwardes Photography (br). **102-103 Paddy Scott. 104-105 Alamy Stock Photo:** Quagga Media. **106-107 Alamy Stock Photo:** J M Barres / agefotostock. **107 Alamy Stock Photo:** Kitchin and Hurst / All Canada Photos (cl). **108 Alamy Stock Photo:** NASA / agefotostock. **109 NASA. 110-111 Getty Images:** Posnov / Moment. **111 Alamy Stock Photo:** van der Meer Marica / Arterra Picture Library (tr). **112 Alamy Stock Photo:** Susan E. Degginger (bl). **112-113 Getty Images:** DEA / Pubbli Aer Foto / De Agostini Editorial. **114 Alamy Stock Photo:** Hitendra Sinkar. **116 Shutterstock.com:** Joanna Rigby-Jones (tr). **116-117 Getty Images:** by wildestanimal / Moment. **118-119 Science Photo Library:** Bernhard Edmaier. **119 Alamy Stock Photo:** Matthijs Wetterauw (br). **120-121 Roberto Zanette. 122-123 Department of Earth Sciences, ETH Zrich. 123 Science Photo Library:** Eth-Bibliothek Zrich (tr). **124-125 EyeEm Mobile GmbH:** Salvatore Paesano. **125 Getty Images:** iGoal.Land.Of.Dreams / Moment (crb). **126-127 NASA:** Earth Observatory images by Robert Simmon and Jesse Allen, using Landsat data from the USGS Earth Explorer.. **126 Dreamstime.com:** Chris Curtis (bl). **128-129 Shutterstock.com:** Jiji Press / Epa-Efe. **131 Science Photo Library:** NASA (cl). **132-133 Anchorage Daily News:** Marc Lester. **132 ESA:** Copernicus data (2014) / ESA / PPO.labs / Norut / COMET-SEOM Insarap study (bl). **134-135 Getty Images / iStock:** RiverNorthPhotography / E+. **135 Dreamstime.com:** Sarit Richerson (cla). **136-137 Getty Images:** Wild Horizon / Universal Images Group

Editorial. **137 Alamy Stock Photo:** Brad Mitchell (clb). **138-139 James Rushforth**. **139 Getty Images:** Ignacio Palacios / Stone (cla). **140-141 Getty Images:** Daiva Baa / EyeEm. **141 Getty Images:** Sebastin Crespo Photography / Moment (cra). **142-143 NASA:** NASA Earth Observatory images by Lauren Dauphin, using Landsat data from the U.S. Geological Survey. **144-145 Getty Images:** Tobias Titz / fStop. **146 SuperStock:** Colin Monteath / age fotostock (tl). **146-147 Getty Images:** Arctic-Images / Stone. **148-149 Getty Images:** Monica Bertolazzi / Moment. **149 Alamy Stock Photo:** Art Publishers / Africa Media Online (tr). **150-151 Getty Images:** Agnes Vigmann / 500px. **151 Science Photo Library:** NASA (tr). **152-153 Alamy Stock Photo:** Dennis Frates. **153 Alamy Stock Photo:** blickwinkel / McPHOTO / TRU (tr). **154-155 Alamy Stock Photo:** Christopher Drabble. **154 Getty Images:** Walter Bibikow / DigitalVision (tr). **156-157 Alamy Stock Photo:** B.A.E. Inc.. **157 Science Photo Library:** Dr Morley Read (cr). **158 Getty Images:** Maxim Blinov / 500px Prime (tl). **158-159 Getty Images:** Daniel Bosma / Moment. **160-161 Shutterstock.com:** travelwild. **161 Alamy Stock Photo:** Daniel Korzeniewski (tr). **162 Getty Images:** Martin Harvey / The Image Bank (cl). **162-163 NASA:** Jesse Allen and Robert Simmon / United States Geological Survey / Landsat 7 / ETM+. **164-165 ESA:** Copernicus Sentinel (2021). **165 Getty Images:** Posnov / Moment (tr). **166-167 Shutterstock.com:** bengharbia. **168 Getty Images:** Amrish Aroonda Manikoth / EyeEm (cl). **168-169 Getty Images:** Westend61. **170-171 Getty Images:** Bernhard Klar / EyeEm. **172-173 Getty Images:** Ignacio Palacios / Stone. **172 Dreamstime.com:** Bennymarty (crb). **174-175 Georg Kantioler**. **175 Alamy Stock Photo:** Danita Delimont Creative (tr). **176-177 Dreamstime.com:** Channarong Pherngjanda. **178-179 Tom Hegen GmbH**. **179 Getty Images / iStock:** mantaphoto (tr). **180-181 Science Photo Library:** Karsten Schneider. **182 Alamy Stock Photo:** Doug Perrine / Nature Picture Library (bc); Eric Grave / Science History Images (cb); Scenics & Science (fcrb). **naturepl.com:** Shane Gross (br); Solvin Zankl (fbr). **Science Photo Library:** Wim Van Egmond (crb). **183 Science Photo Library:** NASA. **184 Science Photo Library:** Ryan Et Al / Geomapapp (cra). **184-185 NASA:** Image courtesy Serge Andrefouet, University of South Florida.. **186 Alamy Stock Photo:** Paul Brady (cla). **186-187 Getty Images:** Streeter Lecka. **188-189 Josef Valenta Photography**. **188 Flying Focus aerial photography:** (bl). **190-191 Getty Images:** Cavan Images. **190 Getty Images:** Fuse / Corbis (tl). **192-193 Getty Images:** Holger Leue / The Image Bank. **193 Getty Images:** Paul Souders / DigitalVision (tr). **194-195 Dreamstime.com:** Rabor74. **195 Science Photo Library:** Planetobserver (br). **196-197 Rob Suisted / Nature's Pic Images**. **196 Getty Images:** Photo 12 / Universal Images Group Editorial (bc). **Julian Hodgson:** (tl). **198-199 Jon Anderson Wildlife Photography**. **198 naturepl.com:** Franco Banfi (bl). **200 Library of Congress, Washington, D.C.:** Berann, Heinrich C. / Heezen, Bruce C. / Tharp, Marie. **201 Alamy Stock Photo:** Granger - Historical Picture Archvie (cl). **Library of Congress, Washington, D.C.:** Berann, Heinrich C. / Heezen, Bruce C. / Tharp, Marie (tr). **202 Alamy Stock Photo:** NOAA (tl). **202-203 Shutterstock.com:** Lovkush Meena. **204 Science Photo Library:** Steve Gschmeissner. **205 Science Photo Library:** Eye Of Science (cla); Eye Of Science (ca); Eye Of Science (cra); Eye Of Science (clb); Eye Of Science (cb); Eye Of Science (crb). **206-207 MARUM- Center for Marine Environmental Sciences, University of Bremen:** CC BY 4.0. **206 Science Photo Library:** OAR / National Undersea Research Program (tl). **208-209 Dreamstime.com:** Fabio Lamanna. **208 Getty Images:** DigitalGlobe / ScapeWare3d / Maxar (tr). **210 Science Photo Library:** Tony & Daphne Hallas (cra). **210-211 Shutterstock.com:** studio23. **212-213 Getty Images:** Elena Pueyo / Moment. **213 Shutterstock.com:** Vladi333 (br). **214-215 G Sharad Haksar:** Eye-light Pictures. **215 Alamy Stock Photo:** JordiStock (crb). **216 Alamy Stock Photo:** Science History Images. **217 NASA:** Albert J. Fleig, Jr. (cl); Robert Simmon (t). **218 Science Photo Library:** NOAA (t). **220 Alamy Stock Photo:** Susan E. Degginger (tl). **220-221 NOAA:** Mike Schuler. **222 Alamy Stock Photo:** Mike Grandmaison / All Canada Photos (bc); Katho Menden (ca). **Ivon Bartholomew:** (cra). **Dreamstime.com:** Intrepix (cr). **223 123RF.com:** Nataliia Kravchuk (bl). **Dreamstime.com:** Iofoto (tl). **Getty Images:** Anadolu Agency / Ozkan Bilgin (cl); Donovan Reese / Photodisc (br). **Michael Johnston:** (c). **225 Science Photo Library:** Bernhard Edmaier (b). **226-227 Alamy Stock Photo:** beyond / Yevgen Timashov / Micheko Productions, Inh. Michele Vitucci. **228-229 Getty Images / iStock:** maximkabb. **228 Conor Regan:** (cb). **230 Science Photo Library:** Kenneth Libbrecht. **230-231 Science Photo Library:** Kenneth Libbrecht. **231 Science Photo Library:** Kenneth Libbrecht (cr, fcr, crb, fcrb, br, fbr). **232-233 Jeremy Holmes**. **233 NOAA:** National Weather Service - Wichita, Kansas (tr). **234-235 James Smart**. **234 Getty Images:** Tasos Katopodis / Stringer (crb). **236-237 Reuters:** Lucas Jackson. **237 Matthew Pasek, University of Southern Florida:** (tr). **238 NOIRLab:** International Gemini Observatory / NOIRLab / NSF / AURA / A. Smith CC BY 4.0 (tl). **238-239 Stephen Hummel**. **240 CSIRO:** Atmospheric Research. **241 U.S. National Ice Core Laboratory:** icecores.org (tr). **The University of Melbourne Archives:** (cl). **242-243 Getty Images:** Mario Tama. **243 NOAA:** map based on satellite data (t). **244-245 Getty Images / iStock:** Nastasic. **246 Science Photo Library:** Dr Ken Macdonald (clb). **248-249 naturepl.com:** Kim Taylor. **249 The Sedgwick Museum of Earth Sciences, University of Cambridge:** (cr). **251 Brendan Burns:** (tl). **252-253 Science Photo Library:** Dirk Wiersma. **253 Dreamstime.com:** Max5128 (cra). **254-255 Dreamstime.com:** Zeytun Images. **254 Science Photo Library:** Dr. Gilbert S. Grant (br). **256 Alamy Stock Photo:** blickwinkel / R. Koenig (tc). **Dreamstime.com:** Alexander Ogurtsov (tl). **257 Arild Hagen. 258-259 Alamy Stock Photo:** McPhoto / Ingo Schulz. **259 James St. John (Ohio State University at Newark):** (bl). **261 Science Photo Library:** Science Stock Photography. **262-263 Dr Sandy Hetherington:** The Hunterian, University of Glasgow. **263 Alamy Stock Photo:** Natural History Museum, London (tr). **264-265 NASA:** DigitalGlobe / Maxar / IKONOS. **266-267 Alamy Stock Photo:** MichaelGrantPlants. **266 Peter Giesen:** (tl). **XU Hong-He, Ph.D:** (tc). **270-271 Dreamstime.com:** Mark Higgins. **270 Science Photo Library:** Dorling Kindersley / UIG (br). **272 Getty Images:** Robert LeBlanc / Moment (cl). **272-273 Alamy Stock Photo:** Album. **274 Shutterstock.com:** Nicholas Toh (cl). **275 Alamy Stock Photo:** Album. **276-277 Gelio Vostok**. **276 Dorling Kindersley:** James Kuether (cla). **278 Dorling Kindersley:** Natural History Museum, London (tl), (cra). **279 Alamy Stock Photo:** Sabena Jane Blackbird (tr). **Science Photo Library:** Dorling Kindersley / UIG (ftr). **280-281 Depositphotos Inc:** Kokhanchikov. **281 Getty Images:** China News Service (tl). **282-283 Vincent Fernandez:** Fernando Abdala, Kristian J Carlson, Della Cook, Adam Yates, Paul Tafforeau, Bruce S Rubidge. **282 Dr. Kawabe Soichiro:** (bl). **283 Science Photo Library:** Philippe Psaila (tr). **284-285 Arild Hagen**. **286 Image Courtesy of the Royal Tyrrell Museum, Drumheller, AB:** (cb). **286-287 Greg Funston**. **288-289 Alamy Stock Photo:** Auk Archive. **288 Alamy Stock Photo:** The Natural History Museum, London (bl). **290-291 Alamy Stock Photo:** Lemaire Stphane / Hemis.fr (bl). **Dorling Kindersley:** Senckenberg Gesellschaft Fuer Naturforschung Museum (tr). **292 Science Photo Library:** Sinclair Stammers (tl). **292-293 Professor Chenyang Cai**. **294-295 Dorling Kindersley:** Senckenberg Gesellschaft Fuer Naturforschung Museum. **294 Alamy Stock Photo:** PD Archive (br). **296 Alamy Stock Photo:** Klaus Honal / agefotostock (br); blickwinkel / R. Koenig (c); blickwinkel / Koenig (cr); blickwinkel / Koenig (bc). **Dr Sonja Wedmann:** Senckenberg Institute (tc, tr). **297 Alamy Stock Photo:** blickwinkel / Koenig (c); tbkmedia.de (cr); blickwinkel / Koenig (bl); Ingi May / Stockimo (bc); blickwinkel / Koenig (br). **Dr Sonja Wedmann:** Senckenberg Institute (tl). **Tiia Monto:** (cl). **298-299 Alamy Stock Photo:** National Geographic Image Collection. **299 Mauricio Anton:** (bc). **300 naturepl.com:** Cyril Ruoso (tl). **300-301 Alamy Stock Photo:** Eric Baccega / agefotostock. **302-303 Florida Museum of Natural History**. **303 Science Photo Library:** Science Source (br). **304-305 Shutterstock.com:** Konrad K / Sipa. **304 Roman Uchytel:** (bl). **306-307 Alamy Stock Photo:** Steven Kazlowski / RGB Ventures / SuperStock. **306 Alamy Stock Photo:** Sanamyan (bl). **308-309 Alamy Stock Photo:** Kersti Lindstrm. **309 Getty Images / iStock:** MEGiordano_ Photography (bl). **310 Altamura National Archaeological Museum:** Ministero della cultura (MIC) / Direzione regionale Musei Puglia / Museo Nazionale Archeologico di Altamura. **311 Arild Hagen. School of Earth Sciences, University of Bristol. 312 Alamy Stock Photo:** Natural History Museum, London (bl). **312-313 Dorling Kindersley:** Natural History Museum, London. **314 Alamy Stock Photo:** Philip Smith (cb). **314-315 Alamy Stock Photo:** keith morris news. **316-317 Getty Images:** Vicki Jauron, Babylon and Beyond Photography / Moment. **317 Alamy Stock Photo:** Malcolm Schuyl (cr). **318-319 Alamy Stock Photo:** Nick Garbutt / Steve Bloom Images. **319 Alamy Stock Photo:** Paul Williams / Nature Picture Library (crb). **320-321 Alamy Stock Photo:** Jeneses Imre. **322 Dorling Kindersley:** Natural History Museum, London (bl). **323 Alamy Stock Photo:** Album. **324 Dorling Kindersley:** Natural History Museum, London (cra), (br). **Dreamstime.com:** Brian Weed (bl); Kevkhiev Yury (fbl). **325 Dorling Kindersley:** Holts Gems (cra); Holts Gems (fbl); Natural History Museum, London (br), (fbr). **326 Dreamstime.com:** Nikand4 (cra); Bjrn Wylezich (fbl); Vvoevale (br); Vvoevale (fbr). **327 Dreamstime.com:** Micropic (bl); Ruslan Minakryn (cra); Ruslan Minakryn (fbr). **Getty Images:** © Santiago Urquijo / Moment (br). **328 Dorling Kindersley:** Natural History Museum, London (bl), (fbr). **Dreamstime.com:** Ruslan Minakryn (br); Vvoevale (fbl). **329 123RF.com:** Dario Lo Presti / porpeller (bl). **Dorling Kindersley:** Natural History Museum, London (fbl). **Dreamstime.com:** Henri Koskinen (cra); Losmandarinas (fbr). **330 Alamy Stock Photo:** PjrRocks (fbr). **Dorling Kindersley:** Natural History Museum, London (bl), (br). **Dreamstime.com:** Simon Zenger / Simarts (cra). **331 Dreamstime.com:** Vvoevale (br). **332 Dorling Kindersley:** Oxford University Museum of Natural History (bl). **Dreamstime.com:** Mohamed El-Jaouhari (fbl). **Getty Images / iStock:** SageElyse (fbr). **333 Dreamstime.com:** Mohamed El-Jaouhari (br). **334 Dreamstime.com:** Rainer Walter Schmied (fbl). **Getty Images / iStock:** jollyphoto (cra). **335 Dreamstime.com:** Bjrn Wylezich (bl). **Getty Images:** Darrell Gulin (cra). **336 Alamy Stock Photo:** Photo Researchers / Science History Images (fbr). **Dreamstime.com:** Annausova75 (cra); Bjrn Wylezich (fbl); Vvoevale (bl). **337 Alamy Stock Photo:** J M Barres / agefotostock (cra). **Dreamstime.com:** Losmandarinas (br). **338 Dorling Kindersley:** Natural History Museum, London (br), (fbr). **Dreamstime.com:** Areg Grigoryan (bl); Bjrn Wylezich (cra). **339 Alamy Stock Photo:** Phil Degginger (fbl); Natural History Museum, London (cra). **Dreamstime.com:** Andrii Popov (br); Bjrn Wylezich (fbr). **341 Getty Images:** DEA / ICAS94. **342 Dreamstime.com:** Rozenn Leard (cra). **343 Alamy Stock Photo:** J M Barres / agefotostock (fbl). **Dorling Kindersley:** Natural History Museum, London (br); Oxford University Museum of Natural History (fbr). **Getty Images:** Pete Rowbottom / Moment (cra). **344 Depositphotos Inc:** siimsepp (cra). **345 Alamy Stock Photo:** Klaus Scholz / mauritius images GmbH (cra); Science History Images / Photo Researchers (bl). **346 Dreamstime.com:** Fokinol (bl). **Getty Images:** Francesco Riccardo Iacomino / Moment (cra). **Shutterstock.com:** Aleksandr Pobedimskiy (fbl). **347 Dreamstime.com:** Keith Wheatley / Kwheatley (cra). **348 Alamy Stock Photo:** Phil Degginger (br). **Depositphotos Inc:** vladislavgajic (cra). **Dorling Kindersley:** Natural History Museum, London (fbr). **349 Alamy Stock Photo:** J M Barres / agefotostock (cra); Phil Degginger (br); Phil Degginger (fbr). **Dorling Kindersley:** Holts Gems (bl). **Siim Sepp:** sandatlas.org (fbl). **351 Alamy Stock Photo:** Album. **352 Alamy Stock Photo:** Matteo Chinellato (bl); durk gardenier (cra); Susan E. Degginger (fbl); Matteo Chinellato (br). **Shutterstock.com:** Matteo Chinellato (fbr). **353 Alamy Stock Photo:** Derek Anderson (cra); Matteo Chinellato (fbr). **Science Photo Library:** Jean-Claude Revy, A. Carion, ISM (bl). **354 Science Photo Library:** Paul D Stewart (cla). **355 Mary Evans Picture Library:** Jean Gabriel Pretre / Florilegius. **356 Alamy Stock Photo:** Sabena Jane Blackbird (fbr). **Science Photo Library:** Dr. Gilbert S. Grant (bl); Sinclair Stammers (cra); Natural History Museum, London (br). **357 Alamy Stock Photo:** FLHC24 (bc). **Science Photo Library:** Sinclair Stammers (bl); Sinclair Stammers (cra); Alan Sirulnikoff (br). **358 Getty Images / iStock:** bobainsworth (cra). **359 Dorling Kindersley:** Swedish Museum of Natural History (fbr). **360 Dorling Kindersley:** Natural History Museum, London (br),(fbr); Swedish Museum of Natural History (bl). **Getty Images / iStock:** Gerald Corsi (cra). **361 Dorling Kindersley:** Natural History Museum, London (fbl); Swedish Museum of Natural History (bl). **362 Dorling Kindersley:** Natural History Museum, London (br),(fbl), (fbr). **363 Dorling Kindersley:** University Museum of Zoology, Cambridge (br); Natural History Museum (cra), (fbl); Swedish Museum of Natural History (fbr). **364 Dorling Kindersley:** Natural History Museum, London (bl), (fbl), (br), (fbr). **365 Dorling Kindersley:** Natural History Museum, London (br), (fbl); Swedish Museum of Natural History (bl); State Museum of Nature, Stuttgart (fbr). **366 Dorling Kindersley:** Senckenberg Nature Museum, Frankfurt (br). **367 Dorling Kindersley:** Trustees of the National Museums Of Scotland (fbr); Hunterian Museum University of Glasgow (cra); Swedish Museum of Natural History (fbl); Swedish Museum of Natural History (bl). **368 123RF.com:** Dave Willman / dcwcreations (br). **Dorling Kindersley:** Courtesy of Dorset Dinosaur Museum (fbr); Natural History Museum, London (cra); Senckenberg Gesellschaft Fuer Naturforschung Museum (fbl). **Getty Images:** Kevin Schafer (bl). **369 Dorling Kindersley:** Natural History Museum, London (br); Senckenberg Gesellschaft Fuer Naturforschung Museum (cra); Swedish Museum of Natural History (fbl); Oxford University Museum of Natural History (bl); Natural History Museum (fbr). **370 Dorling Kindersley:** Natural History Museum, London (cra), (bc); Royal Tyrrell Museum of Palaeontology, Alberta, Canada (br). **Dreamstime.com:** Yezhenliang (bl). **371 Dorling Kindersley:** Natural History Museum (bl); Natural History Museum, London (fbl), (br), (fbr). **372 Dorling Kindersley:** Natural History Museum, London (bl), (bc), (br); Senckenberg Gesellschaft Fuer Naturforschung Museum (cra). **373 Dorling Kindersley:** Natural History Museum, London (bl); Oxford Museum of Natural History (cra); Natural History Museum, London (fbl), (br), (fbr). **374 Dorling Kindersley:** Natural History Museum, London (cra), (fbl), (bl). **375 Dorling Kindersley:** Natural History Museum, London (bl), (fbl), (fbr); Oxford Museum of Natural History (cra). **376 Dorling Kindersley:** Natural History Museum, London (cr), (bl), (br), (fbr); Royal Pavilion & Museums, Brighton & Hove (fbl). **377 Dorling Kindersley:** Natural History Museum, London (cra), (fbl), (bl), (br); Royal Pavilion & Museums, Brighton & Hove (fbr). **378 Alamy Stock Photo:** Victoria Zeffert / Cultura Creative Ltd (br); Robert Francis / robertharding (tr). **379 Alamy Stock Photo:** arkanto (br). **Getty Images:** Richard Eden / Moment (tr). **380 Alamy Stock Photo:** Andrew Sabai (tr); Dominique Braud / Dembinsky Photo Associates (br). **381 Alamy Stock Photo:** NASA / Landsat / Phil Degginger (tr). **Getty Images:** Nick Rains / Corbis Documentary (br). **382 Getty Images:** MB Photography / Moment (tr); Anton Petrus / Moment (br). **383 Dreamstime.com:** Adwo (br). **Getty Images:** SinghaphanAllB / Moment (tr). **384 Getty Images / iStock:** Stefan Tomic / E+ (tr). **Getty Images:** Mark Fox / Moment (br). **385 Ardea:** Yves Bilat (br). **Getty Images:** Natphotos / Photodisc (tr). **386 naturepl.com:** Bruno D'Amicis (br). **Shutterstock.com:** Mikadun (tr). **387 Getty Images:** Bashar Shglila / Moment (tr); Emad Aljumah / Moment (br). **388 Alamy Stock Photo:** Planet Observer / Universal Images Group North America LLC (tr). **Getty Images:** Russell Burden / Photodisc (br). **389 Getty Images:** Martin Harvey / The Image Bank (tr); joSon / Stone (br). **390 Getty Images:** The Image Bank / John Elk III (br). **Shutterstock.com:** Homo Cosmicos (tr). **391 Alamy Stock Photo:** Tom Pfeiffer (br). **Getty Images:** Moment / Daniele Sartori (tr). **392 Alamy Stock Photo:** Aleksander Bolbot (br). **Shutterstock.com:** Mita Stock Images (tr). **393 Getty Images / iStock:** Mikhail Blavatskiy (br). **Getty Images:** DigitalVision / Buena Vista Images (tr). **394 Alamy Stock Photo:** EyeEm / Gregory Michiels (br). **Getty Images:** Moment Open / Anton Petrus (tr). **395 Dreamstime.com:** Chun Guo (br); Anjali Kumari (tr). **396 Alamy Stock Photo:** imageBROKER / Josef Beck (br). **Getty Images:** Moment / DoctorEgg (tr). **397 Dreamstime.com:** Marc Witte / Witteart (br). **Getty Images:** Stone / Peter Harrison (tr). **398 Alamy Stock Photo:** Paul Kingsley (tr). **Getty Images / iStock:** Fyletto (br). **399 Alamy Stock Photo:** Nature Picture Library / Michel Roggo (br). **Getty Images:** Stone / David Clapp (tr)